INTERNATIONAL SERIES OF MONOGRAPHS IN
NATURAL PHILOSOPHY
GENERAL EDITOR: D. TER HAAR

VOLUME 8

FIELD THEORETICAL METHODS IN MANY-BODY SYSTEMS

OTHER TITLES IN THE SERIES IN NATURAL PHILOSOPHY

D. A. KIRZHNITS

Field Theoretical Methods in Many-Body Systems

TRANSLATED BY

A. J. MEADOWS

EDITED BY

D. M. BRINK

PERGAMON PRESS

OXFORD · LONDON · EDINBURGH · NEW YORK

TORONTO · SYDNEY · PARIS · BRAUNSCHWEIG

Pergamon Press Ltd., Headington Hill Hall, Oxford
4 & 5 Fitzroy Square, London W.1

Pergamon Press (Scotland) Ltd., 2 & 3 Teviot Place, Edinburgh 1

Pergamon Press Inc., 44–01 21st Street, Long Island City, New York 11101

Pergamon of Canada, Ltd., 6 Adelaide Street East, Toronto, Ontario

Pergamon Press (Aust.) Pty. Ltd., 20–22 Margaret Street, Sydney, N.S.W.

Pergamon Press S.A.R.L., 24 rue des Écoles, Paris 5^e

Vieweg & Sohn GmbH, Burgplatz 1, Braunschweig

First English edition 1967

This is a translation of the original Russian work
Полевые методы теории многих частиц
(Poleviye metodiy teorii mnogikh chastits)
published by Gostatomizdat, 1963, and contains
corrections and revisions supplied by the author.

Library of Congress Catalog Card No. 66–20567

2750/67

Contents

Contents

Introduction

THIS book describes the basic concepts and methods of present-day microscopic theories of many-body systems.† Methods borrowed from relativistic quantum field theory have been developed during recent years and are widely applied at present to greatly differing fields of physics.

The general problem facing a theory of many-body systems is the necessity of describing both the internal properties of the system and its interaction with external agents; that is, such characteristics of a system as its energy spectrum (the energy of the ground state, the spectrum of the excited states), the mean values of the dynamical variables and their distribution, transition probabilities, thermodynamic and kinetic characteristics, etc. It is supposed that a model of a given system (i. e. the characteristics of the individual particles, the way in which they interact, the external conditions, etc.) is provided.

Many-body theory provides the means for solving a given problem in terms of some approximation, and forms the theoretical foundation of many branches of physical science, e.g. the physics of condensed media (including the solid state), plasma physics, the physics of atoms and atomic nuclei, etc.

The most important aspect of the problems dealt with in many-body theory is the necessity to take into account the interactions

† It is essential to define from the start the meaning of the term "a many-body system", which is normally given a conventional connotation. One- and two-body systems will be excluded from this category, since the calculations are elementary. On the other hand, it is possible, and often convenient, to consider even such a system as a helium atom to be a many-body system. It must be remembered, however, that some aspects of the description—the statistical description, for example—are of no significance for a system composed of only a few particles. Moreover, numerous factors, which simplify the study of many-body systems, are inapplicable to a few particles.

between the particles. The model in which these interactions are ignored (the perfect-gas model) possesses very few properties and they may be calculated by elementary methods. It is just the interaction between particles which accounts for the qualitative differences between the objects considered in the various branches of physics noted above.

To take the interaction in any complicated system accurately into account is, however, exceptionally difficult. Although it now seems possible that electronic computers may be used for this, the scope of many-body theory still remains fairly modest (apart from atomic and molecular theory).

The main cause of these difficulties is that the Schrödinger equation in many variables, which does not separate into independent equations in each variable, is, from the mathematical standpoint, far more complicated than, say, the Schrödinger equation for particles in an external field. It is the non-separable Schrödinger equation which applies to a system of many interacting particles. This mathematically obvious fact is a reflection of the physical situation: due to the continuous field of force between any individual particle and the other particles in the system, the concept of states of particle (i.e. of its wave function, energy, etc.) has no precise meaning. Strictly speaking, one should think only of the state of the system as a whole: hence, when the interactions between particles are taken into account, there is a corresponding change in the object being described—in the perfect-gas model each individual particle may be the object; in any real case it is necessarily the whole system.

It should be clear from what has been said that many-body problems can only be solved approximately.† Numerous approximate methods in many-body theory have been worked out during the past decades. Thus we have perturbation theory, the Hartree–Fock, Thomas–Fermi, Debye–Hückel and Brueckner approximations, the collective–variable method and many others [1]. These methods have undoubtedly played an important part in the development of

† Another reason for the approximate nature of the solution for many of the problems is the lack of information about the interaction between the particles (e.g. nuclear matter). Under these conditions, it is hardly sensible to aspire to extremely accurate solutions of the theoretical equations.

the various branches of physics. In many ways, however, many-body theory has been unsatisfactory until quite recently.

First of all, many of the methods mentioned are very unwieldy. This is particularly true of the "old" perturbation theory which employed Slater–Fock determinants, complicated summations, and so forth. Attempts to extend the calculations beyond the second or, at most, third order usually ran up against serious technical difficulties. Simulaneously, more fundamental problems appeared. For example, in calculating the energy of a homogeneous system, fictitious terms appear which vary with the volume of the system to a higher power than the first. Brueckner's method [2] is likewise highly complicated, requiring very extensive numerical calculations.

Furthermore, these methods have only a limited applicability and—most important of all—it is often far from clear from the way they are formulated under what conditions they are applicable. Questions concerning the conditions which the parameters of a system must satisfy in a given region of applicability, and others concerning the corrections to be applied when these conditions are not satisfied, have as yet received a reasonable answer in only a minority of all cases.

Finally, it is a source of major dissatisfaction that the formulations of the various methods in many-body theory are often of an artificial *ad hoc* character, which varies considerably from one method to the next. As a result, the fundamental relationship between the different methods, which derives from the fact that they are all simply alternative approximations to the accurate Schrödinger equation, has been lost.

The desirability of replacing the various particular methods in many-body theory by one simple, general approach has been felt for a long time: it has been dictated by the growing scope and complications of the theoretical problems. A satisfactory solution, however, has only appeared in the last few years. The methods which have arisen (known, for brevity, as field theoretical methods) derive directly and naturally from the method of second quantization and are therefore both relatively simple and compact.

Green functions, which form the basic element of the field theoretical description, contain considerable physical information about a given system which may be extracted by very simple mathematical operations. If the equations for the Green functions are analysed, due allowance being made for the smallness of the dimensionless parameters characterizing the system, we see that it is possible to simplify them. Here, so-called diagram techniques play an essential part, making it possible to obtain expressions for the terms in perturbation theory to as high an order as desired using only simple rules. An analysis of this sort provides a straightforward and reliable basis for the "old" methods of many-body theory, sheds light on the limits of applicability of these methods, and makes it possible to increase their accuracy.

Turning next to a brief outline of field theoretical methods in many-body theory, we must remark first of all that the method of second quantization, which contains virtually all that is essential for a field theoretical description of a many-body system, was developed when the quantum mechanics of a system of particles was still in its initial stages. In this method the concept of a particle is of minor importance, its place being taken by the concept of a quantized field, $\hat{\psi}(x)$, $\hat{\psi}^+(x)$ (the particle annihilation and creation operators). The particles themselves appear as field quanta. The interaction processes between particles may be interpreted in terms of annihilation of particles in the initial state and their creation in the final state. Generally speaking, the number of particles in an interaction process (as distinct from the difference between the number of particles and the number of holes) is not conserved. The important process of excitation of a system can also be interpreted purely in terms of a field; in the simplest case it leads to the formation of a particle–hole pair, the reverse process whereby the system returns to its initial state then corresponding to the annihilation of this pair. Besides these, the second quantization method can reveal many other aspects of the field theoretical description. During the initial stages in the development of the method, however, the opportunities for a many-body theory were not opened up, and its applications to the quantum mechanics of systems of particles have remained until recently within rather narrow limits (see, however, [3]).

There came a fundamental change in the situation with the development of elementary particle theory in the thirties and forties.

Relativistic quantum field theory was from the very beginning constructed by analogy with many-body theory, that is, using second quantisation methods. The great physical significance of these investigations, and especially the existence of serious fundamental difficulties led to a substantial concentration of effort in this field.

As a result, the necessary formal apparatus of quantum field theory was basically completed by the beginning of the fifties. The techniques, thus developed, could by virtue of their generality hardly fail to outgrow the limits of elementary particle theory. The first papers on the application of field theoretical methods to many-body systems had already begun to appear by the middle of the fifties. This was in effect a return to the source from which they derived. By now, several hundred papers have been published, and we have at our disposal a well-formulated, convenient and flexible method for investigating the properties of many-body systems.

It is particularly important to emphasize that the method seems adequate for application to the most important many-body concept—that of quasi-particles, or elementary excitations. This approach is the main reason for the successes which have been achieved by the theory [4]. The introduction of the idea of a quasi-particle makes it possible to describe many aspects of the problem in single-particle form, i.e. to consider a system of interacting particles as approximating a collection of independent, collectively acting entities—the quasi-particles. Field theoretical methods provide a natural basis for such an approach in many-body theory, and they also establish the limits of its applicability.

This book is based on a series of lectures given by the author in 1960 and 1961 to colleagues at the Lebedev Physical Institute of the Academy of Sciences, U.S.S.R., and at the Physical Energy Institute of the State Committee of the Ministerial Council, U.S.S.R., for the Application of Atomic Energy. It is therefore not intended to be a monograph giving a more or less comprehensive account of contemporary many-body field theory and its numerous applications. It has, instead, rather the nature of an introductory textbook for readers unacquainted with quantum field theory and therefore examines particularly the difficulties encountered in using field theoretical methods.

The limited nature of the task which the author has set himself has, of course, influenced his choice from the material available in the literature. A desire to examine the presented material in greater detail has made it necessary to restrict the number of questions that can be considered.†

In particular the phenomenological aspects of many-body field theoretical theory [7, 24, 25], the relativistic theory of many-body systems [26], and the field theoretical theory of transport phenomena [20] will not be discussed. We limit ourselves to a consideration of the most important case of systems of Fermi particles: thus including multi-electron and multi-nucleon systems. Questions connected with the theory of Bose systems [19, 27, 28], and with superfluid (superconductivity) theory [6–8, 29] will not be considered.

To illustrate the theoretical material presented in the book, a relatively small number of practical problems have been selected from the fields of nuclear theory, plasmas, solid state and atomic theory.

Chapter 1 contains auxiliary material on quantum mechanics and nuclear theory which is essential for the future development [30–37]. It also presents the concept of the pseudopotential, which immediately provides an expression for the effective interaction potential in a system in terms of the particle scattering amplitude. Chapter 1 also gives a general classification of many-body systems in terms of their interaction and degree of condensation. This classification helps to simplify the mathematical apparatus of the theory. Particular attention is devoted in this first chapter to second quantization, which, as has already been emphasized, is essential to the development of field theoretical methods.

The formulation of a many-body theory, as for any other approximate theory, must begin with the choice of an appropriate zero approximation. In this book the Hartree–Fock approximation is chosen since it gives the best description of a system compatible

† The reader, who has obtained an acquaintance with the elements of many-body theory through this book, can find further material of interest in many other works [5–23].

with the concept of single-particle states. This choice also makes it possible to take into account a significant part of the interaction in the initial stage of the calculations, which leads to a considerable simplification in later calculations, especially for dense systems. The perfect-gas model, which is the usual choice for a zero approximation in many problems, seems excessively crude in this case; for example, in atomic problems the screening of the nuclear Coulomb field is of vital importance, yet it is completely ignored in the perfect-gas approach. The perturbation Hamiltonian, describing the so-called dynamical correlation effect, can be represented in terms of a normal operator product, hence reducing considerably the number of Feynman diagrams that need to be considered.

Chapter 2 is devoted to an exposition of many-body theory in the Hartree–Fock approximation. A very convenient operator approach is used which aids the transition to the quasi-classical limiting case of the Hartree–Fock approximation: the Thomas–Fermi approximation. Several problems in the theory of highly compressed matter, and of nuclear matter, are considered as applications. They are further considered in Chapters 3 and 4.

The general field theoretical methods of perturbation theory are worked out in Chapter 3. They are based on the theory of the S-matrix, and include diagram techniques as an important element. The Feynman rules for the formation of the elements of S-matrices of any order are formulated. Relations are also given for determining the density matrix and the ground state energy of a system to any order of perturbation theory. Particular attention is paid in this chapter to the connection between system classification and the selection of those perturbation theory diagrams that are, under these circumstances, of basic importance. The theory of two-electron atoms and nuclear matter including correlation effects are given as examples. Chapters 2 and 3 consider, almost without exception, only the ground state of a system.

Chapter 4 describes the properties of the one- and two-particle Green functions, together with methods of finding what physical information they give about the system considered. Equations are formulated which determine the required Green functions. A detailed consideration is given to the problem of determining the characteristics of the excited states of a system. Spectral representations of the Green functions are deduced. The concept of quasi-particles is then introduced, and the connection between the characteristics of the quasi-particles and the analytical properties of the Green functions is established.

Later in Chapter 4 there is a discussion of the general theory of dilute many-body systems. An expression for the physical quantities in such systems is deduced from a series expansion in terms of the ratio of the scattering amplitude to the average distance between the particles. The properties of the Fermi spectrum of excitations of dilute systems are also investigated.

Dense many-body systems, with both Coulomb and short-range forces, are studied in detail. The Bose-type excitations (plasmons, zero sound) are also investigated. Applications to the atom, and to the problem of plasma oscillations in a homogeneous electron gas (with a compensating positive-charge background) are considered, as are certain applications to nuclear theory. The treatment in these first four chapters is for zero temperature.

Chapter 5, the last in the book, is devoted to investigating the behaviour of a many-body system at finite temperatures, and to formulating the field equations of quantum statistics. The Hartree–Fock equation for "hot" systems is considered: it can be applied to find the equation of state.

Both the technique developed by Matsubara for determining the thermodynamical characteristics of a system, and the use of time-dependent Green functions for finding the excitation spectrum of a "hot" system, are described. Certain applications to plasma theory, and to collective oscillations, are considered.

The book also has several appendices, which contain supplementary mathematical material. Appendix A contains the necessary rules for calculating average values of all operators of physical quantities. Appendix B provides an account of the basic rules and relations for operator calculations. Operator functions of a sum of non-commuting arguments are considered in detail. Appendix C contains essential information on singular integrals, and rules for their calculation.

Dirac's constant $\hbar$ has been taken equal to unity throughout the book. Operators are denoted by a caret ($\wedge$), vector quantities by boldface italics.

The coordinate q represents a combination of the space $\boldsymbol{x}$, spin σ and isotopic spin τ coordinates; the coordinate x combines q and the time t. Thus:

$$\int dq = \int d^3\boldsymbol{x} \sum_{\sigma\tau}, \quad \int d^4x = \int dq \int dt.$$

The functions $\delta(q - q')$ and $\delta^4(x - x')$ denote

$$\delta(q - q') = \delta(\boldsymbol{x} - \boldsymbol{x}')\, \delta_{\sigma\sigma'} \delta_{\tau\tau'},$$

$$\delta^4(x - x') = \delta(q - q')\, \delta(t - t').$$

A numerical symbolism is often used for the coordinates and for the time:

$$(1) = (q_1, t_1), \quad \int d1 = \int d^4x_1, \quad \delta(1 - 2) = \delta^4(x_1 - x_2).$$

The element of momentum space has the following form:

$$d^3\boldsymbol{p} = (2\pi)^{-3}\, dp_x\, dp_y\, dp_z.$$

The Fourier transform of a function is defined by the relations

$$f_{\boldsymbol{p}} = \int d^3\boldsymbol{x}\, f(\boldsymbol{x}) \exp -i(\boldsymbol{p} \cdot \boldsymbol{x}),$$

$$f(\boldsymbol{x}) = \int d^3\boldsymbol{p}\, f_{\boldsymbol{p}} \exp i(\boldsymbol{p} \cdot \boldsymbol{x}).$$

Four-dimensional momenta are also introduced $p = (\boldsymbol{p}, \varepsilon)$:

$$(px) = (\boldsymbol{p} \cdot \boldsymbol{x}) - \varepsilon t,$$

$$d^4p = d^3\boldsymbol{p}\, \frac{d\varepsilon}{2\pi},$$

$$\delta^4(p) = \delta(\boldsymbol{p})\, \delta(\varepsilon).$$

A four-dimensional Fourier transform is defined by the relation

$$f(p) = \int d^4x\, f(x) \exp\,[-i(px)],$$

$$f(x) = \int d^4p\, f(p) \exp\,[i(px)].$$

The commutators and anti-commutators are, respectively,

$$[\hat{a}, \hat{b}]_- = \hat{a}\hat{b} - \hat{b}\hat{a}, \quad [\hat{a}, \hat{b}]_+ = \hat{a}\hat{b} + \hat{b}\hat{a}.$$

The author is deeply indebted to I. E. Tamm and V. L. Ginzburg, on whose initiative, and under whose guidance, the writing of this book was undertaken; to E. L. Feinberg, for reading the manuscript, and for a series of valuable discussions and comments; and to A. S. Davydov, who read the original version of the manuscript. Whilst working on the book, the author received advice and comments from many people. To all of them, particularly to V. N. Aliamovskii, G. M. Vagradov, L. V. Keldysh, E. S. Fradkin and V. V. Shmidt, the author expresses his deep gratitude.

CHAPTER 1

Essential Quantum-Mechanical Data

1. The Schrödinger Equation and the Classification of Many-Body Systems

1.1. In the final analysis, field theoretical methods in many-body quantum theory are simply based on the usual Schrödinger equation:

$$\left(i\frac{\partial}{\partial t} - \hat{H}\right)\Psi(Q, t) = 0, \tag{1.1}$$

where $Q = \{q_1, \ldots, q_N\}$ is the total number of coordinates of a system of N particles, q_i represents both the spatial x_i coordinates and the discrete coordinates (spin σ_i, isospin τ_i) of the ith particle. The Hamiltonian $\hat{H}$ is considered as a given function of the coordinates, of the momentum operators ($\hat{p} = -i\nabla$), and of the matrices acting on the discrete coordinates.

We will suppose that the Hamiltonian does not depend explicitly on time. This makes it possible to introduce the concept of stationary states of the system, corresponding to some energy E,

$$\left.\begin{aligned} \Psi(Q, t) &= \exp(-iEt)\,\Psi(Q), \\ (\hat{H} - E)\,\Psi(Q) &= 0. \end{aligned}\right\} \tag{1.2}$$

The state with the lowest energy is called the ground state. We will be considering almost exclusively stationary states.

The wave function of a system of identical fermions must be an anti-symmetric function of the coordinates. By altering the equations, it is possible to extend this property to different particles so long as we accept the restriction that they possess similar characteristics (such as mass, laws of interaction, etc.). The most important example of this type is a system composed of protons and neutrons, if we neglect electromagnetic interactions. The antisymmetry of the wave function is ensured in this case by introducing an additional, isotopic spin degree of freedom for the particle, so that both proton and neutron may be treated as different states of a single particle—the nucleon.†

1.2. The Hamiltonian of a system can be represented as the sum of two components $\hat{H} = \hat{H}_F + \hat{H}_I$. The free Hamiltonian $\hat{H}_F$ is the sum of the operators representing the kinetic energy of the particles and their energy due to interaction with an external field U:

$$\left.\begin{aligned} \hat{H}_F &= \sum_{i=1}^{N} \hat{T}_i, \\ \hat{T} &= \hat{p}^2/2M + U(q). \end{aligned}\right\} \tag{1.3}$$

The interaction Hamiltonian $\hat{H}_I$ contains terms which represent the interactions between any two particles, any three particles, and so on. Pair interaction is usually the most important:

$$\hat{H}_I = \tfrac{1}{2} \sum_{i,j=1}^{N}{}' \hat{V}_{ij} \tag{1.4}$$

(the dash after the summation sign indicates that terms for which $i = j$ are omitted).

In many-body theory the interaction potential $\hat{V}$, is taken to be a given function of the coordinates, spins, momenta, etc., of the particles.‡ The form of the function is derived either from theoretical considerations (electromagnetic interactions), or directly from experiment.

It is helpful in some cases to introduce an effective interaction potential which may be obtained by a separate calculation. For example, suppose we have a system containing two kinds of particles (say a and b) in which the a particles are not themselves

† For details, see [37] or [143]. These books also discuss the various isotopic spin equations which we will later need to use.

‡ Since the particle motions are non-relativistic, we will ignore the electromagnetic retardation effect in the interaction. Should it be necessary to allow for the retardation to first-order terms in $1/c^2$, the Breit potential, which is concerned with momentum-dependent interactions, may be used [38].

of interest, but only their influence on the motion of the b particles. If this influence is taken into account in such a way that the dynamical variables for the a particles are eliminated, the result is effectively a change in the interaction potential between the b particles.

Such a situation is encountered in the superconductivity of metals, where one is considering the system of electrons and the atoms in a lattice. The effective interaction potential between the electrons, when the lattice vibrations are allowed for, can differ radically from the original Coulomb potential. This in fact forms the basis of contemporary superconductivity theory [8, 29].

1.3. In the simplest cases the interaction potential is a function of the particle coordinates alone. The Coulomb interaction is an example of this:

$$V_{ij} = \frac{e^2}{r_{ij}}, \tag{1.5}$$

where e is the charge of the particle and $r_{ij} = |\boldsymbol{x}_i - \boldsymbol{x}_j|$.

The interaction potential may also depend on the momentum operators, spin operators, etc. This implies from the physical standpoint that the nature of the interaction between the particles (its magnitude and sign) depends upon the particular state of the particles at any time. This is true of interactions between complicated particles—atoms, molecules, nuclei—and also of the interaction between nucleons. It is this latter case in particular that we will be studying in this, and subsequent, sections.

The interaction potential thus includes a series of typical combinations of operators. The most important of these are the exchange operators for the spatial $\hat{\mathscr{P}}_{ij}^{(x)}$, spin $\hat{\mathscr{P}}_{ij}^{(\sigma)}$ and isotopic spin $\hat{\mathscr{P}}_{ij}^{(\tau)}$ coordinates, which act on the wave function to produce an interchange in the corresponding coordinates of the ith and jth particles. We may write explicitly [30],

$$\hat{\mathscr{P}}_{ij}^{(\sigma)} = \frac{1 + (\boldsymbol{\sigma}_i \cdot \boldsymbol{\sigma}_j)}{2}, \quad \hat{\mathscr{P}}_{ij}^{(\tau)} = \frac{1 + (\boldsymbol{\tau}_i \cdot \boldsymbol{\tau}_j)}{2}, \tag{1.6}$$

where $\boldsymbol{\sigma}$ and $\boldsymbol{\tau}$ are the Pauli matrices, which act on the corresponding coordinates. The operators $\hat{\mathscr{P}}^{(x)}$ may be expressed in explicit terms by means of the momentum operator for the particles. If electromagnetic effects are neglected, the operator $\hat{\mathscr{P}}^{(x)}$ reduces to $-\hat{\mathscr{P}}^{(\sigma)}\hat{\mathscr{P}}^{(\tau)}$ (due, in fact, to the antisymmetry of the wave function)

$$\hat{\mathscr{P}}\Psi = \hat{\mathscr{P}}^{(x)}\hat{\mathscr{P}}^{(\sigma)}\hat{\mathscr{P}}^{(\tau)}\Psi = -\Psi.$$

The general expression for the potential of the exchange forces can be written in the form

$$\hat{V} = V_W(r) + V_B(r)\,\hat{\mathscr{P}}^{(\sigma)} + V_H(r)\,\hat{\mathscr{P}}^{(\tau)} - V_M(r)\,\hat{\mathscr{P}}^{(\sigma)}\,\hat{\mathscr{P}}^{(\tau)}, \tag{1.7}$$

where the individual terms are called the Wigner force, the Bartlett force, the Heisenberg force and the Majorana force, respectively. One may distinguish especially the Serber forces which contain the characteristic combination

$$\hat{S} = \tfrac{1}{2}(1 + \hat{\mathscr{P}}^{(x)}) = \tfrac{1}{2}(1 - \hat{\mathscr{P}}^{(\sigma)}\hat{\mathscr{P}}^{(\tau)}). \tag{1.8}$$

These forces differ from zero only for states in which the orbital momentum of the relative particle motion has an even value.

Besides the specific nuclear forces which do not depend on the nature of the interacting nucleons, it is essential to take into account the Coulomb interactions occurring only between protons. To do this, it is necessary to introduce the operators $\zeta_{(p)}$ and $\zeta_{(n)}$, which specify the proton and neutron states:

$$\zeta_{(p)} = \tfrac{1}{2}(1 + \tau_3), \quad \zeta_{(n)} = \tfrac{1}{2}(1 - \tau_3), \tag{1.9}$$

where τ_3 is the appropriate Pauli matrix. The potential for the Coulomb interaction between nucleons has the following form:

$$V_{(C)ij} = \zeta_{(p)i}\zeta_{(p)j}\frac{e^2}{r_{ij}}. \tag{1.10}$$

The nucleon-interaction potential also contains terms, corresponding to tensor and spin–orbit forces, which depend on the spin operators and on the operators for the angular momentum. We will not require explicit expressions for the potentials of these forces.

1.4. It must be remembered that any direct data on the interaction potential between the members of a multi-nucleon system are, of necessity, absent. That is, a theoretical solution of this problem can only be obtained from some meson theory for such systems and at the present no suitable theory exists. Furthermore, experimental data from nuclear physics on nucleon interactions refer almost exclusively to an isolated pair of particles (i.e. nucleon–nucleon scattering, deuteron data, etc.).

Since the distance between particles in any real multi-nucleon system is small, the supposition that the nucleons in such a system have the same interaction potential as isolated nucleons seems, from an *a priori* viewpoint, improbable. One might expect multi-particle interaction forces between nucleons to appear. It should be added, however, that there are no convincing, quantitative estimates of the part played by multi-particle forces at present, and the whole question still remains open.

It must be mentioned in this connection that, besides the phenomenological approach to nuclear theory, there is also another possible approach based on the assumption that multi-particle forces are small. In this case the characteristics of multi-nucleon systems may be calculated using an empirical potential for the interaction between isolated particles. If the results of the calculations are compared with the experimental data, it is possible, in principle, to determine the limits within which the assumption applies.

Comparisons of this sort have been made for many nuclear characteristics: they have not yet led to any obvious contradictions. We may suppose, therefore, that a microscopic theory of the nucleus, based on an empirical value for the interaction potential of a pair, can claim to provide a qualitative, and often also a quantitative, description of the nucleus (at least for low energies).

Hence, in our future consideration of a multi-nucleon system, we will use an empirical interaction. To simplify the calculations, the potential is chosen to have the following approximate form [39]:†

$$\hat{V} = \hat{V}_{(a)} + \hat{V}_{(c)} + \hat{V}_{(C)}. \tag{1.11}$$

† More detailed information on the general form of nuclear forces can be found in [37, 40, 143]. A summary of various semi-empirical expressions for $\hat{V}$ is given in [41, 42, 73].

Here $\hat{V}_{(a)}$ is the main part of the attraction describing the inter-particle interaction. It is assumed to be a Serber force:

$$\hat{V}_{(a)} = V_{(a)}(r)\,\hat{S}.$$

The function $V_{(a)}(r)$ can be approximated by a square well potential:

$$V_{(a)}(r) = -V_0 \quad (c < r < a); \quad V_{(a)}(r) = 0 \quad (r < c, r > a).$$

Numerically,

$$V_0 = \frac{\pi^2}{4M(a-c)^2} \approx 28 \text{ MeV},$$

$$a \approx 2{\cdot}3 \text{ fm}, \quad c \approx 0{\cdot}4 \text{ fm} \quad (1 \text{ fm} = 1 \text{ fermi} = 10^{-13} \text{ cm}).$$

Intense Wigner-type ("hard-core") repulsion forces come into play at small distances [43]:

$$V_{(c)}(r) = \infty \quad (r < c); \quad V_{(c)}(r) = 0 \quad (r > c).$$

Finally, $\hat{V}_{(C)}$ is the Coulomb interaction between protons [see (1.10)].

This expression for $\hat{V}$ fails to represent several of the peculiarities of real nuclear forces, such as the difference in the interaction for singlet and triplet spin states (for the proton–neutron system with even values of the orbital momentum l), the finite value of the long-range potential for odd l, the important role of tensor and spin–orbit forces, etc.

Nevertheless, one should note that these effects are important, generally speaking, only for the two-body problem. When the integral characteristics of a multi-nucleon system are considered, the averaging process seems to reduce the significance of the effects. The potential (1.11), which is both simple and suitable for calculations, may reflect, therefore, certain averaged properties of the nuclear forces, and is hence of use in considering a whole range of problems in the theory of nuclear matter. The effects enumerated above can in fact be calculated without major complication by the methods to be discussed below.

1.5. The potential $V_{(c)}$ contains an explicit infinity, which represents a major, formal objection to the expression (1.11). Although this gives rise to no physical problems (since it simply represents the impossibility of nucleons approaching one another to less than some distance c), the corresponding calculations become highly complicated. If the relative momentum k of the colliding nucleon is sufficiently small,

$$kc \ll 1, \tag{1.12}$$

it seems possible to replace the potential $V_{(c)}$ by some other equivalent potential which does not contain an explicit infinity [44, 45].

With this end in view, we will consider several general problems on scattering between two isolated particles, where the interaction potential $V(r)$ between them differs from zero only in the region $r < c$ (this potential may, in particular, coincide with $V_{(c)}$). The Schrödinger equation in the centre-of-mass system may be written in the form

$$(\nabla^2 + k^2)\,\psi(\boldsymbol{r}) = MV(\boldsymbol{r})\,\psi(\boldsymbol{r}), \tag{1.13}$$

where M is the mass of the particle. Its solution in the region $r > c$ satisfies the Schrödinger equation for a free particle and can be written in the form

$$\psi(\boldsymbol{r}) = \exp i(\boldsymbol{k}\cdot\boldsymbol{r}) + \left[a_0(k) + a_1(k)\frac{(\boldsymbol{k}\cdot\nabla)}{ik^2} + \cdots\right]\frac{\exp(ikr)}{r}. \tag{1.14}$$

Here $a_l(k)$ is the partial wave amplitude for scattering with an angular momentum l; we are considering only the s- and p-waves which are of predominant importance when the requirement (1.12) is satisfied.

When $r < c$, the solution of eqn. (1.13) will obviously differ radically from (1.14). However, if (1.12) is obeyed, the exact details of what happens to the wave function within the small region where the forces act cannot be of great significance. From the uncertainty principle, the effective distance between the particles is of the order $1/k \gg c$. The result is that the particles spend only a relatively short time in the region we are considering. It may therefore be supposed that (1.14) is true not only for $r > c$, but also for $r < c$. Similarly, one may look for an interaction potential $\hat{V}_{\text{eff}}$ which produces an expression of the type (1.14) applicable over the whole range of r.

If the operator $\nabla^2 + k^2$ is applied to (1.14), then, remembering that $\nabla^2 r^{-1} = -4\pi\delta(\boldsymbol{r})$, the left-hand side of eqn. (1.13) may be written as

$$(\nabla^2 + k^2)\,\psi(\boldsymbol{r}) = -4\pi\left(a_0 + a_1\frac{(\boldsymbol{k}\cdot\nabla)}{ik^2} + \cdots\right)\delta(\boldsymbol{r}). \quad (1.15)$$

If this is compared with the right-hand side of (1.13), the required form of the potential $\hat{V}_{\text{eff}}$ can be found.

We will start by considering *s*-scattering, which is the most important. Using the easily verified relationship,

$$\delta(\boldsymbol{r})\frac{\partial}{\partial r}(r\psi(\boldsymbol{r})) = (1 + ika_0)\,\delta(\boldsymbol{r}),$$

and comparing the right-hand sides of eqns. (1.15) and (1.13), it is not difficult to arrive at the following expression for the required potential:

$$\hat{V}_{\text{eff}}(\boldsymbol{r}) = -\frac{4\pi}{Mk}\tan\delta_0(k)\,\delta(\boldsymbol{r})\left(1 + r\frac{\partial}{\partial r}\right), \quad (1.16)$$

where $\delta_0(k)$ represents the phase of the *s*-scattering

$$a_0 = \frac{1}{2ik}[\exp(2i\delta_0) - 1].$$

For the potential $V_{(c)}$ the phase $\delta_0(k) = -kc$ and, to the first order in kc, we may write

$$\hat{V}_{\text{eff}}(\boldsymbol{r}) = \frac{4\pi c}{M}\delta(\boldsymbol{r})\left(1 + r\frac{\partial}{\partial r}\right). \quad (1.17)$$

The potential, defined in this equation, is called the pseudopotential, and is much more convenient for purposes of calculation than $V_{(c)}$. The operator $1 + r(\partial/\partial r)$, which it contains, reduces to unity when acting on functions regular at $r = 0$, and to zero when acting on a function with a singularity of the type $1/r$. Owing to this $1 + r(\partial/\partial r)$ operator, the pseudopotential of (1.17) can be used for higher order calculations of perturbation theory, where as the Fermi pseudopotential only applies to calculations in the first Born approximation.

It should be emphasized that (1.16) was deduced only on the basis of (1.12), which limits the radius of action of the forces as compared with the wave length of the relative particle motion, $1/k$. It was not assumed that the scattering amplitude is small relative to $1/k$. The expression (1.16) may, therefore, also be used to describe the situation at resonance, when $\delta_0 \approx \frac{1}{2}\pi$.

We will now determine the corrections to the pseudopotential, eqn. (1.17), of order $(kc)^2$ (it seems that linear terms are absent in this case). We must first consider the next term in powers of (kc) of the factor $\tan \delta_0(k)$ in (1.16). To the required approximation, we may substitute $k^2 \to -\nabla^2$, and omit the operator $1 + r(\partial/\partial r)$ (in agreement with (1.15)).† This gives

$$\delta_0 \hat{V}_{\text{eff}}(\boldsymbol{r}) = -\frac{4\pi c^3}{3M}\,\delta(\boldsymbol{r})\,\nabla^2. \tag{1.18}$$

Here, and later,

$$\nabla = \frac{\nabla_{x_i} - \nabla_{x_j}}{2}.$$

Besides this, correction terms of the order considered may arise from p-scattering. Substituting the expression

$$\begin{aligned}(\boldsymbol{k}\cdot\nabla)\,\delta(\boldsymbol{r}) &= -i\sum_\alpha \nabla_\alpha[\delta(\boldsymbol{r})\,\nabla_\alpha \exp i(\boldsymbol{k}\cdot\boldsymbol{r})]\\ &\simeq -i\sum_\alpha\,[\nabla_\alpha\delta(\boldsymbol{r})\,\nabla_\alpha + \delta(\boldsymbol{r})\,\nabla^2]\,\psi(\boldsymbol{r})\end{aligned}$$

into eqn. (1.15), and remembering that $a_1(k) = -k^2c^3$ for the potential $V_{(c)}$, we have

$$\delta_1 \hat{V}_{\text{eff}}(\boldsymbol{r}) = -\frac{4\pi c^3}{M}\left[\sum_\alpha \nabla_\alpha\delta(\boldsymbol{r})\cdot\nabla_\alpha + \delta(\boldsymbol{r})\,\nabla^2\right]. \tag{1.19}$$

Finally, we must discuss the error involved in assuming that (1.14) may be substituted for the accurate wave function. It is obvious that the error will only affect the terms of higher order in kc. An analysis [46] shows that the relative correction is at least of order $(kc)^3$, and may therefore be ignored.

† Equation (1.18) will in the future be used to the first Born approximation, where this operator acts on a function which is regular at $r = 0$.

1.6. We can combine the quantities which enter into the Schrödinger equation, in the general case, into two independent, dimensionless parameters. One—the interaction parameter α—is defined as the ratio of the average interaction energy of a pair of particles to their average kinetic energy. The other—the condensation parameter η—is the ratio of the effective radius of action of the forces to the average distance between particles.

Many-body systems may be classified in terms of the magnitudes of these parameters; that is, by the interaction force between the particles, and by the degree of condensation of the system. Interaction is large, or small, if α is large, or small, compared with unity. Similarly, a system is classified as dense, or dilute, if η is respectively large, or small. This sort of classification is very useful when it comes to choosing a suitable approximate method of describing a system (see § 16).

It must be emphasized that a low value for α in dense systems in no way signifies that interaction effects as a whole are small throughout the system. The point is that in dense systems many particles will lie within the radius of action of the forces, so the total interaction energy may reach an appreciable value, and can even exceed the total kinetic energy.

We will now consider the two parameters we have introduced in specific cases. Systems with Coulomb and with short-range forces will require separate investigations.

A system of particles held together by short-range forces is characterized by the mass of the particles M, the average value of the interaction potential V_0† and the range R of the forces. One should add also the average number density of the particles $\bar{\varrho}$ or the related average distance between particles $d = (3/4\pi\bar{\varrho})^{1/3}$. If the system is in a gas-like state, and is subject to a constraining compressive force, then d is fixed independently, e.g. by the boundary conditions. If the system is in a self-condensed state (e.g. the atomic nucleus), then d is, in principle, expressible in terms of M, V_0 and R.

† The "hard-core" potential constitutes a special case, which will be considered later.

A system with Coulomb interactions is characterized by fewer parameters: M, d and e (the charge of the particle). The concept of a force-range disappears for a Coulomb force. We may note, incidentally, that an effective radius of interaction can be introduced via the Debye screening length (see §§ 16 and 27) [4, 47]:

$$R_0 \sim v_0/\omega_L, \tag{1.20}$$

where v_0 is the characteristic particle velocity; $\omega_L = (4\pi\bar{\varrho}e^2/M)^{1/2}$ is the plasma frequency.

We will consider throughout that a system with short-range forces is at zero temperature.†

On the other hand, for a system with Coulomb forces, the alternative limiting case of high temperatures is also of considerable interest.

1.7. We will consider first a system with short-range forces. As we have remarked, the condensation parameter η is determined by the ratio

$$\eta = R/d. \tag{1.21}$$

Suppose the interaction potential contains two forces with differing ranges, then the system may be dense relative to one of the ranges, and dilute relative to the other. This occurs, for example, in the nuclear force potential (see § 1.4), where the attraction forces have a range $a \approx 2{\cdot}3$ fm, and the repulsion forces have $c \approx 0{\cdot}4$ fm. The average distance between the particles in the nucleus is approximately 1·3 fm. The atomic nucleus is thus a fairly condensed system as regards the attractive forces, and fairly dilute as regards the repulsive ones.

We will now make a few general remarks about models of a system with short-range repulsive forces and long-range attractive forces. This model can describe not only nuclear matter, but also certain atomic and molecular systems, e.g. liquid He^3 (see the paper by Brueckner in ref. [14]). If there are no attractive forces, the parameter η for the repulsive forces can be arbitrarily small. The

† "Hot" systems are of no particular interest here, since their thermal energy is of little significance. The total temperature of an excited nucleus, for example, is only of the order 1 MeV.

attractive forces in the model we are considering, however, remove the arbitrariness of this parameter. This is because in the self-condensed state (which we are now discussing), the average distance between particles is determined by the equilibrium between the attractive and repulsive forces, and it decreases as the range of the repulsive forces decreases. The parameter η for these forces cannot, therefore, be considered as necessarily very small. At the same time, the corresponding parameter for the attractive forces can be arbitrarily large.

We will now estimate the interaction parameter α. In dense systems the average interaction energy between pairs of particles is of the order V_0, their average kinetic energy is of the order $p_0^2/2M \sim (Md^2)^{-1}$, where $p_0 \sim 1/d$ is some characteristic value of the momentum. Hence

$$\alpha \sim MV_0 d^2. \tag{1.22}$$

In dilute systems the estimate of the kinetic energy remains the same, but the effective interaction energy is $(d/R)^3$ times smaller. This is because the particles are within the range of the forces for only a small part of the time. Therefore,

$$\alpha \sim MV_0 R^3 d^{-1}. \tag{1.23}$$

In dilute systems, another parameter α' is of basic importance, where

$$\alpha' \sim MV_0 R^2 \sim \alpha/\eta. \tag{1.24}$$

This appears in the description of two-particle scattering and determines the applicability of the Born approximation [31].

If α' is not small relative to unity, another relation for α may be used which makes it possible to consider the "hard-core" potential, and which reduces to (1.23) as $\alpha' \to 0$. Writing V_0R^3 in the form $\int V(\mathbf{r})d^3\mathbf{r}$, and substituting the expression for the pseudopotential† [see (1.16)], we find

$$\alpha \sim l/d, \tag{1.23'}$$

† The requirement that a system be relatively dilute is, in fact, a restatement of (1.12), since $k \sim 1/d$; thus the transition to the pseudopotential is justified in this case.

where $l = -\tan \delta(k)/k$ is the scattering length. For the hard-core potential $l \approx c$, and

$$\alpha \sim \eta \ll 1. \tag{1.25}$$

For resonance scattering, $\delta \approx \frac{1}{2}\pi$, $l \gg d$, and we have $\alpha \gg 1$; interaction between particles is now of considerable importance, even though the system is dilute.

We will now analyse the nuclear force potential in terms of the relations we have derived. The relation (1.22), when applied to the attractive forces between nucleons in the atomic nucleus, gives a value for α close to unity. This interaction must therefore be considered as of intermediate intensity. A system of nucleons compressed-due to some external constraint (such as is considered in astrophysics) may already be classified as a system with weak interactions. For example, a system with an density of 10^{14} g/cm^3 has $\alpha \sim \frac{1}{2}$.

The hard-core nuclear potential also belongs to the class of weak interactions [see (1.25)]; however, if the system is compressed, the corresponding value of α tends to unity.

A special case is presented by very dilute multi-nucleon systems, where the average distance between the particles exceeds the radius of action of the attractive forces. The interaction in such systems is of a resonance type: the presence of a low energy level (real or virtual) results in a large scattering length ($l \sim 20$ fm). The corresponding α is therefore larger. There is some reason to think that there is a similar situation of interest for neutron matter (see §§ 7 and 27) [48].

1.8. We will next consider a system with Coulomb interactions at zero temperature.

It is convenient to introduce from the start the radius of the Bohr orbit: $a_0 = (Me^2)^{-1}$. Putting $p_0 \sim Mv_0 \sim 1/d$, we obtain the following relation for the Debye length from (1.20):

$$R_0 \sim (a_0 d)^{1/2}, \tag{1.26}$$

whence the condensation parameter is

$$\eta \sim (a_0/d)^{1/2}. \tag{1.27}$$

A system may thus be considered dense, or dilute, depending on the ratio of the average distance between particles to the radius of the Bohr orbit.

An expression for α for dense Coulomb systems can be obtained from (1.22), with e^2/d substituted for V_0. This gives

$$\alpha \sim \frac{d}{a_0} \sim \frac{e^2}{v_0} \sim (a_0 p_0)^{-1}. \tag{1.28}$$

The expression $\alpha \sim e^2/v_0$ is frequently used in the literature.

We will not consider dilute Coulomb systems with $d \gg a_0$. The Coulomb interaction is so large in such systems that a transition to a condensed state occurs [49, 50]. From a physical standpoint this may be related to the fact that the virial theorem cannot be fulfilled in the gaseous state as the kinetic energy is much smaller than the Coulomb energy. The transition to the condensed state restores the energy balance, since the oscillations of the particles about their equilibrium positions necessarily lead to an increase in their kinetic energy. Hence, an increase in the distance between the particles actively encourages the appearance of short-range order in the system.

We will next estimate the value of α for several real systems. The atomic nucleus, for example, is a system with weak Coulomb interactions; d is approximately equal to 1 fm, and the Bohr radius of the proton to 30 fm. For the conduction electrons of a metal, $\alpha \sim 1$. In an atom with Z electrons, $\alpha \sim Z^{-2/3}$, so that it is small for large Z (see § 5).

It can be seen from what has been said that α and η are not independent for a Coulomb system: thus a dense system is simultaneously one with weak interactions. This is also the case at high temperatures. Here, the kinetic energy corresponds to kT, and $v_0 \sim (kT/M)^{1/2}$, whence

$$R_0 \sim \left(\frac{kTd^3}{e^2}\right)^{1/2}, \quad \eta \sim \left(\frac{kTd}{e^2}\right)^{1/2}, \quad \alpha \sim \left(\frac{e^2}{dkT}\right). \tag{1.29}$$

It may be noted from the definition of η that an increase in the distance between the particles corresponds to the system becoming not more dilute, but more dense. This somewhat paradoxical conclusion is due to the fact that the Debye length grows rapidly as the distance between the particles increases.

2. Basic Ideas of Representation Theory

2.1. Other methods, or other representations besides the one described in § 1, may be used to describe a quantum system. It is, of course, essential that any change to a new representation keeps the observed quantities—the eigenvalues of the operators, the average values, the transition probabilities, etc.—invariant.

The transition from one representation to another is made via a unitary transformation operator $\hat{U}$. The corresponding change in the wave function is given by

$$\Psi' = \hat{U}\Psi \tag{2.1}$$

and the change in the operator of any physical quantity $\hat{\alpha}$ by

$$\hat{\alpha}' = \hat{U}\hat{\alpha}\hat{U}^{-1}. \tag{2.2}$$

The transformed quantities are denoted by a dash.

It can easily be seen that the matrix elements of an operator do not, in general, change during the transformation to a new representation:

$$\langle\Psi_1'|\hat{\alpha}'|\Psi_2'\rangle = \langle\Psi_1|\hat{\alpha}|\Psi_2\rangle. \tag{2.3}$$

As a result, the observable quantities, which we have previously enumerated, also remain invariant.

2.2. We can look for a new representation in the first place by changing the way in which the wave function and the operators depend on the time. It is possible to transfer this dependence, in part or in full, from the wave function to the operators. In the initial equation (1.1), the wave function depended on the time, but the operators did not. This is called the Schrödinger representation.

In the Heisenberg representation, on the other hand, the wave function is not time dependent, but the operators are. We will denote quantities in this representation by the suffix H. Equating the time derivative of $\Psi_H = \hat{U}\Psi$ to zero, and using eqn. (1.1), we obtain

$$i\,\partial\hat{U}/\partial t = -\hat{U}\hat{H}.$$

Since the two representations should coincide at $t = 0$, we have $\hat{U} = \exp(i\hat{H}t)$, so that

$$\Psi_H = \exp(i\hat{H}t)\,\Psi,$$

$$\hat{\alpha}_H = \exp(i\hat{H}t)\,\hat{\alpha}\,\exp(-i\hat{H}t).$$

If we put $\hat{\alpha} = \hat{H}$ in the second equation, we find $\hat{H}_H = \hat{H}$. The equation of motion for an operator in the Heisenberg representation may be readily obtained by differentiating the relation for $\hat{\alpha}_H$.

$$i\frac{\partial \hat{\alpha}_H}{\partial t} = [\hat{\alpha}_H, \hat{H}_H]_- = [\hat{\alpha}_H, \hat{H}]_- \tag{2.4}$$

It follows that $\partial \hat{H}_H/\partial t = 0$, hence the operator $\hat{H}_H$ can be taken at any time. If the operator $\hat{U}$ is applied to the wave function for a stationary state, it is easy to see that Ψ_H then coincides with the time-independent part of the wave function $\Psi(Q)$ [see eqn. (1.2)].

We need, for future use, a general expression for the matrix element of an operator between the stationary states of a given system, i.e. the quantity:

$$M_{(mn)} = \langle \Psi_{H(m)} | \hat{\alpha}_H | \Psi_{H(n)} \rangle$$

where $\Psi_{H(n)}$ is an eigenfunction of the Hamiltonian $\hat{H}\Psi_{H(n)} = E_{(n)}\Psi_{H(n)}$. The explicit time-dependence of $M_{(mn)}$ can be found without difficulty. We differentiate $M_{(mn)}$ with respect to t and use (2.4) to obtain

$$i\frac{\partial M_{(mn)}}{\partial t} = \langle \Psi_{H(m)} | [\hat{\alpha}_H, \hat{H}]_- | \Psi_{H(n)} \rangle = (E_{(n)} - E_{(m)})\, M_{(mn)}.$$

We have, finally,

$$\langle \Psi_{H(m)} | \hat{\alpha}_H | \Psi_{H(n)} \rangle = \langle \Psi_{H(m)} | \hat{\alpha} | \Psi_{H(n)} \rangle \exp\left[-i(E_{(n)} - E_{(m)})\, t\right]. \tag{2.5}$$

Here $\hat{\alpha}$ is the operator in the Schrödinger representation.

2.3. If the time dependence is partially transferred from the wave function to the operator, we are led to the interaction representation. We will express the Hamiltonian $\hat{H}$ as a sum of $\hat{H}_0 + \hat{H}'$, where $\hat{H}_0$ is to be identified with the free Hamiltonian $\hat{H}_F$ and $\hat{H}'$ with the interaction Hamiltonian $\hat{H}_I$ (we will employ a different division of $\hat{H}$ into $\hat{H}_0$ and $\hat{H}'$ in Chapter 3).

We will require that the time dependence of the wave function is determined by the operator $\hat{H}'$, and that $\hat{H}_0$ determines the time variation of the operators. As the corresponding equations show, this is equivalent to a representation which is a Heisenberg representation with respect to $\hat{H}_0$ and a Schrödinger representation with respect to $\hat{H}'$. If there is no interaction, then we obviously have a pure Heisenberg representation.

Indicating quantities in this interaction representation by the suffix "Int", the Schrödinger equation may be written in the form:

$$i\frac{\partial \Psi_{\text{Int}}}{\partial t} = \hat{H}'_{\text{Int}} \Psi_{\text{Int}}. \tag{2.6}$$

Using eqns. (2.1) and (2.2) together with (1.1), we have

$$i\frac{\partial \hat{U}}{\partial t} = -\hat{U}\hat{H}_0,$$

whence

$$\hat{U} = \exp(i\hat{H}_0 t).$$

Therefore

$$\left.\begin{aligned} \Psi_{\text{Int}} &= \exp(i\hat{H}_0 t)\,\Psi \\ \hat{\alpha}_{\text{Int}} &= \exp(i\hat{H}_0 t)\,\hat{\alpha}\,\exp(-i\hat{H}_0 t) \\ i\frac{\partial \hat{\alpha}_{\text{Int}}}{\partial t} &= [\hat{\alpha}_{\text{Int}}, \hat{H}_0]_- \end{aligned}\right\} \tag{2.7}$$

We next find the explicit time dependence of the matrix element:

$$\langle \Psi_{0(m)} | \hat{\alpha}_{\text{Int}} | \Psi_{0(n)} \rangle.$$

Repeating the line of reasoning that led us to (2.5), we obtain

$$\langle \Psi_{0(m)} | \hat{\alpha}_{\text{Int}} | \Psi_{0(n)} \rangle = \langle \Psi_{0(m)} | \hat{\alpha} | \Psi_{0(n)} \rangle \exp[-i(E_{0(n)} - E_{0(m)})\,t], \tag{2.8}$$

where $\Psi_{0(n)}$ is the solution of the equation

$$\hat{H}_0 \Psi_{0(n)} = E_{0(n)} \Psi_{0(n)}.$$

It is also easy enough to find an operator to effect the transition from the interaction representation to the Heisenberg representation. Performing the intermediate transformation to the Schrödinger representation, we have†

$$\hat{U} = \exp(i\hat{H}t)\exp(-i\hat{H}_0 t). \tag{2.9}$$

† The exponentials in this equation may not be combined since the operators $\hat{H}_0$ and $\hat{H}$ do not necessarily commute.

The S-matrix (see Chapter 3) provides another expression for this operator.

It is useful in transforming from the Schrödinger representation to the Heisenberg and interaction representations to remember that the transform of a product of operators is equal to the product of the transformed operators

$$\hat{U}\hat{a}\hat{b}\ldots\hat{z}\hat{U}^{-1} = \hat{U}\hat{a}\hat{U}^{-1}\hat{U}\hat{b}\hat{U}^{-1}\ldots$$

or

$$(\hat{a}\hat{b}\ldots\hat{z})' = \hat{a}'\hat{b}'\ldots\hat{z}'. \tag{2.10}$$

In particular if we transform a relationship of the form $[\hat{a}, \hat{b}]_+ = c$, where c is, in general, a number, not an operator (a c-number) we find that

$$[\hat{a}'(t), \hat{b}'(t)]_+ = \hat{U}(t)\, c\hat{U}(t)^{-1} = c, \tag{2.11}$$

where the transformed operators refer to the same instant of time. Thus a relation of this type (called a commutation relation) preserves its form in all representations, so long as the operators are calculated at the same time.

2.4. Another possible change in the method of describing a system can be brought about by transforming from the Q-coordinates to new dynamical variables λ. We will suppose $\Phi_\lambda(Q)$ to be a complete orthonormal set of functions, corresponding to particular values of these variables.

We will expand the wave function $\Psi(Q, t)$ in this system:

$$\Psi(Q, t) = \sum_\lambda \Psi(\lambda, t)\, \Phi_\lambda(Q). \tag{2.12}$$

The coefficients in this expansion

$$\Psi(\lambda, t) = \int dQ\, \Phi_\lambda^*(Q)\, \Psi(Q, t) \tag{2.13}$$

play the part of a wave function in the new representation. In particular the square of its modulus corresponds to the probability of observing a given value λ of the new dynamical variable. The symbol $\int dQ$ signifies integration over all spatial coordinates, and summation over all discrete coordinates.

An explicit form for the corresponding transformation operator $\hat{U}$ might be obtained by comparing (2.13) with (2.1). It is simpler, however, to start directly from (2.3), which provides an expression for the matrix element of a transformed operator:

$$\langle\lambda'|\hat{\alpha}'|\lambda\rangle = \int dQ\, \Phi^*_{\lambda'}(Q)\, \hat{\alpha}\Phi_\lambda(Q). \tag{2.14}$$

2.5. The new dynamical variables may be chosen from the characteristic properties of particles in a system: such as their coordinates, momenta, energies, etc. We are thus led towards the so-called configuration representation (using this term with the widest possible connotation).

The function $\Phi_\lambda(Q)$ can then be represented as the product of single-particle functions $\chi_{\lambda_i}(q_i)$, each of which refers to one of the particles in the system (λ_i is the value of the corresponding dynamical variable for the ith particle). In particular if the λ-variables denote the coordinates of a particle, then $\chi_{\lambda_i}(q_i) = \delta(q_i - q_{0i})$ is the wave function of a state for a specific value of the coordinate equal to q_{0i}. In the momentum representation, with $\lambda_i = \boldsymbol{p}_i$,†

$$\chi_{p_i}(\boldsymbol{x}_i) = \frac{1}{\sqrt{\Omega}} \exp i\,(\boldsymbol{p}_i \cdot \boldsymbol{x}_i)$$

The transformed wave function $\Psi(\boldsymbol{p}_1 \ldots \boldsymbol{p}_N, t)$ is the Fourier transform of the function $\Psi(Q, t)$.

There is another possible choice of the dynamical variables which leads to the important occupation number (or second quantization) representation. This will be considered in greater detail in the next section.

For a transformation to this representation it is necessary to give an arbitrary set (basis) of single-particle wave functions $\chi_{\lambda_1}(q) \ldots, \chi_{\lambda_\nu}(q) \ldots$, where the numerical suffix denotes the number of the state in this set, and λ, as before, indicates the operator which has the functions χ as its eigenfunctions. The dynamical variable here is not the quantity λ but the number of particles in each of the states of the chosen basis. So long as the particles in the system are identical, this approach will give a complete description.

† Discrete coordinates have been omitted here for simplicity. Ω is the normalizing volume.

From what has been said, it can be seen that a change to a new representation affects different aspects of the description of a system. In particular the character of the time dependence of the quantities considered may be altered, thus giving rise to the Schrödinger, Heisenberg and interaction representations. There also exist various possibilities for specifying the nature of the dynamical variables. Thus the configuration representation may use those quantities which characterize the particles in the given system, whilst the occupation-number representation uses the occupation number of the basic states.

Finally, there are various ways of choosing the particle characteristic (denoted above by λ) which determines the state of the system (in the configuration representation) or the basis of states (in the occupation-number representation). We may thus differentiate between the coordinate, momentum and other representations.

The description of a quantum system should therefore begin with the choice of some definite combination of these three types of representation. For example, a combination of the coordinate and configuration representations, and of the Schrödinger representation was used in the first section; in Chapter 3 a combination of the momentum, interaction and occupation-number representations will be applied extensively.

3. The Occupation-Number Representation

3.1. The dynamical variables in this representation are the occupation numbers n_ν,† which are determined by the number of particles in each state, χ_ν, of the chosen basis set. The latter is fixed by the choice of the arbitrary, complete, orthonormal set of single-particle functions $\chi_\nu(q)$ ($\nu = 1, 2, \ldots$). In particular χ_ν may be chosen to be the set of eigenfunctions of the operator $\hat{T}$ [see (1.3)], with eigenvalues ε_ν.

$$\hat{T}\chi_\nu = \varepsilon_\nu \chi_\nu. \tag{3.1}$$

This choice corresponds to the energy representation.

† Similarly, we will write $\Phi_n(Q)$ instead of $\Phi_\lambda(Q)$.

The transition to this new representation should begin with the construction of a set of functions $\Phi_n(Q)$, $(n = n_1, n_2, \ldots)$, corresponding to n_1 particles in the state χ_1, n_2 particles in the state χ_2, etc. Examining all possible combinations of whole numbers n_ν which satisfy the conditions (for Fermi particles), $n_\nu = 0, 1$, and $\sum_\nu n_\nu = N$, we arrive at the following complete orthonormal set of functions:

$$\Phi_n(Q) = \left(\frac{n_1!\, n_2! \ldots}{N!}\right)^{1/2} S\chi_{\nu_1}(q_1) \ldots \chi_{\nu_N}(q_N), \tag{3.2}$$

where, amongst the suffixes $\nu_1 \leqq \nu_2 \leqq \nu_3 \ldots$ there are n_1 with $\nu = 1$, n_2 with $\nu = 2$, etc. The symbol S signifies antisymmetrization in the q-coordinates.

In particular, for $N = 2$,

$$\Phi_{2,0,0\ldots} = \Phi_{0,2,0\ldots} = 0,$$

$$\Phi_{1,1,0\ldots} = \frac{1}{\sqrt{2}} [\chi_1(q_1)\, \chi_2(q_2) - \chi_1(q_2)\, \chi_2(q_1)], \quad \text{etc.}$$

3.2. It is easy to obtain an expression for the wave function in the new representation using the general equation (2.13):

$$\Psi(n, t) = \int dQ\, \Phi_n^*(Q)\, \Psi(Q, t), \tag{3.3}$$

which acts as the probability amplitude for observing n_1 particles in the state χ_1, n_2 particles in the state χ_2, etc. In particular if we take $\Psi(Q, t)$ in this equation to be the function $\Phi_{n_0}(Q)$, $(n_0 = n_{01}, n_{02}, \ldots)$ itself, we arrive at the wave function of a state with values of the occupation numbers equal to n_0:

$$\Psi_{n_0}(n) = \int dQ\, \Phi_n^*(Q)\, \Phi_{n_0}(Q) = \delta_{n, n_0}. \tag{3.4}$$

The real wave function (3.3) can always be represented as a combination of the functions (3.4) with time-dependent coefficients:

$$\Psi(n, t) = \sum_{n_0} C(n_0, t)\, \Psi_{n_0}(n),$$

where

$$C(n_0, t) = \Psi(n_0, t).$$

The vacuum state, corresponding to a complete absence of particles, plays a special part in this theory. The wave function of this state, for which $n_{01} = n_{02} = \cdots = 0$, is denoted by $\Psi_{\text{vac}}(n)$. It is normalized to unity like the other functions in (3.4) which correspond to fixed occupation numbers.

3.3. Before formulating the rules for the construction of operators in this new representation (which must act on functions of the occupation numbers), we must introduce the so-called annihilation and creation operators, $\hat{A}_\nu$ and $\hat{A}_\nu^+$. Their action on any arbitrary function of the occupation numbers is defined in the following way:

$$\left.\begin{aligned} \hat{A}_\nu \Psi(\ldots, n_\nu, \ldots) &= (-1)^\alpha \sqrt{(1+n_\nu)}\, \Psi(\ldots, n_\nu+1, \ldots), \\ \hat{A}_\nu^+ \Psi(\ldots, n_\nu, \ldots) &= (-1)^\alpha \sqrt{(2-n_\nu)}\, \Psi(\ldots, n_\nu-1, \ldots). \end{aligned}\right\} \quad (3.5)$$

An occupation number n_λ, with $\lambda \neq \nu$, is not altered at all; the index α is equal to $\sum_{\lambda=1}^{\nu-1} n_\lambda$.

It follows from this definition that $\hat{A}_\nu$ is an annihilation operator and $\hat{A}_\nu^+$ a creation operator for a particle in the state χ_ν. The action of these operators on the wave function $\Psi_{n_0}(n)$ produces expressions which are correspondingly proportional to†

$$\delta_{n_1, n_{01}} \cdots \delta_{n_\nu+1, n_{0\nu}} \cdots \equiv \delta_{n_1, n_{01}} \cdots \delta_{n_\nu, n_{0\nu}-1} \cdots$$

$$\delta_{n_1, n_{01}} \cdots \delta_{n_\nu-1, n_{0\nu}} \cdots \equiv \delta_{n_1, n_{01}} \cdots \delta_{n_\nu, n_{0\nu}+1} \cdots$$

Thus

$$\hat{A}_\nu \Psi_{n_{01}, \ldots, n_{0\nu}, \ldots} \sim \sqrt{(n_{0\nu})}\, \Psi_{n_{01}, \ldots, n_{0\nu}-1 \ldots}$$

$$\hat{A}_\nu^+ \Psi_{n_{01}, \ldots, n_{0\nu}, \ldots} \sim \sqrt{(1-n_{0\nu})}\, \Psi_{n_{01}, \cdots, n_{0\nu}+1, \ldots}$$

In particular the annihilation operator, when acting on a vacuum, reduces to zero:

$$\hat{A}_\nu \Psi_{\text{vac}} = 0, \quad \Psi^*_{\text{vac}} \hat{A}_\nu^+ = 0. \quad (3.6)$$

The second equality is the conjugate of the first.

† The operators $\hat{A}$ and $\hat{A}^+$ are usually defined in terms of their effect on the index of the state n_0 rather than on the dynamical variable n. The definition (3.5) therefore differs from the normal, but it seems more consistent in terms of the general concept of an operator in quantum theory.

The following anti-commutation rules for the operators $\hat{A}$ and $\hat{A}^+$ may be obtained without difficulty:

$$\left.\begin{aligned} [\hat{A}_\mu, \hat{A}_\nu^+]_+ &= \delta_{\mu\nu}, \\ [\hat{A}_\mu, \hat{A}_\nu]_+ &= [\hat{A}_\mu^+, \hat{A}_\nu^+]_+ = 0. \end{aligned}\right\} \tag{3.7}$$

From (3.7) we obtain

$$(\hat{A}_\nu)^2 = (\hat{A}_\nu^+)^2 = 0.$$

Using these operators, it is easy to construct the state $\Psi_{n_0}(n)$ from the vacuum state wave function:

$$\Psi_{n_0}(n) = (\hat{A}_1^+)^{n_{01}} (\hat{A}_2^+)^{n_{02}} \ldots \Psi_{\text{vac}}. \tag{3.8}$$

From the previous equations, we arrive at the inequality $n_{0\nu} \leqq 1$, which expresses the Pauli principle.

The definition of the operators $\hat{A}$ and $\hat{A}^+$ are seldom, if ever, needed. It is usually necessary to calculate the matrix element of a product of these operators, and this can be done immediately via the relations (3.6) to (3.8) and the vacuum normalization condition. Thus, for example,

$$\langle \Psi_{\text{vac}} | \hat{A}_\nu \hat{A}_\mu^+ | \Psi_{\text{vac}} \rangle = \langle \Psi_{\text{vac}} | \delta_{\mu\nu} - \hat{A}_\mu^+ \hat{A}_\nu | \Psi_{\text{vac}} \rangle = \delta_{\mu\nu}.$$

This is also true for the field operators (see § 3.4).

3.4. Besides the operators $\hat{A}_\nu$ and $\hat{A}_\nu^+$ which represent the annihilation and creation of a particle in a state ν, much theoretical use is made of the operator functions (or field operators), $\hat{\psi}(q)$ and $\hat{\psi}^+(q)$, which describe the annihilation and creation of a particle at a point q. These functions are defined in the following way:

$$\hat{\psi}(q) = \sum_\nu \hat{A}_\nu \chi_\nu(q), \quad \hat{\psi}^+(q) = \sum_\nu \hat{A}_\nu^+ \chi_\nu^*(q). \tag{3.9}$$

We will now show that, when the operator $\hat{\psi}(q_0)$ acts on the vacuum, a state of the particle localized at q_0 is, indeed, obtained. We have

$$\hat{\psi}^+(q_0)\, \Psi_{\text{vac}} = \sum_\nu \chi_\nu^*(q_0)\, \delta_{n_1,0\ldots}\, \delta_{n_\nu,1\ldots},$$

where unity corresponds to the occupied νth state. Going over to the configuration representation [see (2.12)], and remembering that $\Phi_n(Q)$ is simply equal to $\chi_\nu(q)$ for a single-particle state, we find, since the functions $\chi_\nu(q)$ form a complete set, that†

$$\Psi(Q) = \sum_\nu \chi_\nu^*(q_0)\, \chi_\nu(q) = \delta(q - q_0),$$

which proves the above result.

Using this same completeness condition, commutation rules for the field operators may easily be derived. From the anti-commutation rules (3.7), we find that

$$\left.\begin{aligned} [\hat{\psi}(q),\ \hat{\psi}^+(q')]_+ &= \delta(q - q'), \\ [\hat{\psi}, \hat{\psi}]_+ = [\hat{\psi}^+, \hat{\psi}^+]_+ &= 0. \end{aligned}\right\} \tag{3.10}$$

We can then write down an expression similar to (3.6):

$$\hat{\psi}(q)\, \Psi_{\text{vac}} = 0, \quad \Psi^*_{\text{vac}} \hat{\psi}^+(q) = 0. \tag{3.11}$$

3.5. We can show directly from (3.5) that the operator

$$\hat{n}_\nu = \hat{A}_\nu^+ \hat{A}_\nu \tag{3.12}$$

satisfies the relation

$$\hat{n}_\nu \Psi_{n_0} = n_{0\nu} \Psi_{n_0}$$

and may therefore be called the occupation-number operator. The operators $\hat{n}_\nu$ commute with one another.

The operator for the total number of particles, $\hat{N} = \sum_\nu \hat{n}_\nu$, satisfies not only the relation $\hat{N}\Psi_{n_0} = N\Psi_{n_0}$, but also the stricter condition

$$\hat{N}\Psi(n, t) = N\Psi(n, t) \tag{3.13}$$

connected with the conservation of the total number of particles in the system. In other words the operator $\hat{N}$ commutes with the total Hamiltonian of the system.

The operator $\hat{N}$, can also be expressed in terms of the field operators:

$$\hat{N} = \int dq\, \hat{\psi}^+(q)\, \hat{\psi}(q). \tag{3.14}$$

From the commutation relations (3.10), we obtain the useful relationship

$$\hat{N}\hat{\psi}(q) = \hat{\psi}(q)\,(\hat{N} - 1). \tag{3.15}$$

† $\delta(q - q_0)$ signifies the product of the δ-function for the spatial coordinates, and of the δ-symbol for the discrete coordinates.

3.6. We will now formulate the rules for constructing an operator for any arbitrary, physical quantity $\hat{\alpha}$ in the occupation-number representation. The required operator may be expressed in terms of the creation and annihilation operators, and has the following simple form ($\hat{\alpha}$ acts on the coordinates q_i)

$$\hat{\alpha}' = \frac{1}{N!}\int dQ\, \hat{\psi}^+(q_N) \dots \hat{\psi}^+(q_1)\, \hat{\alpha}\hat{\psi}(q_1) \dots \hat{\psi}(q_N). \tag{3.16}$$

Together with (2.14), all the proof that is necessary is that the equality

$$\langle \Psi_{n_0'}|\hat{\alpha}'|\Psi_{n_0}\rangle = \int dQ\, \Phi^*_{n_0'}(Q)\, \hat{\alpha}\Phi_{n_0}(Q)$$

is fulfilled. This comes from the relation

$$\frac{1}{\sqrt{N!}}\hat{\psi}(q_1) \dots \hat{\psi}(q_N)\, \Psi_{n_0} = \Phi_{n_0}(Q)\, \Psi_{\text{vac}}$$

which is, in turn, easily shown to be true. In fact the product of the operators $\hat{\psi}$ annihilates all particles that are present in Ψ_{n_0}, and transforms this to the vacuum function. Each act of annihilation is accompanied, according to (3.9), by the appearance of a function $\chi_\nu(q)$. Moreover, the product of these functions, thus obtained, will be antisymmetric as was the product of the operators initially. It is, therefore, simply a matter of confirming that the numerical coefficients are correct.

The norm of the right-hand side of this last relation equals unity; the norm of the left-hand side is equal to

$$(N!)^{-1} \int dQ\, \langle \Psi_{n_0}|\hat{\psi}^+(q_N) \dots \hat{\psi}^+(q_1)\, \hat{\psi}(q_1) \dots \hat{\psi}(q_N)|\Psi_{n_0}\rangle.$$

If we take out the operator $\hat{N} = \int dq_1\, \hat{\psi}^+(q_1)\, \hat{\psi}(q_1)$ and transfer it to the right, using (3.15), and repeat the same procedure for $\hat{N} = \int dq_2\, \hat{\psi}^+(q_2)\, \hat{\psi}(q_2)$, etc., then we finally arrive at an expression equal to unity:

$$(N!)^{-1} \langle \Psi_{n_0}|\hat{N}(\hat{N}-1) \dots |\Psi_{n_0}\rangle.$$

If $\hat{\alpha}$ is an n-particle operator

$$\hat{\alpha}_n = \frac{1}{n!}\sum_{i_1 \dots i_n = 1}^{N}{}' \hat{a}_{i_1 \dots i_n}, \tag{3.17}$$

where the dash indicates omission of all terms for which at least one pair of indices coincide, then (3.16) may be simplified. To do this, it is necessary for the operators $\hat{N}$ to be introduced for those coordinates which do not enter into $\hat{a}_{i_1 \dots i_N}$, and then for them to be transferred to the right in accordance with (3.15) and (3.13). This gives the well-known expression

$$\hat{\alpha}'_n = \frac{1}{n!} \int dq_1 \dots dq_n \, \hat{\psi}^+(q_n) \dots \hat{\psi}^+(q_1) \, \hat{a} \hat{\psi}(q_1) \dots \hat{\psi}(q_n), \tag{3.18}$$

where there is no longer a summation over particles.

3.7. The rules which have been formulated permit the construction of a Hamiltonian in the occupation number representation for a system with pair interactions. Starting from (1.3) and (1.4), together with (3.17), and (3.18), we obtain

$$\hat{H} = \int dq \, \hat{\psi}^+(q) \, \hat{T} \hat{\psi}(q) + \tfrac{1}{2} \int dq \, dq' \, \hat{\psi}^+(q) \, \hat{\psi}^+(q') \, \hat{V}(q, q') \, \hat{\psi}(q') \, \hat{\psi}(q). \tag{3.19}$$

The Schrödinger equation itself has the following form in this representation

$$\left(i \frac{\partial}{\partial t} - \hat{H} \right) \Psi(n, t) = 0.$$

The Hamiltonian (3.19) can also be expressed directly in terms of the operators $\hat{A}$, $\hat{A}^+$. Inserting (3.9) into (3.19), and introducing the notation

$$\langle \nu | \hat{T} | \mu \rangle = \int dq \, \chi^*_\mu(q) \, \hat{T} \chi_\nu(q),$$

$$\langle \nu\mu | \hat{V} | \sigma\lambda \rangle = \int dq \, dq' \chi^*_\sigma(q) \, \chi^*_\lambda(q') \, \hat{V} \chi_\mu(q') \, \chi_\nu(q),$$

we obtain

$$\hat{H} = \sum_{\mu\nu} \langle \nu | \hat{T} | \mu \rangle \, \hat{A}^+_\mu \hat{A}_\nu + \tfrac{1}{2} \sum_{\mu\nu\lambda\sigma} \langle \nu\mu | \hat{V} | \sigma\lambda \rangle \, \hat{A}^+_\sigma \hat{A}^+_\lambda \hat{A}_\mu \hat{A}_\nu. \tag{3.20}$$

If the base of the representation $\chi_\nu(q)$ satisfies (3.1), then the first term of (3.20) can be written in the form

$$\hat{H}_F = \sum_\nu \varepsilon_\nu \hat{n}_\nu. \tag{3.21}$$

3.8. We have, up to the present, been considering only systems of identical particles. A system of protons and neutrons, excluding electromagnetic interactions, can also be described by the same equations, so long as q represents the spatial, spin and isotopic coordinates and ν the corresponding indices labelling the states. In particular ν must include the third component of the isotopic spin t, with values $t = \frac{1}{2}$ and $t = -\frac{1}{2}$ corresponding to the proton and neutron states, respectively.

The field operator $\hat{\psi}(q)$, satisfying (3.10), can be expressed in terms of the operators of the proton and neutron fields ($\tilde{q} = \boldsymbol{x}, \sigma$):

$$\hat{\psi}(q) = \sum_{\nu} \hat{A}_{\nu}\chi_{\nu}(q) = \hat{\psi}_{(p)}(\tilde{q})\,\delta_{\tau,\frac{1}{2}} + \hat{\psi}_{(n)}(\tilde{q})\,\delta_{\tau,-\frac{1}{2}}. \qquad (3.22)$$

The operators $\hat{\psi}_{(p)}$, $\hat{\psi}^{+}_{(p)}$, and also $\hat{\psi}_{(n)}$, $\hat{\psi}^{+}_{(n)}$, satisfy the anti-commutation rules (3.10). The field operators of the different particles anti-commute with each other. This does not contradict the distinguishability of the particles (in the sense that there is no Pauli prohibition), since the operators of the physical quantities, which are bilinear in $\hat{\psi}$ and $\hat{\psi}^{+}$, commute with one another.

The Hamiltonian of a system of protons and neutrons in the occupation-number representation may be written in the form (3.19), where $\hat{\psi}(q)$ is defined by eqn. (3.22), and $\hat{V}$ is the potential (1.11) (without the Coulomb term).† Substituting the expansion (3.22) in $\hat{H}$, we may write

$$\begin{aligned}\hat{H} = \int d\tilde{q}\,\{&\hat{\psi}^{+}_{(p)}(\tilde{q})\,\hat{T}\hat{\psi}_{(p)}(\tilde{q}) + \hat{\psi}^{+}_{(n)}(\tilde{q})\,\hat{T}\hat{\psi}_{(n)}(\tilde{q})\}\\ &+ \tfrac{1}{2}\int d\tilde{q}\,d\tilde{q}'\{\hat{\psi}^{+}_{(p)}(\tilde{q})\,\hat{\psi}^{+}_{(p)}(\tilde{q}')\,\hat{V}_{(pp)}\hat{\psi}_{(p)}(\tilde{q}')\,\hat{\psi}_{(p)}(\tilde{q})\\ &+ \hat{\psi}^{+}_{(n)}(\tilde{q})\,\hat{\psi}^{+}_{(n)}(\tilde{q}')\,\hat{V}_{(nn)}\hat{\psi}_{(n)}(\tilde{q}')\,\hat{\psi}_{(n)}(\tilde{q})\\ &+ 2\hat{\psi}^{+}_{(p)}(\tilde{q})\,\hat{\psi}^{+}_{(n)}(\tilde{q}')\,\hat{V}^{(1)}_{(pn)}\hat{\psi}_{(n)}(\tilde{q}')\,\hat{\psi}_{(p)}(\tilde{q})\\ &+ 2\hat{\psi}^{+}_{(p)}(\tilde{q})\,\hat{\psi}^{+}_{(n)}(\tilde{q}')\,\hat{V}^{(2)}_{(pn)}\hat{\psi}_{(p)}(\tilde{q}')\,\hat{\psi}_{(n)}(\tilde{q})\}. \end{aligned} \qquad (3.23)$$

The remaining terms, which contain an odd number of the indices (p) and (n), disappear because

$$\sum_{\tau} \delta_{\tau,\frac{1}{2}}, \delta_{\tau,-\frac{1}{2}} = 0.$$

The interaction potentials between similar particles are of the form

$$\hat{V}_{(pp)} = \hat{V}_{(nn)} = V_{(c)} + V_{(a)}\frac{1 - \hat{\mathscr{P}}^{(\sigma)}}{2}, \qquad (3.24)$$

† See § 7, for the inclusion of the Coulomb interaction in the nucleus.

The interaction potential between protons and neutrons consists of two parts. The first,

$$\hat{V}^{(1)}_{(pn)} = V_{(c)} + \frac{V_{(a)}}{2}, \tag{3.25}$$

does not change the interacting particles; the second,

$$\hat{V}^{(2)}_{(pn)} = -V_{(a)}\frac{\mathscr{P}^{(\sigma)}}{2}, \tag{3.25'}$$

leads to the transformation of a proton into a neutron, and vice versa (cf. (1.8)).

When we require the Hamiltonian of the system, we will therefore use in future the general expression (3.19), which is suitable for both multi-electron and multi-nucleon systems.

3.9. We have so far, in this section, been considering multiparticle systems in the Schrödinger representation. In our future work we will also need the equation of motion for the field operators in the Heisenberg representation. To obtain them, we use the general relation (2.4), replacing $\hat{\alpha}_H$ by the operator $\hat{\psi}_H$, or $\hat{\psi}^+_H$, and $\hat{H}$ by the Hamiltonian (3.19). It will in fact prove more convenient to use the operator $\hat{H}_H$, referred to the same moment of time t as the field operators, rather than $\hat{H}$. We may write from (2.10):

$$\hat{H}_H(t) = \int dq\, \hat{\psi}^+_H(q,t)\,\hat{T}\hat{\psi}_H(q,t) + \tfrac{1}{2}\int dq\, dq'\, \hat{\psi}^+_H(q,t)\,\hat{\psi}^+_H(q',t) \times \hat{V}\hat{\psi}_H(q',t)\,\hat{\psi}_H(q,t). \tag{3.26}$$

So, when calculating the commutator in (2.4), only operators referring to the same moment of time are encountered. The corresponding commutation rules coincide with the analogous rules for operators in the Schrödinger representation (3.10) [see (2.11)]:

$$[\hat{\psi}_H(q,t), \hat{\psi}^+_H(q',t)]_+ = \delta(q-q'),$$

$$[\hat{\psi}_H(q,t), \hat{\psi}_H(q',t)]_+ = [\hat{\psi}^+_H(q,t), \hat{\psi}^+_H(q',t)]_+ = 0. \quad (3.27$$

If the times considered are not coincident, the commutation rules take on a very complicated form.

Using the simple relationships,

$$[\hat{a}, \hat{b}\hat{c}]_- = [\hat{b}, \hat{a}]_+ \hat{c} - \hat{b}[\hat{c}, \hat{a}]_+ ,$$

$$[\hat{a}, \hat{b}\hat{c}\hat{d}\hat{e}]_- = [\hat{b}, \hat{a}]_+ \hat{c}\hat{d}\hat{e} - \hat{b}[\hat{c}, \hat{a}]_+ \hat{d}\hat{e} + \hat{b}\hat{c}[\hat{d}, \hat{a}]_+ \hat{e} - \hat{b}\hat{c}\hat{d}[\hat{e}, \hat{a}]_+ ,$$

where $\hat{a}$, $\hat{b}$, $\hat{c}$, $\hat{d}$ and $\hat{e}$ are arbitrary operators, we finally obtain the following equation in the Heisenberg representation:

$$\left(i\frac{\partial}{\partial t} - \hat{T}\right)\hat{\psi}_H(q, t) = \int dq' \, \hat{\psi}_H^+(q', t) \, \hat{V}\hat{\psi}_H(q', t) \, \hat{\psi}_H(q, t). \quad (3.28)$$

The analogous equation for $\hat{\psi}_H^+(q, t)$ is the Hermitian conjugate of this equation.

CHAPTER 2

Many-Body Systems in the Hartree–Fock Approximation

4. The Hartree–Fock Approximation

4.1. The mathematical difficulties inherent in constructing a theory of a system of interacting particles suggests that one should begin by choosing some suitable zero (initial) approximation. There are two essential requirements. Firstly, the approximation should be reasonably simple mathematically. Secondly, it should nevertheless reflect as closely as possible the properties of real, many body-systems.

As has already been pointed out in the introduction, the basic mathematical difficulties in any theory of many interacting particles derive from the fact that the idea of each particle in the system having its own individual states (as distinct from states of the system as a whole) no longer has a precise meaning, and even under the best circumstances can only be considered as a more or less happy approximation to reality. It is natural, nevertheless, to require that the zero approximation we choose should be compatible with the concept of individual particle states. We will call this the single-particle approximation in many-body theory.

Thus, to satisfy the second of our requirements, we must look for the best† of the single-particle approximations. This would seem to be the Hartree–Fock (or self-consistent field) approximation. In many cases this approximation by itself provides a sufficiently accurate description of a system of interacting particles.

We will use the Schrödinger representation throughout this chapter.

† See § 4.6 for the meaning of the term "the best approximation".

4.2. The problem of constructing a zero approximation really reduces to the finding of some approximate Hamiltonian $\hat{H}_0$ to replace the exact Hamiltonian $\hat{H}$. The most general expression for $\hat{H}_0$, corresponding to the single-particle approximation may be written in the form†

$$\hat{H}_0 = \int dq\, \hat{\psi}^+(q) (\hat{T} + \hat{W}) \hat{\psi}(q) + C, \tag{4.1}$$

where C is some constant, and $\hat{W}$ is an as yet unknown operator.

The operator $\hat{T} + \hat{W}$ represents the effective Hamiltonian of the particle in the approximation considered. The solution of the equation

$$(\hat{T} + \hat{W}) \chi_\nu(q) = \varepsilon_\nu \chi_\nu(q) \tag{4.2}$$

gives the wave functions and energies of the individual states of the particles in the system. If there is no interaction, the operator $\hat{W}$ reduces to zero and equation (4.2) is transformed to (3.1). The explicit form of $\hat{W}$ and C will be derived below.

It is convenient to choose the set of functions χ_ν as the base of the occupation-number representation. In this case, expression (4.1) can be rewritten in the form

$$\hat{H}_0 = \sum_\nu \varepsilon_\nu \hat{n}_\nu + C, \tag{4.3}$$

where $\hat{n}_\nu$ are the occupation-number operators (see § 3.7). Since these operators commute with one another, one has

$$[\hat{H}_0, \hat{n}_\nu]_- = 0. \tag{4.4}$$

In other words the occupation numbers of the individual levels are integrals of motion. This obviously results from the single-particle character of the approximation.

Introducing the eigenfunction Ψ_0 of the operator $\hat{H}_0$,

$$\hat{H}_0 \Psi_0 = E_0 \Psi_0, \tag{4.5}$$

eqn. (4.4) may be rewritten in the following form:

$$\hat{n}_\nu \Psi_0 = n_\nu \Psi_0. \tag{4.6}$$

† There are other, more complicated forms of the single-particle Hamiltonian. They may, however, be related to (4.1) by a canonical transformation.

Thus the wave function of a stationary state of a system in the single-particle approximation belongs to the class of functions with given occupation of the levels.

4.3. The study of many-body systems in the single-particle approximation is considerably simplified by introducing such quantities as the density matrix, the occupation operator and the distribution function. These quantities are directly connected with the physical characteristics of the system.

The single-particle density matrix for the state Ψ_0 of the system is defined by

$$R(q, q') = \langle \Psi_0 | \hat{\psi}^+(q') \, \hat{\psi}(q) | \Psi_0 \rangle = \sum_\nu n_\nu \chi_\nu^*(q') \, \chi_\nu(q) \qquad (4.7)$$

(see Appendix A). This provides a comprehensive picture of the single-particle characteristics of the system—the average values of the single-particle operators,† the density distribution of the corresponding quantities, etc. In particular the number-density distribution for the particles is given by

$$\varrho(\boldsymbol{x}) = \mathrm{Tr}_{\sigma\tau} \, R(\boldsymbol{x}, \sigma, \tau; \boldsymbol{x}, \sigma', \tau'), \qquad (4.8)$$

where $\mathrm{Tr}_{\sigma\tau}$ is the trace for the discrete variables.

Similarly, we can introduce higher order density matrices which describe the two-particle characteristics of the system, etc. In the single-particle approximation these matrices reduce to $R(q, q')$ (see Appendix A). This latter density matrix may be considered as the matrix element of an operator $\hat{\varrho}$ (the occupation operator), and may be calculated using the wave functions $\delta(q - q_0)$, which correspond to some definite value of the coordinate q:‡

$$R(q, q') = (q|\hat{\varrho}|q') = \int dq_0 \, \delta(q_0 - q) \, \hat{\varrho}_{q_0} \delta(q_0 - q')$$
$$= \hat{\varrho}_q \delta(q - q'). \qquad (4.9)$$

† By a single-particle operator we understand an operator of the type $\int dq \, \hat{\psi}^+(q) \, \hat{a}_1 \, \hat{\psi}(q)$, which, in the configuration representation takes the form $\sum_{i=1}^{N} \hat{a}_i$. Analogously, a two-particle operator is of the type $\frac{1}{2} \int dq \, dq' \hat{\psi}^+(q) \times \hat{\psi}^+(q') \, \hat{a}_2 \hat{\psi}(q') \, \hat{\psi}(q)$, and so on (see § 3).

‡ The operator index denotes the variable on which the operator is acting.

If we expand the δ-function in this expression in terms of the set of functions χ_ν and remember that they must be complete

$$\sum_\nu \chi_\nu^*(q')\,\chi_\nu(q) = \delta(q-q'),$$

then we find from a comparison of this equation with (4.7) that

$$\hat{\varrho}\chi_\nu(q) = n_\nu\chi_\nu(q). \tag{4.9'}$$

The operator $\hat{\varrho}$, whose eigenvalues are the occupation numbers of the system, thus depends on the state of the system under consideration. Since $n_\nu^2 = n_\nu$ ($n_\nu = 0,1$), $\hat{\varrho}$ satisfies the condition

$$\hat{\varrho}^2 = \hat{\varrho}. \tag{4.10}$$

The explicit form of $\hat{\varrho}$ may be found in the following way. The occupation numbers n_ν depend on the energy of the state ε_ν, and some other quantities (spin components, angular momentum, etc.) which together with the energy provide a complete description of the state χ_ν. We will limit our consideration to those systems for which the level occupation is determined solely by the energy. We will suppose, that is, that all the states corresponding to a given energy (generally degenerate) are either all filled or all empty. These are usually referred to as closed shell systems (see § 9).

We will therefore assume that $n_\nu = n(\varepsilon_\nu)$. From eqn. (4.2) we can write† $n_\nu\chi_\nu = n(\varepsilon_\nu)\,\chi_\nu = n(\hat{T}+\hat{W})\,\chi_\nu$, whence

$$\hat{\varrho} = n(\hat{T}+\hat{W}). \tag{4.11}$$

Thus the operator $\hat{\varrho}$ can be obtained by a formal substitution of the Hamiltonian $\hat{T}+\hat{W}$ for the energy ε_ν in the expression for the occupation numbers. The operator $\hat{\varrho}$ evidently commutes with $\hat{T}+\hat{W}$, and corresponds to a time-independent distribution, conforming with the relation (4.4).

4.4. The mean value of the single-particle operator, $\hat{\alpha}_1 = \int dq \times \hat{\psi}^+(q)\,\hat{a}_1\hat{\psi}(q)$ may be calculated easily using $\hat{\varrho}$.

$$\langle\Psi_0|\hat{\alpha}_1|\Psi_0\rangle = \mathrm{Tr}\,(\hat{a}_1\hat{\varrho}). \tag{4.12}$$

† Functions of an operator have the same set of eigenfunctions as the original operator, and eigenvalues which are the same functions of the original eigenvalues (see Appendix B).

Similarly, for a two-particle operator

$$\hat{\alpha}_2 = \tfrac{1}{2} \int dq\, dq'\, \hat{\psi}^+(q)\, \hat{\psi}^+(q')\, \hat{a}_2 \hat{\psi}(q')\, \hat{\psi}(q)$$

and we have

$$\langle \Psi_0 | \hat{\alpha}_2 | \Psi_0 \rangle = \tfrac{1}{2} \operatorname{Tr} [\hat{a}_2 (1 - \hat{\mathscr{P}}_{12})\, \hat{\varrho}_{q_1} \hat{\varrho}_{q_2}]. \qquad (4.13)$$

Here $\hat{\mathscr{P}}_{12}$ is the coordinate exchange operator (see § 1.3); Tr is the trace for all the variables. The normalization condition takes the form

$$\operatorname{Tr} \hat{\varrho} = N, \qquad (4.14)$$

where N is the total number of particles in the system. (See Appendix A for the derivation of (4.12) and (4.13): it contains the necessary rules for calculating the trace.) In calculating quantities connected with the density it is convenient to introduce the so-called distribution function, $f(\boldsymbol{x}, \boldsymbol{p})$, which can be obtained by a Fourier transformation of $R(q, q')$ in terms of the difference of the spatial coordinates, $\boldsymbol{x} - \boldsymbol{x}'$:

$$R(q, q') = \int d^3\boldsymbol{p}\, f(\boldsymbol{x}, \boldsymbol{p}) \exp [i(\boldsymbol{p} \cdot \boldsymbol{x} - \boldsymbol{x}')]. \qquad (4.15)$$

The quantity $f(\boldsymbol{x}, \boldsymbol{p})$, which is a matrix in the discrete coordinates, if integrated over the momenta, gives the coordinate probability distribution

$$\varrho(\boldsymbol{x}) = \operatorname{Tr}_{\sigma\tau} \int d^3\boldsymbol{p} f(\boldsymbol{x}, \boldsymbol{p}). \qquad (4.16)$$

Similarly, the momentum probability distribution is given by

$$\varrho(\boldsymbol{p}) = \operatorname{Tr}_{\sigma\tau} \int d^3\boldsymbol{x} f(\boldsymbol{x}, \boldsymbol{p}). \qquad (4.17)$$

We can easily relate the distribution function to the operator $\hat{\varrho}$ which we have introduced. Inserting (4.9) in the left-hand side of (4.15), and expanding $\delta(\boldsymbol{x} - \boldsymbol{x}')$ in a Fourier integral, we have

$$f(\boldsymbol{x}, \boldsymbol{p}) = \langle \hat{\varrho} \rangle_{\boldsymbol{p}}, \qquad (4.18)$$

where we have introduced the notation†

$$\langle \hat{a} \rangle_{\boldsymbol{p}} \equiv \exp [-i(\boldsymbol{p} \cdot \boldsymbol{x})]\, \hat{a} \exp i(\boldsymbol{p} \cdot \boldsymbol{x}). \qquad (4.19)$$

† If we use the eigenfunctions of some other operator than the momentum operator $\hat{\boldsymbol{p}}$ in this relation, we arrive at the distribution function for that quantity [51].

4.5. The simplest of the single-particle approximations is the perfect-gas approximation, in which the interaction between particles is completely ignored. In this case the operator $\hat{H}_0$ coincides with the free Hamiltonian $\hat{H}_F$, and the quantities C and $\hat{W}$ are equal to zero.

The individual wave functions χ_ν for the perfect gas are determined by eqn. (3.1.). In the simplest case, when there is no external field, we may choose the solution of this equation in the form:

$$\chi_\nu(q) = \Omega^{-1/2}\delta_{\sigma s}\delta_{\tau t} \exp i(\boldsymbol{p} \cdot \boldsymbol{x}), \qquad (4.20)$$

which corresponds to a particle state with known values for the momentum, ordinary spin component s and isotopic spin t.

The energy of the system E_0 is obtained by averaging the Hamiltonian (3.21) over the state Ψ_0 and has the form

$$E_0 = \sum_\nu \varepsilon_\nu n_\nu,$$

where the ε_ν are the energies of the individual levels. It is easy, using this relation, to determine the characteristics of the ground state of the system. The minimum energy E_0 will obviously occur when

$$n_\nu = \theta(\varepsilon_F - \varepsilon_\nu), \qquad (4.21)$$

where the quantity ε_F, which is called the Fermi energy, determines the upper limit to the filled part of the spectrum, and is found from the normalization condition. The function $\theta(x)$ is defined in the following way:

$$\theta(x) = \begin{cases} 1, & x > 0 \\ 0, & x < 0. \end{cases}$$

The wave function for the ground state of the system can be written as

$$\Psi_0 = \Big(\prod_{\varepsilon_\nu \leqq \varepsilon_F} \hat{A}_\nu^+\Big)\Psi_{\text{vac}}, \qquad (4.22)$$

which contains creation operators referring to all filled levels. If one transforms this expression to the configuration representation, the usual Slater–Fock determinant is obtained. In accordance with (4.11) the occupation operator in this case may be taken as

$$\hat{\varrho} = \theta(\varepsilon_F - \hat{T}). \qquad (4.23)$$

If there is no external field, the ground state of the system may be characterized by the limiting momentum p_0, which is defined by $\varepsilon_F = p_0^2/2M$. Hence, since $\hat{T} = \hat{p}^2/2M$, we obtain

$$\hat{\varrho} = \theta(p_0^2 - \hat{p}^2)$$

and the distribution function becomes†

$$f(\boldsymbol{x}, \boldsymbol{p}) = \theta(p_0^2 - p^2), \tag{4.24}$$

which does not depend on the coordinates in this case.

The number density of the particles, which is determined by the limiting momentum, may easily be found using (4.16):

$$\varrho = \frac{g p_0^3}{6\pi^2}. \tag{4.25}$$

Here $g = \mathrm{Tr}_{\sigma\tau} 1$ is the statistical weight equal to the number of substates with different values of the discrete variables ($g = 2$ for electrons; $g = 4$ for nucleons).

The energy of the ground state in the absence of any external field may be determined directly from (4.12) (see Appendix A, § A.2):

$$E_0 = \mathrm{Tr}\, \hat{T}\hat{\varrho} = \mathrm{Tr}_{\sigma\tau} \int d^3\boldsymbol{x}\, d^3\boldsymbol{p} \left\langle \frac{\hat{p}^2}{2M} \cdot \theta(p_0^2 - \hat{p}^2) \right\rangle_p = \frac{\Omega g p_0^5}{20\pi^2 M}.$$

The quantity E_0 is proportional to the total number of particles $N = \Omega\varrho$, and is equal to $E_0 = \frac{3}{5}(p_0^2/2M)\, N$. We note that the derivative $(\partial E_0/\partial N)_\Omega$, which is the chemical potential of the system μ coincides with the limiting Fermi energy:

$$\mu = \frac{p_0^2}{2M} = \varepsilon_F. \tag{4.26}$$

This relation is quite general (see § 23).

4.6. There exist, besides the perfect-gas approximation, several other single-particle approximations which take into account interaction between particles. The best of these approximations, as will become apparent, is the Hartree–Fock approximation. The approximate Hamiltonian $\hat{H}_0$, should differ as little as possible from $\hat{H}$, which

† The momentum operator $\hat{p}$, in the preceding equation, when acting on $\exp i(\boldsymbol{p} \cdot \boldsymbol{x})$ in (4.19), changes simply to $\boldsymbol{p}$.

requires that the mean of the square of the difference of these operators $I = \langle \Psi_0 | (\hat{H} - \hat{H}_0)^2 | \Psi_0 \rangle$ be a minimum. Here Ψ_0 is the wave function, in the single-particle approximation, of the state in which we are interested.

If we carry out a formal variation of I with $\hat{H}_0$, we obtain

$$\delta I = \langle \Psi_0 | [\hat{H}_0 - \hat{H}, \delta \hat{H}_0]_+ | \Psi_0 \rangle = 0,$$

where $\delta \hat{H}_0 = \int dq\, \hat{\psi}^+(q)\, \delta \hat{W}\, \hat{\psi}(q) + \delta C$. Equating the coefficient of δC to zero, we find

$$\langle \Psi_0 | (\hat{H}_0 - \hat{H}) | \Psi_0 \rangle = 0.$$

This, together with (4.12) and (4.13), gives

$$C = -\operatorname{Tr} \hat{W} \hat{\varrho} + \tfrac{1}{2} \operatorname{Tr} [\hat{V}(1 - \hat{\mathscr{P}}_{12})\, \hat{\varrho}_{q_1} \hat{\varrho}_{q_2}]. \qquad (4.27)$$

Equating to zero the part of δI containing $\delta \hat{W}$, and using (4.27) and the results of Appendix A (see § A.5), we may write

$$\operatorname{Tr} \{[\hat{W}_{q_1} - \operatorname{Tr}_{q_2} (\hat{V}(q_1, q_2)\,(1 - \hat{\mathscr{P}}_{12})\, \hat{\varrho}_{q_2})] \times (\hat{\varrho}_{q_1} - \hat{\varrho}_{q_3})^2\, \hat{\mathscr{P}}_{13}\, \delta \hat{W}_{q_3}\} = 0.$$

Whence we obtain the required expression for the operator $\hat{W}$:

$$\hat{W}_q = \operatorname{Tr}_{q'} [\hat{V}(q, q')\,(1 - \hat{\mathscr{P}}_{qq'})\, \hat{\varrho}_{q'}]. \qquad (4.28)$$

Substituting this expression into (4.27), we find

$$C = -\tfrac{1}{2} \operatorname{Tr} [\hat{V}(q, q')\,(1 - \hat{\mathscr{P}}_{qq'})\, \hat{\varrho}_q \hat{\varrho}_{q'}], \qquad (4.28')$$

which solves the problem of finding the Hamiltonian $\hat{H}_0$.

We require, for future use, an expression for the operator $\hat{H}_0$ in terms of the density matrix. If we insert the last two relations into (4.1), and use the results of Appendix A (see §§ A.2 and A.3), we find†

$$\hat{H}_0 = \int dq\, \hat{\psi}^+(q)\, \hat{T} \hat{\psi}(q)$$
$$+ \int_{q'' \to q'} dq\, dq'\, \hat{\psi}^+(q)\, \hat{V}(q, q')\,(1 - \hat{\mathscr{P}}_{qq'})\, R(q', q'')\, \hat{\psi}(q)$$
$$-\tfrac{1}{2} \int_{\substack{q'' \to q \\ q''' \to q'}} dq\, dq'\, \hat{V}(q, q')\,(1 - \hat{\mathscr{P}}_{qq'})\, R(q, q'')\, R(q', q''').$$

† In the equations cited, one first operates on the density matrix, and then goes to the limit $q'' \to q'$, etc.

If the potential $\hat{V}$ is a function of the coordinates only, then this simplifies to

$$\hat{H}_0 = \int dq\, \hat{\psi}^+(q)\, \hat{T}\hat{\psi}(q) + \int dq\, dq'\, V(q, q')\, \{R(q', q')\, \hat{\psi}^+(q)\, \hat{\psi}(q) - R(q, q')\, \hat{\psi}^+(q)\, \hat{\psi}(q') - \tfrac{1}{2}[R(q, q)\, R(q', q') - R(q, q')\, R(q', q)]\}. \quad (4.29)$$

4.7. We will now consider the transformation of the expression for the operator $\hat{W}$. It is usually written as a difference, $\hat{W} = \hat{B} - \hat{A}$, where $\hat{B} = \mathrm{Tr}_{q'}[\hat{V}(q, q')\, \hat{\varrho}_{q'}]$ is the direct interaction operator, and $\hat{A} = \mathrm{Tr}_{q'}[\hat{V}(q, q')\, \hat{\mathscr{P}}_{qq'}\hat{\varrho}_{q'}]$ is the self-consistent exchange interaction operator.

Using eqn. (A.20), we can write†

$$\hat{B} = \mathrm{Tr}_{\sigma'\tau'} \int d^3x'\, d^3p'\, V[\boldsymbol{x} - \boldsymbol{x}', -\tfrac{1}{2}(i\boldsymbol{p}' + \nabla_{x'})]\, f(\boldsymbol{x}', \boldsymbol{p}' + \hat{\boldsymbol{p}}). \quad (4.30)$$

Thus, in the general case, the operator $\hat{B}$ is a function of the momentum operator.

If $\hat{V}$ is independent of the momentum operator, then [52]

$$\hat{B} = \mathrm{Tr}_{\sigma'\tau'} \int d^3x'\, d^3p'\, \hat{V}(q, q')\, f(\boldsymbol{x}', \boldsymbol{p}'). \quad (4.30')$$

If $\hat{V}$ is a function of the coordinates only, then one can obtain, with the help of eqn. (4.16), the well-known expression

$$B(x) = \int d^3x'\, V(\boldsymbol{x} - \boldsymbol{x}')\, \varrho(\boldsymbol{x}'). \quad (4.30'')$$

The operator $\hat{B}$ is then a function of the coordinates only.

For the operator $\hat{A}$, we find from (A.20) that

$$\hat{A} = \mathrm{Tr}_{\sigma'\tau'} \int d^3x'\, d^3p' \exp[i(\boldsymbol{p}' \cdot \boldsymbol{x} - \boldsymbol{x}')]\, \hat{\mathscr{P}}_{\sigma\sigma'}\hat{\mathscr{P}}_{\tau\tau'} \times V[\boldsymbol{x} - \boldsymbol{x}', \tfrac{1}{2}(i\boldsymbol{p}' + \nabla_x)]\, f(\boldsymbol{x}, \boldsymbol{p}' + \hat{\boldsymbol{p}}). \quad (4.31)$$

Here ∇_x acts on the distribution function. If $\hat{V}$ is independent of the momentum operator, then, introducing the Fourier transform of the potential in terms of the difference of the spatial coordinates,

$$\hat{V}(q, q') = \int d^3p\, \hat{v}(\boldsymbol{p}) \exp[i(\boldsymbol{p} \cdot \boldsymbol{x} - \boldsymbol{x}')], \quad (4.32)$$

† Here $\hat{V} = V[\boldsymbol{x} - \boldsymbol{x}', \tfrac{1}{2}(\nabla_x - \nabla_{x'})]$.

where $\hat{v}$ is a matrix in the discrete variables, and we obtain

$$\hat{A} = (2\pi)^3 \, \mathrm{Tr}_{\sigma'\tau'} \int d^3p' \hat{v}(p') f(x, p' + \hat{p}) \hat{\mathscr{P}}_{\sigma\sigma'} \hat{\mathscr{P}}_{\tau\tau'}. \tag{4.33}$$

Finally, if $\hat{V}$ is a function of the coordinates only

$$\hat{A} = (2\pi)^3 \int d^3p' v(p') f(x, p' + \hat{p}) \tag{4.33'}$$

and is necessarily dependent on the momentum operator.

Two remarks should be made about the relations obtained in this section. In the first place the momentum operator in the distribution function must be understood in the sense that, when it acts on a plane wave, $\exp i(p \cdot x)$

$$f(x, p' + \hat{p}) \to f(x, p' + p).$$

Moreover, (4.30″) and (4.33′) were based on the assumption that $f(x, p)$ is diagonal in the discrete variables, which is characteristic of systems with closed shells.

We now consider eqn. (4.2) which describes the individual states of the particles. If we insert (4.28) and calculate the trace for the set of functions χ_ν, we find, using eqn. (A.19), that

$$(\hat{T} + \hat{W})\chi_\nu(q) = \Big[\hat{T} + \sum_\mu n_\mu \int dq' \chi_\mu^*(q') \hat{V} \chi_\mu(q')\Big] \chi_\nu(q)$$

$$- \sum_\mu n_\mu \int dq' \chi_\mu^*(q') \hat{V} \chi_\nu(q') \chi_\mu(q) = \varepsilon_\nu \chi_\nu(q).$$

These are simply the well-known Hartree–Fock equations [53].†

The operator $\hat{W}$, which corresponds to the effective interaction between the particles, is similar in form to the external field potential, but depends on the state of the system.‡ This interaction is called self-consistent, since the operator $\hat{W}$ depends on wave functions which it, in turn, determines. Physically speaking, $\hat{W}$ corresponds to the mean potential at a given particle due to all the remaining particles in the system.

† As is evident from these equations, the term corresponding to $\mu = \nu$ disappears automatically.

‡ Despite this dependence, the Hamiltonian $\hat{T} + \hat{W}$ is the same for all states χ_ν. Hence, the quantum-mechanical requirements, especially that the set of functions χ_ν be complete and orthogonal, must still be satisfied.

4.8. An expression for the distribution function may be found from the operator relation (4.18) [51, 54, 55]. From (4.11) and (4.28), we can write

$$f(\boldsymbol{x}, \boldsymbol{p}) = \langle n(\hat{T} + \hat{W}) \rangle_{\boldsymbol{p}}. \tag{4.34}$$

If (4.34) is combined with the definitions of the operators $\hat{T}$ and $\hat{W}$ in eqns. (1.3), (4.30) to (4.33), the operator form of the Hartree–Fock equations is obtained. It must be emphasized that the apparent simplicity of these equations is illusory. The reason is that (4.34) contains a function of a sum of non-commuting operators. In fact the commutation rules

$$[\hat{p}, F(\boldsymbol{x})]_- = -i\nabla F(\boldsymbol{x})$$

indicate that the operators $\hat{T}$, $\hat{A}$ and $\hat{B}$ do not commute. The rules for dealing with these functions are quite complicated (see Appendix B). Thus the transition to the operator form simply corresponds to changing the emphasis of the calculations: from solving the Schrödinger equation to calculating a function of non-commuting arguments.

Generally speaking, both of these problems are equally complicated. However, the operator approach has undoubted advantages in the quasi-classical case considered in § 5, and also for investigations of a number of general properties.

4.9. We will now determine the energy of a system in the Hartree–Fock approximation. According to (4.5), we may write

$$E_0 = \langle \Psi_0 | \hat{H}_0 | \Psi_0 \rangle.$$

From (4.3) and the relations (A.14) and (A.15) derived in Appendix A, we have

$$E_0 = \sum_{\nu} \varepsilon_\nu n_\nu - \tfrac{1}{2} \sum_{\mu\nu} n_\mu n_\nu \langle \mu\nu | \hat{V}(1 - \hat{\mathscr{P}}) | \mu\nu \rangle.$$

This expression may be used to determine how the levels are occupied in the ground state of the system. In most cases this is the same as for a perfect gas.† Hence (4.21) and (4.22) remain

† If there were no second term in E_0 this would be necessarily true. In principle, however, we must include the possibility that the lowest value of E_0 does not correspond to an occupation of the lowest levels ε_ν. It isnecessary, in this case, that the corresponding increase in the first term in the expression should be compensated by an opposite change in the second term.

valid, provided we remember that states are determined by (4.2) and not by (3.1). The expression (4.34) now assumes the form

$$f(\mathbf{x}, \mathbf{p}) = \langle \theta(\varepsilon_F - \hat{T} - \hat{W}) \rangle_p . \tag{4.34'}$$

The expression we have introduced for E_0 is not very suitable for practical calculation. From (4.12), (4.13), (4.28) and (4.28'), we may write

$$E_0 = \operatorname{Tr} \hat{\varrho}\hat{\tau}, \tag{4.35}$$

where $\hat{\tau} = \hat{T} + \hat{W}/2$. Using (A.16') to calculate the trace, and denoting by $\mathscr{E}$ the energy per unit volume ($\mathscr{E} = E_0/\Omega$), we have

$$\mathscr{E} = \operatorname{Tr}_{\sigma\tau} \int d^3\mathbf{p} \, \overline{\langle \theta(\varepsilon_F - \hat{T} - \hat{W}) \hat{\tau} \rangle_p}, \tag{4.36}$$

where the bar signifies an average over the volume, $\bar{a} = \Omega^{-1} \int d^3\mathbf{x}\, a$.

We will write the operator $\hat{\tau}$ in the following form:

$$\hat{\tau} = \hat{\tau}_1 + \hat{\tau}_2,$$

$$\hat{\tau}_1 = \tfrac{1}{2}(\hat{T} + \varepsilon_F),$$

$$\hat{\tau}_2 = \tfrac{1}{2}(\hat{T} + \hat{W} - \varepsilon_F).$$

From eqn. (4.16), the part of the energy corresponding to $\hat{\tau}_1$ can be rewritten as

$$\mathscr{E}_1 = \tfrac{1}{2}(\mathscr{E}_k + \varepsilon_F \bar{\varrho} + \overline{\varrho U}).$$

U is the external field, and $\mathscr{E}_k$ is the mean kinetic energy per unit volume:

$$\mathscr{E}_k = \operatorname{Tr}_{\sigma\tau} \int d^3\mathbf{p} \frac{p^2}{2M} \overline{f(\mathbf{x}, \mathbf{p})}.$$

The energy $\mathscr{E}_2$ has the form

$$\mathscr{E}_2 = -\tfrac{1}{2} \operatorname{Tr}_{\sigma\tau} \int d^3\mathbf{p} \, \overline{\langle (\varepsilon_F - \hat{T} - \hat{W}) \theta(\varepsilon_F - \hat{T} - \hat{W}) \rangle_p}.$$

A formal differentiation of this expression with respect to ε_F,† together with the equalities

$$\frac{\partial \theta(x)}{\partial x} = \delta(x) \quad \text{and} \quad x\delta(x) = 0$$

† The differentiation is only for that argument of ε_F which enters explicitly into the expression under the integral sign ($\hat{W}$ is also implicity dependent on ε_F). This also refers to integration with respect to ε_F in subsequent equations.

leads to

$$\frac{\partial \mathscr{E}_2}{\partial \varepsilon_F} = -\frac{1}{2}\bar{\varrho}.$$

Remembering also that $\mathscr{E}_2$ must tend to zero as $\varepsilon_F \to -\infty$ (when there are, in general, no particles present), we finally find [51]

$$\mathscr{E} = \frac{1}{2}\left\{\mathscr{E}_k + \varepsilon_F\bar{\varrho} + \overline{\varrho U} - \int_{-\infty}^{\varepsilon_F} \bar{\varrho}\, d\varepsilon_F\right\}. \tag{4.37}$$

To determine the total energy, it is therefore necessary to find the kinetic energy and the density as explicit functions of the Fermi energy. As a result of this transformation, we can avoid calculating complicated traces which contain the operator $\hat{A}$. If the exchange term can be neglected, we can also use the foregoing procedure to avoid calculating $\mathscr{E}_k$. To see this, we compare (4.37) with (4.36) for $\hat{A} = 0$, and find

$$\mathscr{E} = \mathscr{E}_k + \overline{\varrho U} + \tfrac{1}{2}\overline{\varrho B}.$$

Eliminating $\mathscr{E}_k$, we then find

$$\left.\begin{aligned} \mathscr{E}_k &= -\int_{-\infty}^{\varepsilon_F} \bar{\varrho}\, d\varepsilon_F + \varepsilon_F\bar{\varrho} - \overline{\varrho(U+B)}, \\ \mathscr{E} &= -\int_{-\infty}^{\varepsilon_F} \bar{\varrho}\, d\varepsilon_F + \varepsilon_F\bar{\varrho} - \tfrac{1}{2}\overline{\varrho B}. \end{aligned}\right\} \tag{4.38}$$

5. The Thomas–Fermi Approximation

5.1. When the Hartree–Fock approximation is applied to real systems, we find in most cases that the numerical calculations are quite complicated. Moreover, the results are not universally applicable; for example, the solution of a particular type of problem for two different atoms requires two separate calculations.

The problem is easier if the particle motion under the action of the self-consistent potential $\hat{W}$ can be considered as quasi-classical, at least over most of space. The quasi-classical approximation to the Hartree–Fock equations may be referred to, very generally, as the Thomas–Fermi approximation [56].

The condition that the particle motion should be quasi-classical is provided by the usual inequality:

$$\xi = \left|\frac{d\lambda(x)}{dx}\right| \ll 1, \tag{5.1}$$

where $\lambda(x)$ is the quasi-classical de Broglie wavelength of the particle. This is determined in the following way. We make a formal replacement of the momentum operator in $\hat{T} + \hat{W}$ by some function $p(x)$ (the quasi-classical momentum), which is determined by

$$(\hat{T} + \hat{W})|_{\hat{p}\to p(x)} = \varepsilon_\nu .$$

The wavelength is connected with $p(x)$ by the relation: $\lambda(x) = 1/p(x)$. In particular if we consider the ground state of the system, and put $\varepsilon_\nu = \varepsilon_F$, we arrive at

$$(\hat{T} + \hat{W})|_{\hat{p}\to p_0(x)} = \varepsilon_F . \tag{5.2}$$

The quantity $p_0(x)$, thus defined, may, from the analogy with the uniform perfect gas, be called the limiting momentum of the quasi-classical system. It is, however, distinguished from the former by the fact that it is a function of the coordinates. If $\hat{W}$ is independent of the momentum, then

$$p_0(x) = [2M(\varepsilon_F - U - W)]^{1/2} .$$

The physical significance of (5.1) is that the wavelength must be small compared with the characteristic size of inhomogeneities which occur. In other words since we can say $\lambda \sim 1/p_0$, both the limiting momentum and the effective potential must be slowly varying functions of the coordinates.

5.2. So long as ξ is sufficiently small, gradients of the limiting momentum and of the potential may be ignored in the zero approximation. All commutators of terms in the Hamiltonian $\hat{T} + \hat{W}$ go to zero. As a result, (4.34) and (4.34′) become ordinary functions; the momentum operator they contain now acts only on the function $\exp i(\boldsymbol{p}\cdot\boldsymbol{x})$ [see (4.19)], and can be replaced simply by the number $\boldsymbol{p}$. We then arrive at the following expression for the quasi-classical distribution function of the ground state of the system:†

$$f(\boldsymbol{x}, \boldsymbol{p}) = \theta[\varepsilon_F - \varepsilon_p(x)], \tag{5.3}$$

where $\varepsilon_p(x) = (\hat{T} + \hat{W})|_{\hat{p}\to p}$ is the quasi-classical value for the energy of a particle possessing a momentum $\boldsymbol{p}$.

In this, and subsequent sections, we will consider only systems with Coulomb interactions. We will, in fact, limit ourselves to a system of electrons; the isotopic terms will, consequently, be neglected. From (4.30″) and (4.33′) we have

$$\varepsilon_p(x) = \frac{p^2}{2M} + U(x) + 2e^2 \int \frac{d^3x'}{|\boldsymbol{x} - \boldsymbol{x}'|} \int d^3\boldsymbol{p}' f(\boldsymbol{x}', \boldsymbol{p}')$$

$$- (2\pi)^3\, 4\pi e^2 \int \frac{d^3\boldsymbol{p}'}{(p')^2} f(\boldsymbol{x}, \boldsymbol{p}' + \boldsymbol{p}). \tag{5.4}$$

Ordinarily, the function ε_p increases monotonically with the momentum, so that eqn. (5.2) has only one root. From the properties of the function θ, we may obviously write

$$f(\boldsymbol{x}, \boldsymbol{p}) = \theta[p_0^2(x) - p^2]. \tag{5.3'}$$

The quasi-classical properties of the particle motion are clearly apparent in this expression. Indeed the distribution function (multiplied by a factor of $2/(2\pi)^3$) may be interpreted as the number density of the particles in phase (coordinate–momentum) space. As the last equation indicates, each cell in phase space with a volume $(2\pi)^3$ is occupied by two particles in the region $p^2 < p_0^2(x)$.‡ It is well known that this is a specific characteristic of a quasi-classical system.

The density and energy of a quasi-classical system may be obtained without difficulty from the expression for the distribution

† This expression cannot be applied to the small region round the Fermi surface whose width is about $\xi^{1/2}$ [57].

‡ There are no particles at all in the region $p^2 > p_0^2(x)$. We also note that the region of phase space with $p_0^2(x) < 0$ is not occupied by particles, as is immediately evident from (5.3′).

function. From (4.16) and (4.38), we find

$$\varrho(x) = \frac{p_0^3(x)}{3\pi^2}, \tag{5.5}$$

$$\mathscr{E}_k = \frac{\overline{p_0^5(x)}}{10\pi^2 M}. \tag{6.6}$$

The method discussed in § 4.9 may be applied to the calculation of the total energy. Starting with the relation:

$$\int_{-\infty}^{\varepsilon_F} \bar{\varrho}\, d\varepsilon_F = 2 \int \overline{d^3\boldsymbol{p}\, (\varepsilon_F - \varepsilon_p)\, \theta(\varepsilon_F - \varepsilon_p)}$$

and using (5.4), we can write

$$\mathscr{E} - \mathscr{E}_k = \overline{\varrho U} + \frac{1}{2}\overline{\varrho B} - \frac{e^2}{4\pi^3}\overline{p_0^4}, \tag{5.7}$$

where the potential B is given by (4.30″). The three terms on the right-hand side of (5.7) correspond physically to the energy from the external field and the energies of the direct and the exchange interaction.

These equations contain the unknown function $p_0(x)$. We may determine it by means of an appropriate, self-consistent procedure. If we insert (5.3′) in (5.2), and use the corresponding expressions for $\hat{T}$ and $\hat{W}$, we obtain the Thomas–Fermi equation in integral form:

$$\frac{p_0^2(x)}{2M} + U(x) + \frac{e^2}{3\pi^2} \int \frac{d^3\boldsymbol{x}'}{|\boldsymbol{x} - \boldsymbol{x}'|} p_0^3(x') - \frac{e^2}{\pi} p_0(x) = \varepsilon_F. \tag{5.8}$$

The Fermi energy occurs as a parameter in this equation, and must be determined from the normalization condition:

$$\int d^3\boldsymbol{x}\, \varrho(x) = N.$$

5.3. The simplest example of a quasi-classical system is a uniform system whose characteristics (particularly the Fermi momentum) are independent of the coordinates. The quasi-classical condition (5.1) is then fulfilled in a trivial way, since $\xi \equiv 0$. Hence, there is no difference between the Hartree–Fock and the Thomas–Fermi approximations for a uniform system.

Uniform systems are obtained when the external field $U(x)$ does not distinguish between different points in space. Uniform systems of similarly charged particles with Coulomb interactions must have a $U(x)$ which is, in particular, not equal to zero. Indeed if we put $p_0 = \text{const.}$ in (5.4), (5.7) and (5.8), and suppose that $U = 0$, we find that divergent space integrals $\int d^3x/x$ appear. This reflects the fact that a uniform (and, consequently, spatially unbounded) system of similarly charged particles cannot be in equilibrium. If the problem under consideration is to have any physical significance, it is necessary to add a uniform external background field of the opposite sign which neutralizes the total charge of the system. The potential $U(x) = -e^2\varrho \int d^3x'/|x - x'|$ (ϱ is the number density of particles in the system), which is created by this background, is itself infinite, but its presence completely removes the divergence in (5.4) and (5.8). The latter reduces simply to a relation between the constants ε_F and p_0:

$$\frac{p_0^2}{2M} - \frac{e^2 p_0}{\pi} = \varepsilon_F. \tag{5.9}$$

Equation (5.4) now becomes

$$\varepsilon_p = \frac{p^2}{2M} - \frac{e^2 p_0}{\pi}\left\{1 + \frac{p_0^2 - p^2}{2pp_0}\ln\left|\frac{p + p_0}{p - p_0}\right|\right\}. \tag{5.10}$$

Whence it can be seen that the dispersion law for a particle (i.e. the variation of energy with momentum) is changed by the interaction. The change is smaller, the larger the momentum of the particle. In the converse case of low momenta, the dispersion relation can be approximated by expanding ε_p as a power series in p and retaining only terms up to p^2. The true mass of the particle M is now, however, changed to an effective mass M_{eff} ($\neq M$). These qualitative characteristics of the dispersion law are true in general.

To obtain a finite expression for the energy, we have to add not only the interaction energy of the particle with the background to eqn. (5.7), but also the interaction energy of the background with itself. This leads us to the following simple expression:

$$\mathscr{E} = \frac{p_0^5}{10\pi^2 M} - \frac{e^2 p_0^4}{4\pi^3}. \tag{5.11}$$

The second term is the exchange term.†

In an unbounded system with Coulomb interactions the compensating background is not necessarily uniform. When one considers the electrons in a solid, for example, the compensating charge is provided by the ionic lattice (see § 6).

Similarly charged systems, which are spatially bounded by forces of non-electric nature, can also be in equilibrium in the absence of any compensating charge, e.g. the atomic nucleus.

5.4. We will now consider inhomogeneous systems with Coulomb interactions which can be described by the Thomas–Fermi equation (5.8). We start with the final (exchange) term of the left-hand side of this equation. Compared with the first term, it is of order $e^2 M/p_0 \sim (a_0 p_0)^{-1}$, i.e. the same as the interaction parameter (see § 1).

As will be shown later (see § 5.5), there are also corrections to the Thomas–Fermi equation of the same order of magnitude due to the inaccuracy of the quasi-classical approximation (such terms will be called in future the quantum corrections). If these are not to be included in the equation, the exchange term, too, should be omitted. This leads to the normal Thomas–Fermi equation:‡

$$\frac{p_0^2(x)}{2M} = \varepsilon_F - U(x) - \frac{e^2}{3\pi^2} \int \frac{d^3 x'}{|x - x'|} p_0^3(x'). \tag{5.12}$$

† It is evident from the foregoing equations that the Hartree–Fock approximation for uniform systems is equivalent to considering only terms of the zero and first order in the interaction parameter $(a_0 p_0)^{-1}$. On the other hand, an inhomogeneous system (e.g. an atom) may, on this approximation be accounted for to all orders of the parameter. The reason is that, in the latter case, the wave function, $\chi_\nu(q)$ of an individual state of the particle is a complicated function of the quantity $(a_0 p_0)^{-1}$. In the uniform case the functions $\chi_\nu(q)$ may be chosen to be plane waves (as for a perfect gas) owing to the translational symmetry.

‡ If the exchange terms are retained in the quasi-classical equation (5.8) (which is then called the Thomas–Fermi–Dirac equation), this means that a certain number of terms with a given order of magnitude are being retained, whilst others of the same order are omitted. It seems more logical to consider exchange and quantum corrections on the same footing [58].

Applying the Laplace operator to both sides of this equation, we obtain the Thomas–Fermi equation in differential form

$$\nabla^2 p_0^2(x) = \frac{8\pi}{a_0} [\varrho(x) - \sigma(x)], \tag{5.13}$$

where $\varrho(x)$ is the number density of particles in the system (5.5); $\sigma = \nabla^2 U/4\pi$ is the charge density of the sources providing the external field (divided by e).

5.5. We next turn to the relative significance of the exchange and quantum corrections. The order of the latter is defined by the parameter ξ^2 (if any physical quantity is expanded in ξ, the odd terms vanish identically [4]). In the most important case when one can ignore the region immediately adjacent to a point source in the external field (the nucleus), the exchange and quantum corrections are of the same order of magnitude. If we compare the left-hand side and the first term on the right-hand side of (5.13), we arrive at the order of magnitude estimate

$$\nabla^2 p_0^2 \sim p_0^2/x_0^2 \sim p_0^3/a_0,$$

whence

$$(x_0 p_0)^{-2} \sim (a_0 p_0)^{-1}. \tag{5.14}$$

Here x_0 is a characteristic parameter, with the dimensions of length, which represents the distance over which p_0 varies appreciably. As can be seen from (5.1), the left-hand side of (5.14) coincides with ξ^2, thus confirming that the quantum and exchange corrections are of the same order of magnitude.

It follows from (5.14) that the size of the quantum (and exchange) corrections diminishes as the system increases in density. Indeed if we substitute the density from (5.5) into this relation, we have

$$\xi^2 \sim (a_0 \varrho^{1/3})^{-1}. \tag{5.14'}$$

In the region close to a nucleus these estimates are inappropriate. In this region

$$U(r) \sim -\frac{Ze^2}{r}, \quad p^2(x) \sim \frac{2Z}{a_0 r},$$

where r is the distance from the nucleus. Hence,

$$\xi^2 \sim \left|\frac{\partial(1/p_0)}{\partial r}\right|^2 \sim a_0/Zr,$$

$$(a_0 p_0)^{-1} \sim (r/a_0 Z)^{1/2}.$$

Thus in the region $r \leqq a_0/Z$, the quantum corrections are large, whilst the exchange corrections are small.

5.6. The smallness of the parameter (5.14) indicates that the interaction energy of a pair of particles is small compared with their kinetic energy. However, it is not possible to deduce from this that the total interaction energy of a particle in the system is small compared with its total kinetic energy (see § 1.6).

To examine this question, we estimate the magnitude of the potential $U + B$. A particle in a system interacts only with those particles which are less than some distance x_1 away. For systems which are spatially bounded (the atom, the nucleus), this distance coincides with the size of the system. If, on the other hand, we consider an infinitely extended system, then a given particle is subject to screening due to the electrostatic potential of the compensating background, with x_1 representing the screening length. In the solid state, for example, x_1 is of the order of the distance between the lattice atoms. The potential can be estimated by substituting x_1 into the appropriate expression:

$$U + B \sim e^2 p_0^3 x_1^2.$$

The ratio of this quantity to the kinetic energy is $p_0 x_1^2/a_0 \sim (p_0 x_1)^2 (a_0 p_0)^{-1}$. This parameter, which also gives the ratio of the total interaction energy to the kinetic energy, can be considerably larger than $(a_0 p_0)^{-1}$. For example, in a compressed solid state, $x_1 \sim Z^{1/3}/p_0$ and $(p_0 x_1)^2 \sim Z^{2/3}$; for an atom, $p_0 \sim Z^{2/3}/a_0$, $x_1 \sim x_0 \sim a_0 Z^{-1/3}$, $(a_0 p_0)^{-1} \sim \xi^2 \sim Z^{-2/3}$ [59], and this ratio is of the order of unity.

The above result does not apply to the uniform case. Here, $x_0 \to \infty$, and eqn. (5.13) reduces to the identity $0 \equiv 0$. Since U and B completely compensate each other, the parameter x_1, on the other hand, goes to zero.† As a result, only one parameter is left to characterize the particle interaction, namely $(a_0 p_0)^{-1}$.

The Coulomb interaction between particles in an inhomogeneous system can thus be of basic importance, even if the parameter $(a_0 p_0)^{-1}$ is small compared with unity. This is due to the long-range nature of the Coulomb force, which has the effect that, for sufficiently large values of the parameter $(p_0 x_1)^2$, a large number of particles interact with any given particle.

5.7. If we wish to examine the limits within which the Thomas–Fermi equation applies, and also find the necessary corrections to this equation, we must consider terms to the order of $\xi^2 \sim (a_0 p_0)^{-1}$ in the Hartree–Fock equations. To this approximation, quantum and exchange effects may be considered independently of each other. We will begin with the exchange effects.

If we apply the Laplace operator to both sides of eqn. (5.8), we find

$$\nabla^2 p_0^2 - \frac{8\pi}{a_0}\left(\frac{p_0^3}{3\pi^2} - \hat{\sigma}\right) = \frac{2}{\pi a_0}\nabla^2 p_0 .$$

The exchange term, which should be considered small, is on the right-hand side. We will therefore divide p_0^2 into two parts: one corresponding to eqn. (5.12) and one corresponding to an exchange correction, $p_0^2 \to p_0^2 + \delta_1 p_0^2$. Since the latter is small, we can write

$$\left(\nabla^2 - \frac{4}{\pi a_0} p_0\right)\delta_1 p_0^2 = \frac{2}{\pi a_0}\nabla^2 p_0 . \tag{5.15}$$

The solution of this equation is chosen so as to conform with the normalization condition. In other words the equality

$$\overline{\delta_1 \varrho} = \frac{1}{2\pi^2}\overline{p_0 \delta_1 p_0^2} = 0 \tag{5.16}$$

must be satisfied.

We obtain the correction to the total energy due to exchange forces by substituting $p_0^2 \to p_0^2 + \delta_1 p_0^2$ in (5.6) and in the first two

† This conclusion only holds if correlation effects are ignored. The inclusion of these leads to Debye screening.

terms of (5.7). The last term in (5.7), denoted by $\mathscr{E}_1$, is of a purely exchange nature, and must therefore be taken into account completely. Hence

$$\delta_1\mathscr{E} = \frac{1}{4\pi^2 M}\overline{p_0^3\delta_1 p_0^2} + \overline{(U+B)\,\delta_1\varrho} + \mathscr{E}_1. \quad (5.17)$$

If we take into account (5.12) and the normalization condition, the first two terms in (5.17) cancel, and we finally find

$$\delta_1\mathscr{E} = \mathscr{E}_1 = -\frac{e^2}{4\pi^3}\overline{p_0^4}. \quad (5.18)$$

5.8. We must now determine the quantum corrections to the Thomas–Fermi approximation [51, 54], due to the inaccuracy of the quasi-classical approximation. The concept we have used of cells in phase space is evidently inaccurate, since we have, to a great degree, ignored the uncertainty principle. A formal source of the quantum effects may thus be traced to the non-commutation of the coordinate and momentum operators.

When we obtained the Thomas–Fermi equation (5.8), we based our argument on expression (5.3) for the distribution function. This expression is, however, inaccurate, since we did not allow for the non-commutation of the operators in the exact equation (4.34′). We will write the operator function in (4.34′) in the form $\theta(\hat{a} + \hat{b})$, where $\hat{a} = -\hat{p}^2 = \nabla^2$ and $\hat{b} = p_0^2(x)$; the exchange terms are omitted.

In fact (5.3) is the zero term in the expansion of the distribution function in ξ. The terms of higher order originate in the commutators of the operators $\hat{a}$ and $\hat{b}$; the power of ξ for each term in the expansion is determined by the number and type of the commutators in it.

If we limit ourselves to terms of second order in ξ, we need consider only those commutators and their products which require not more than two commutation operations. In fact in the transition to commutators of a more complicated type, each new commutation operation leads to at least one additional differentiation of the function $p_0^2(x)$, i.e. at least one extra power of ξ.

A general formula is developed in Appendix B for expanding an arbitrary function, $f(\hat{a} + \hat{b})$, in terms of the commutators to the desired degree of accuracy [see (B.17)]:

$$f(\hat{a} + \hat{b}) = f(a + \hat{b}) - \tfrac{1}{2}f''(a + \hat{b})\,[\hat{b}, \hat{a}]_- + \tfrac{1}{6}f'''(a + \hat{b})$$
$$\times \{[\hat{b},[\hat{b}, \hat{a}]_-]_- - [\hat{a},[\hat{b}, \hat{a}]_-]_-\} + \tfrac{1}{8}f^{\mathrm{IV}}(a + \hat{b})\,[\hat{b}, \hat{a}]_-^2 + \cdots \quad (5.19)$$

It is assumed here that the operator $f(\hat{a} + \hat{b})$ is acting on an eigenfunction of the operator $\hat{a}$ with an eigenvalue a. This is precisely the situation in the case we are considering: the function $\theta[p_0^2(x) - \hat{p}^2]$ acts on $\exp i(\boldsymbol{p}\cdot\boldsymbol{x})$; $a + \hat{b}$ in the foregoing equation corresponds to $p_0^2(x) - p^2$.

To the required order of accuracy, the commutators figuring in (5.19) are

$$[\hat{b}, \hat{a}]_- = -\nabla^2 p_0^2 - 2(\nabla p_0^2)\cdot\nabla,$$
$$[\hat{b},[\hat{b}, \hat{a}]_-]_- = 2(\nabla p_0^2)^2,$$
$$[\hat{a},[\hat{b}, \hat{a}]_-]_- = -4(\nabla_i\nabla_k p_0^2)\,\nabla_i\nabla_k,$$
$$[\hat{b}, \hat{a}]_-^2 = 4(\nabla_i p_0^2)\,(\nabla_k p_0^2)\,\nabla_i\nabla_k.$$

Using (4.34′), we arrive at the following expression for the distribution function with quantum corrections:

$$f(\boldsymbol{x}, \boldsymbol{p}) = \theta(p_0^2 - p^2) + \tfrac{1}{2}[(\nabla^2 p_0^2 + 2i(\boldsymbol{p}\cdot\nabla p_0^2)]\delta'(p_0^2 - p^2)$$
$$+ \tfrac{1}{3}\,[(\nabla p_0^2)^2 - 2(\boldsymbol{p}\cdot\nabla)^2\,p_0^2]\,\delta''(p_0^2 - p^2)$$
$$- \frac{(\boldsymbol{p}\cdot\nabla p_0^2)^2}{2}\,\delta'''(p_0^2 - p^2). \quad (5.20)$$

5.9. If this expression is substituted into (4.16), a corrected expression for the number density of the particles is obtained [54, 55, 58, 60, 61]:

$$\left.\begin{aligned} \varrho &= \frac{p_0^3}{3\pi^2} + \varrho_2, \\ \varrho_2 &= \frac{-1}{96\pi^2 p_0^3}\,[(\nabla p_0^2)^2 - 4p_0^2\nabla^2 p_0^2]. \end{aligned}\right\} \quad (5.21)$$

It is convenient to look for an expression for the kinetic energy by means of the method described in § 4. We may note that, in agreement with (5.12), only p_0^2 depends on ε_F (not ∇p_0^2, $\nabla^2 p_0^2$, etc.), and the dependence is linear. The integration of (5.21) in terms of ε_F may, therefore, be done very simply. An application of (4.38) leads to the following expression for the kinetic energy:

$$\left.\begin{aligned} \mathscr{E}_k &= \frac{\overline{p_0^5}}{10\pi^2 M} + \mathscr{E}_{k2}, \\ \mathscr{E}_{k2} &= -\frac{1}{192\pi^2 M p_0}\,[3\overline{(\nabla p_0^2)^2} + 4\overline{p_0^2\,\nabla^2 p_0^2}]. \end{aligned}\right\} \quad (5.22)$$

If we now use (5.21) to introduce ϱ as the argument, we find

$$\mathscr{E}_k = \frac{3}{10M}(3\pi^2)^{2/3}\,\overline{\varrho^{5/3}} + \frac{1}{72M}\left(\frac{\overline{(\nabla\varrho)^2}}{\varrho} - 6\overline{\nabla^2\varrho}\right). \quad (5.23)$$

The second term of this expression is similar in form to the familiar Weizsäcker correction [62], but has a coefficient which is only one-ninth as large [54, 58].

The fourth-order correction terms in ξ can be determined by a similar, though more tedious, method. Omitting the intermediate steps [51, 57], we have the final expression:

$$\begin{aligned} \varrho_4 = \frac{1}{30720\pi^2 p_0^9}\{&-64p_0^6\,\nabla^4 p_0^2 + 80p_0^4(\nabla^2 p_0^2)^2 + 192p_0^4(\nabla p_0^2\cdot\nabla\nabla^2 p_0^2) \\ &+ 64p_0^4(\nabla_i\nabla_k p_0^2)^2 - 200\nabla^2 p_0^2(\nabla p_0^2)^2 \\ &- 240p_0^2\nabla_i\nabla_k p_0^2\nabla_i p_0^2\nabla_k p_0^2 + 175(\nabla p_0^2)^4\}. \end{aligned} \quad (5.24)$$

The corresponding expression for $\mathscr{E}_{k4}$ differs from (5.24) by an additional factor $p_0^2/6M$, and by the fact that the coefficients within the parantheses are replaced by -576, 400, 960, 320, -840, -1008 and 675, respectively.

These expressions for the correction terms indicate clearly how rapidly the numerical coefficients decrease when a physical quantity is expanded in terms of ξ.

5.10. The quantity $p_0^2(x)$ which enters into the above equations is not the same as the quasi-classical term, since it also contains the quantum corrections. To determine these, we substitute (5.21) into (5.13), and put $p_0^2 \to p_0^2 + \delta_2 p_0^2$:

$$\left(\nabla^2 - \frac{4}{\pi a_0} p_0\right) \delta_2 p_0^2 = -\frac{1}{12\pi p_0^3 a_0} [(\nabla p_0^2)^2 - 4p_0^2 \nabla^2 p_0^2].$$

Using the identity

$$\frac{(\nabla p_0^2)^2}{p_0^3} \equiv \frac{2\nabla^2 p_0^2}{p_0} - 4\nabla^2 p_0$$

and substituting $\nabla^2 p_0^2 \to (8/3\pi a_0)\, p_0^3$,† we may easily confirm that there is a simple connection between the quantum $(\delta_2 p_0^2)$ and the exchange $(\delta_1 p_0^2)$ corrections to p_0^2 (see § 5.7):

$$\delta_2 p_0^2 = \frac{2}{9} \delta_1 p_0^2 - \frac{1}{9\pi a_0} p_0. \tag{5.25}$$

The quantum correction to the density has, from eqn. (5.21), the following form:

$$\delta_2 \varrho = \frac{1}{2\pi^2} p_0\, \delta_2 p_0^2 + \varrho_2 = \frac{1}{9\pi^2} p_0 \delta_1 p_0^2 + \frac{1}{24\pi^2} \nabla^2 p_0 .$$

Since the last term disappears on integration, it follows that, so long as the condition (5.16) is fulfilled, the quantum corrections to p_0^2 will be correctly normalized. We thus see that the exchange and quantum corrections are of the same order of magnitude and, moreover, have the same functional dependence on the coordinates.

Finally, we must determine the quantum correction to the energy of the system. The calculations are similar to those performed in § 5.7, replacing the index 1 by 2 in (5.17), and interpreting $\mathscr{E}_2$ as the quantity $\mathscr{E}_{k2}$ in (5.22). Using the Thomas–Fermi equation and the normalization condition, we have

$$\delta_2 \mathscr{E} = \mathscr{E}_2 - \overline{\varrho_2 p_0^2 / 2M},$$

whence

$$\delta_2 \mathscr{E} = -\frac{1}{96\pi^2 M} \left(\overline{\frac{(\nabla p_0^2)^2}{p_0} + 4 p_0 \nabla^2 p_0^2} \right). \tag{5.26}$$

† We omit the term $-8\pi\sigma$ since it occurs in combination with p_0^{-1}. If, as is supposed here, we are considering only point sources for the external field, this combination reduces to zero.

6. Applications to the Theory of Strongly Compressed Matter

6.1. The theory of the foregoing paragraphs can be used successfully to describe the properties of strongly compressed substances. A strongly compressed substance is one subject to a high external pressure P which obeys the condition

$$P \gg \mathscr{E}_0, \tag{6.1}$$

where $\mathscr{E}_0$ represents the order of magnitude of the energy density in the uncompressed state of the system. $\mathscr{E}_0$ may also be interpreted as the "internal" pressure of the electron gas in the substance due to the kinetic energy of the electrons. In an uncompressed substance this pressure is compensated by the Coulomb forces.

Substances, which are subject to low pressures only, possess extremely diverse properties. Their properties have a strong irregular dependence on the chemical composition. When these substances are subject to compression there is a clear tendency for these individual properties to be smoothed out, so that, for pressures corresponding to (6.1), the dependence becomes comparatively smooth and monotonic.† The theory is therefore considerably simplified.

At zero temperature, we may assume that the substance is in a solid (crystalline) state [50, 63] (in fact we know that this statement can be true for highly compressed substances which have very high temperatures, such as are found in stellar interiors‡). We may note, in this connection, that liquid helium solidifies under a pressure of a few tens of atmospheres.

6.2. We can estimate $\mathscr{E}_0$ in (6.1) very easily in the following way. If we denote the electron energy by ε, the number of particles by N and some characteristic length by L, we can write $\mathscr{E}_0 \sim N\varepsilon/L^3$.

† When (6.1) is fulfilled, the outer electron shells of the atoms in the substance are stripped off, and it is these shells which are the cause of the non-monotonic behaviour.

‡ As analysis [50] shows the increase in oscillation amplitude of nuclei, $A \sim a_0(R/a_0)^{3/4}$, with increase in density can lead to fusion of a solid substance when $A \sim R$ (where R is the average distance between nuclei; $a_0 = (MZ^2e^2)^{-1}$ is the Bohr radius of the nucleus). The corresponding compression is so high, that, for lower pressures than these, the transition to a neutron state due to electron capture by nuclei must occur [4].

For simplicity, we will consider a substance made up of one type of atom only, with a nuclear charge $Z \gg 1$. We must distinguish three characteristic scale sizes for $\mathscr{E}_0$.

In the peripheral region of the atom, $\varepsilon \sim e^2/a_0$, $N \sim 1$, $L \sim a_0$, whence the internal pressure has the following order of magnitude,

$$\mathscr{E}_0^{(1)} \sim \frac{e^2}{a_0^4}.$$

In the central region of the atom the number of electrons is of the order of the total number of particles $N \sim Z$, the scale length $L \sim a_0 Z^{-1/3}$ and the particle energy $\varepsilon \sim p_0^2/M \sim e^2/a_0 Z^{4/3}$ (see § 5.5). Hence

$$\mathscr{E}_0^{(2)} \sim Z^{10/3} \frac{e^2}{a_0^4}.$$

Finally, in the inner part of the atom, which has dimensions of the same order as the K-shell, we have $L \sim a_0 Z^{-1}$, $\varepsilon \sim Z^2 e^2/a_0$ and $N \sim 1$. Therefore

$$\mathscr{E}_0^{(3)} \sim Z^5 \frac{e^2}{a_0^4}.$$

The $\mathscr{E}_0$ in (6.1) should correspond to the smallest of these parameters, namely $\mathscr{E}_0^{(1)}$. Thus the lower limit of the region containing strongly compressed substances should be given by pressures of the order $e^2/a_0^4 \sim 10^8$ atmos.†

It is convenient to divide the region where strongly compressed substances are found into three parts:

$$\text{(I) } \mathscr{E}_0^{(1)} < P < \mathscr{E}_0^{(2)}, \quad \text{(II) } \mathscr{E}_0^{(2)} < P < \mathscr{E}_0^{(3)}, \quad \text{(III) } \mathscr{E}_0^{(3)} < P.$$

In region I the external pressure can only strip off the outer electrons. The inner electron shells are compressed, and their electron density distribution has a comparatively slow spatial variation. In region II a considerable number of electrons have been removed, which move freely through most of the space and

† Up to the moment, pressures of the order of 10^7 atmos. have been attained under laboratory conditions. In nature, strongly compressed substances occur in the cores of many astronomical objects. Especially high pressures ($\sim 10^{17}$ to 10^{20} atmos.) are expected in the stars known as "white dwarfs".

have a practically uniform distribution. Finally, in region III the electrons and nuclei form distinct entities. A substance in this region is a nuclear lattice permeated by a nearly perfect electron gas.†

6.3. The most important property of a strongly condensed substance is its equation of state, i.e. the variation of density with external pressure. To determine the exact relation, we can use the familiar equation:

$$P = -\left(\frac{\partial E}{\partial \Omega}\right)_N, \tag{6.2}$$

where E is the total energy of the system, Ω is its volume and N is the total number of electrons. The average density of the latter is $\bar{\varrho} = N/\Omega$.

Since the ratio of the electron mass to the nuclear mass is small, we may suppose that the lattice configuration is fixed. We will therefore consider the field created by the nuclei to be an external field:

$$U = -Ze^2 \sum_n \frac{1}{|\boldsymbol{x} - \boldsymbol{x}_n|},$$

where the summation is taken over all the lattice points. The equation of state is comparatively little changed for the range of pressures under consideration, if the oscillations of the lattice points are also included.

The total energy of the system may be conveniently written in the form

$$E_k = \langle \Psi | \int dq\, \hat{\psi}^+(q) \frac{\hat{p}^2}{2M} \hat{\psi}(q) | \Psi \rangle,$$

$$E - E_k = \left\langle \Psi \left| \int dq\, \hat{\psi}^+(q)\, U \hat{\psi}(q) + \frac{e^2}{2} \int \frac{dq\, dq'}{|\boldsymbol{x} - \boldsymbol{x}'|} \hat{\psi}^+(q)\, \hat{\psi}^+(q') \right.\right.$$

$$\left.\left. \times\, \hat{\psi}(q')\, \hat{\psi}(q) \right| \Psi \right\rangle + \frac{Z^2 e^2}{2} {\sum_{n,m}}' \frac{1}{|\boldsymbol{x}_n - \boldsymbol{x}_m|}. \tag{6.3}$$

† We will be limiting our consideration to pressures which produce only non-relativistic electrons. That is, we impose the condition $P \ll 10^{17}$ atmos.

The last term in the expression for $E - E_k$ takes into account the interaction of the nuclei with one another (less their self-interaction).

We will introduce in (6.3) a new dimensionless variable $\boldsymbol{x}' = \Omega^{-1/3}\boldsymbol{x}$. The normalization condition now gives $\hat{\psi}(\boldsymbol{x}) = \Omega^{-1/2} \times \hat{\psi}(\boldsymbol{x}')$; the momentum operator $\hat{\boldsymbol{p}} = -i\nabla_x = -i\Omega^{-1/3}\nabla_{x'}$. Hence it is evident that E_k is a homogeneous function of Ω of degree $-\frac{2}{3}$; $E - E_k$ is of degree $-\frac{1}{3}$. Applying Euler's theorem for homogeneous functions, $x\,\partial f_n/\partial x = nf_n$ (where n is the degree of the homogeneous function), we find eventually

$$P = \frac{1}{3\Omega}(2E_k + E - E_k) = \frac{1}{3}(\mathscr{E} + \mathscr{E}_k). \tag{6.4}$$

This is a very useful expression, since from it, the equation of state may be determined immediately from the kinetic and total energies of the substance.

6.4. As we will see in Chapter 4, a strongly compressed substance may be described with sufficient accuracy by the Hartree–Fock approximation. It seems, moreover, that, so long as (6.1) holds, the quantum (and therefore the exchange) corrections are small. This follows at once from an estimate of the size of ξ^2 [see (5.14′). When (6.1) is fulfilled, the electron density ϱ is necessarily larger than a_0^{-3} (see § 6.2), which also leads to $\xi^2 \ll 1$. It should be emphasized in this connection that the region near the nucleus, which needs separate consideration, only affects the equation of state for the highest compressions. Section 6.7 is devoted to an estimate of the quantum effects, which appear to be unimportant under these conditions.

Thus the Thomas–Fermi approximation may be used to deduce the equation of state of a highly compressed substance. If we substitute (5.6) and (5.7) (the latter without an exchange term) into (6.4), we obtain

$$P = \frac{\overline{p_0^5}}{15\pi^2 M} - \frac{Ze^2}{3}\sum_n \frac{\overline{\varrho(x)}}{|\boldsymbol{x} - \boldsymbol{x}_n|} + \frac{e^2}{6}\overline{\varrho B} + \frac{Z^2e^2}{6}\sum_{n,m}{}' \frac{1}{|\boldsymbol{x}_n - \boldsymbol{x}_m|}, \tag{6.5}$$

where ϱ and B are given by (5.5) and (4.30″). P must be determined by solving the Thomas–Fermi equation (5.12) for $p_0^2(x)$ with the correct boundary conditions.

6.5. Equation (5.12) has a particularly simple solution in the regions II and III, i.e. when the condition $P \gg Z^{10/3}e^2/a_0^4$ is fulfilled. As we have observed, the electrons in this case may generally be thought of as free and their density distribution is practically uniform.

We will look for a solution of eqn. (5.12) in the form

$$p_0^2(x) = (3\pi^2\bar{\varrho})^{2/3} + \alpha(x),$$

where the first term represents a uniform distribution, and the second is a small correction for inhomogeneity. Substituting this into (5.12) gives

$$\alpha(x) = 2M[\delta\varepsilon_F - U(x) - B(x)],$$

where $B(x) = e^2\bar{\varrho}\int d^3x'/|x - x'|$ and $\delta\varepsilon_F = \varepsilon_F - (3\pi^2\bar{\varrho})^{2/3}/2M$. We find from the normalization condition that $\delta\varepsilon_F = \overline{U} + \overline{B}$. We next introduce a dimensionless variable, $y = (\bar{\varrho}/Z)^{1/3}\,x$, equal to the ratio of x to the average distance between nuclei. So, finally

$$p_0^2(x) = (3\pi^2\bar{\varrho})^{2/3}\left\{1 + \frac{2}{(3\pi^2)^{2/3}}\cdot\frac{Z^{2/3}}{a_0\bar{\varrho}^{1/3}}\,[Q(y) - \overline{Q}]\right\}, \quad (6.6)$$

where

$$Q(y) = \int\frac{d^3y'}{|y - y'|} - \sum_n \frac{1}{|y - y_n|}.$$

The second term in the brackets in (6.6) is the correction for inhomogeneity, and is of the order $Z^{2/3}/(a_0\bar{\varrho}^{1/3})$. We can determine the equation of state to the same accuracy. If (6.6) is substituted in the first term of (6.5), we obtain the term $(5M)^{-1}\,(3\pi^2)^{2/3}\,\bar{\varrho}^{5/3}$ (owing to the normalization condition, corrections for inhomogeneity do not occur). The remaining terms in the expression for P are already at the limits of the accuracy to which we are working. We may therefore calculate them by putting $p_0^2(x) = (3\pi^2\bar{\varrho})^{2/3}$, which, on transformation to the variable y, gives

$$\frac{e^2}{6\Omega}Z^{5/3}\,\bar{\varrho}^{1/3}\left[\int\frac{d^3y\,d^3y'}{|y - y'|} - 2\sum_n\int\frac{d^3y'}{|y' - y_n|} + \sum_{n,m}{}'\frac{1}{|y_n - y_m|}\right].$$

The quantity in square brackets is proportional to the total number of particles, and will be denoted by $-(6N/Z)\,\theta$.

We finally obtain the following equation of state

$$P = \frac{1}{5M}(3\pi^2)^{2/3}\,\bar{\varrho}^{5/3} - \theta e^2 \bar{\varrho}^{4/3} Z^{2/3}. \tag{6.7}$$

The coefficient θ is a dimensionless quantity of the order of unity, which must be calculated from the electrostatic configuration of the lattice. Its physical significance is similar to that of the Madelung constant [64].

6.6. The feasibility of finding a solution of the Thomas–Fermi equation for low pressures is limited by the complicated geometry of the problem (due to the large number of sources of the external field, i.e. the nuclei of the lattice). Frequent use is therefore made of a simplified cell model [59, 65].†

It is supposed in this model that a substance consists of neutral spherically symmetric cells, each of which contains one nucleus. The cells are subject to the boundary condition $\partial p_0^2(x)/\partial x = 0$, which ensures their neutrality. In this case the cells have no electrostatic affect on each other, and the problem reduces to the consideration of a single, isolated cell.

The quantities Q and θ may be easily calculated on this model. The radius of a neutral cell is determined by $4\pi R^3\bar{\varrho}/3 = Z$; in dimensionless units this corresponds to $(3/4\pi)^{1/3}$. Elementary calculations show that

$$Q(y) = \frac{2\pi}{3}\left[3 - \left(\frac{4\pi}{3}\right)^{2/3} y^2 + 2\left(\frac{3}{4\pi}\right)^{1/3}\frac{1}{y}\right], \tag{6.8}$$

where y is the distance from the centre of the shell. The constant θ is given by

$$\theta = \frac{3}{10}\left(\frac{4\pi}{3}\right)^{1/3}. \tag{6.9}$$

† Several calculations have been made with more realistic models which give better results at comparatively low pressures [66, 67]. The model we have adopted cannot describe the different kinds of phase transitions which are observed when a substance is compressed.

These results refer to regions II and III. To deduce the equation of state for a substance in region I, we must integrate the Thomas–Fermi equation. For the cell model, this takes the form

$$\nabla^2 p_0^2(x) = \frac{8}{3\pi a_0} p_0^3(x)$$

with the boundary conditions

$$x p_0^2(x)\Big|_0 = \frac{2Z}{a_0}, \quad \frac{\partial p_0^2(x)}{\partial x}\Big|_R = 0.$$

The corresponding calculations have been made by Latter [68]. The results for iron are given in Fig. 1 (curve 1).

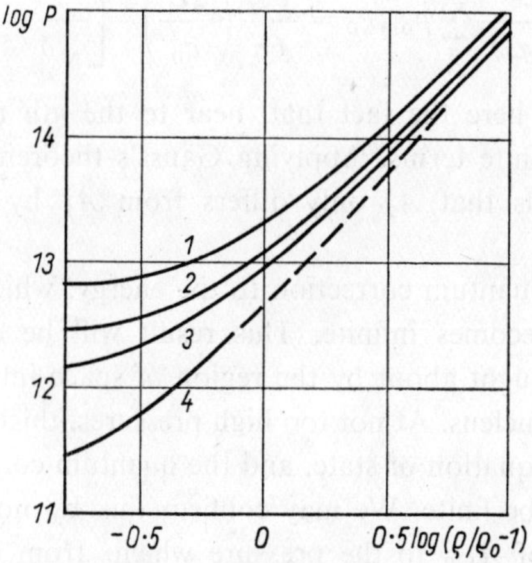

FIG. 1

6.7. We now proceed to a calculation of the exchange and quantum corrections to the pressure; these, once known, enable us to examine the limits within which the quasi-classical equation of state applies.†

† The literature contains a vast amount of numerical material concerned with the quasi-classical equation of state for substances subject to a wide range of pressures [68, 69]. A fair amount of these results are, however, incorrect, because the highly important quantum effects have been ignored.

We will begin by calculating the quantum correction. Using

$$\frac{(\nabla p_0^2)^2}{p_0} = -2p_0 \nabla^2 p_0^2 + \frac{4}{3} \nabla^2 p_0^3$$

and eqn. (5.13), we can reduce (5.26) to the form

$$\delta_2 \mathscr{E} = -\frac{e^2}{18\pi^3} \overline{p_0^4} + A_1 + A_2, \tag{6.10}$$

where

$$A_1 = \frac{e^2}{6\pi} \overline{\sigma p_0}; \quad A_2 = -\frac{1}{72\pi^2 M} \overline{\nabla^2 p_0^3}.$$

Inserting the expression for $\sigma = Ze \sum_n \delta(\boldsymbol{x} - \boldsymbol{x}_n)$, we can write for $A_1 (\delta \equiv |\boldsymbol{x} - \boldsymbol{x}_n|)$:

$$A_1 = \frac{Ze^2}{6\pi\Omega} \sum_n p_0|_{\delta \to 0} = \frac{e^2 \bar{\varrho}}{6\pi} \left(\frac{2Z}{a_0}\right)^{1/2} \left[\frac{1}{\sqrt{\delta}} + O(\sqrt{\delta})\right]$$

We are using here the fact that, near to the nth nucleus, $p_0^2(x) = 2Z/a_0\delta$ + finite terms. Applying Gauss's theorem, it is immediately obvious that A_2 only differs from A_1 by a numerical coefficient.

Hence the quantum correction to the energy, which is of lower order in ξ, becomes infinite. This result will be considered in § 6.8. It is brought about by the region of space immediately adjacent to the nucleus. At not too high pressures, this region has no affect on the equation of state, and the quantum correction to the pressure may be finite. We may confirm this by noting that the contribution of $A_{1,2}$ to the pressure which, from (6.2) is equal to $-[\partial(A_{1,2}\Omega)/\partial\Omega]_N$, goes identically to zero ($\bar{\varrho}\Omega = N$).

The quantum correction to the pressure may thus be written as

$$\delta_2 P = \frac{e^2}{18\pi^3} \cdot \frac{\partial(\overline{p_0^4}\Omega)}{\partial\Omega}. \tag{6.11}$$

The exchange correction to the pressure can be written in an analogous way, using (5.18):

$$\delta_1 P = \frac{e^2}{4\pi^3} \cdot \frac{\partial(\overline{p_0^4}\Omega)}{\partial\Omega}.$$

We are led to conclude that the ratio of the quantum and exchange effects to the pressure is constant, and equal to 2/9 [70].

$$\frac{\delta_2 P}{\delta_1 P} = \frac{2}{9}. \tag{6.12}$$

Using these results, we can find directly the quantum and exchange corrections to the pressure in region II if we substitute $(3\pi^2 \bar{\varrho})^{1/3}$ for p_0:

$$\delta_1 P = -\frac{e^2}{4\pi}(3\pi^2)^{1/3}\,\bar{\varrho}^{4/3}, \tag{6.13}$$

$$\delta_2 P = -\frac{e^2}{18\pi}(3\pi^2)^{1/3}\,\bar{\varrho}^{4/3}. \tag{6.14}$$

Comparing these expressions with (6.7), we see that they differ from the correction for inhomogeneity only by the very small factor $Z^{-2/3}$.

6.8. At lower pressures (region I), the exchange and quantum corrections to the pressure are most conveniently determined from eqn. (6.4). Putting $\delta_1 \mathscr{E}_k = \overline{p_0^3 \delta_1 p_0^2}/4\pi^2 M$, we find from eqn. (5.18)

$$\delta_1 P = \frac{1}{12\pi^2 M}\overline{p_0^3\left(\delta_1 p_0^2 - \frac{p_0}{\pi a_0}\right)}. \tag{6.15}$$

The quantum correction can be determined from (6.12).

Corresponding calculations have been made for the cell model [71].† The results are given in Fig. 1.

Curve 1 corresponds to the quasi-classical equation of state; curve 2 to the Thomas–Fermi–Dirac model; curve 3 to the equation with quantum and exchange corrections; curve 4 is taken from the experimental results [72]. All the data refer to iron ($Z = 26$).‡

It is evident that there is a general tendency for the calculated and empirical curves to approximate to one another. In particular

† Actually, a slightly different method was used, but one that was fully equivalent to the one discussed here.

‡ The quasi-classical pressure variation with Z is given by $P = Z^{10/3} f(\bar{\varrho}/Z^2)$; the corresponding variation of the quantum and exchange corrections by $\delta P = Z^{8/3}\varphi(\bar{\varrho}/Z^2)$.

the negative contribution to the pressure of the exchange and quantum corrections is in good agreement with experiment.

The pressures so far obtained are too small for a substance to be considered highly compressed in the sense of § 6.1. At the maximum pressures which have so far been achieved, the quantum and exchange corrections are still of the same order as the pressure.

6.9. We must still find the quantum and exchange corrections to the pressure in region III, corresponding to a pressure $P > Z^5e^2/a_0^4$, and a density $\bar{\varrho} > Z^3/a_0^3$. The exchange correction in this region is given, as before, by (6.13), but the quantum correction requires a special consideration.

It was remarked in § 5.5 that, near the nucleus, quantum effects become important. In this region quantum corrections of all orders in ξ are essential. Thus, the low-order correction to the energy which we encountered in § 6.7 tends to infinity.†

We will write the expression for p_0^2 near the nucleus in the form

$$p_0^2(x) = 2M\varepsilon_F + \frac{2Z}{a_0\delta},$$

where δ is the distance to the nearest nucleus; the potential of the remaining nuclei and electrons is ignored. The limiting energy ε_F may be written, with sufficient accuracy, as $(3\pi^2\bar{\varrho})^{2/3}/2M$. Comparing the two terms in p_0^2, we see that the region of the atom unaffected by the external pressure (where the second term predominates) corresponds to a distance $\delta < x_2 \equiv Z/a_0\bar{\varrho}^{2/3}$.

On the other hand, we can find easily enough the size of the atomic region for which quantum corrections of all orders are important, i.e. $\xi \gtrsim 1$. From the definition $\xi = (1/p_0^3) \cdot (dp_0^2(x)/dx)$, we find the radius of this region to be:

$$x_3 \sim \left(\frac{Z}{a_0\bar{\varrho}}\right)^{1/2}.$$

It is immediately obvious that in region III

$$\frac{x_2}{x_3} \sim \left(\frac{Z^3}{a_0^3\bar{\varrho}}\right)^{1/6} < 1.$$

† The energy is a non-analytic function of ξ in this region, which prevents its being expanded as a series in ξ.

Thus the pressure is appreciably changed in this region by that part of the atom for which higher-order quantum corrections are of importance. It is therefore impossible, in this instance, to investigate the quantum effects by the methods used above.

The quantum effects in region III may, nevertheless, be determined simply from the fact that the interaction of electrons, both with one another and with their nuclei, is in this case small compared with their kinetic energy. This is true here even for the most tightly bound inner electrons. We may therefore use perturbation theory, considering $E - E_k$ in (6.3) as a small correction. Since we require an accuracy to the order $(a_0 p_0)^{-1}$, we will use only the lowest order perturbation theory, so that in (6.3) we replace Ψ by Ψ_0—the wave function for a perfect electron gas. If we carry out calculations similar to those in § 4, we obtain

$$\mathscr{E} - \mathscr{E}_k = -3\theta e^2 \bar{\varrho}^{4/3} Z^{2/3} - \frac{e^2}{4\pi} (3\pi^2)^{1/3} \bar{\varrho}^{4/3},$$

$$\mathscr{E}_k = \frac{3}{10M} (3\pi^2)^{2/3} \bar{\varrho}^{5/3}.$$

If we now consider the pressure, we find that, to an accuracy $(a_0 p_0)^{-1}$, it is a combination of (6.7) and (6.13). The term corresponding to the quantum correction is absent to this approximation

$$\delta_2 P = 0. \tag{6.16}$$

Thus the quantum corrections are compensated to a different order in ξ.

We see that the total (quantum and exchange) correction to the pressure is $\frac{11}{9}\delta_1 P$ in regions I and II, and $\delta_1 P$ in region III This latter quantity is determined from (6.15).

7. Applications to the Theory of the Atomic Nucleus

7.1. We will consider in this section some applications of the methods we have developed to the theory of multi-nucleon systems: the atomic nucleus and neutron matter. Unlike high-density matter, these systems cannot be considered purely in terms of the

Hartree–Fock approximation. We will be using the results we obtain here as a zeroth approximation in later chapters. Nevertheless, some of the characteristics of multi-nucleon systems can be calculated within the limits of the Hartree–Fock approach.

We will use (1.11) for the interaction potential between nucleons. When the limiting momentum for the particles satisfies

$$p_0 c \ll 1, \tag{7.1}$$

the requirement (1.12) is fulfilled and the potential $V_{(c)}$ in (1.11) may be replaced by the pseudopotential.

The distribution function for a multi-nucleon system, $f(\boldsymbol{x}, \boldsymbol{p})$, is represented by a matrix in the isotopic coordinates. One can always write it in the form

$$f(\boldsymbol{x}, \boldsymbol{p}) = f_{(p)}(\boldsymbol{x}, \boldsymbol{p})\,\zeta_{(p)} + f_{(n)}(\boldsymbol{x}, \boldsymbol{p})\,\zeta_{(n)}, \tag{7.2}$$

where $\zeta_{(p,n)}$ are the matrices which distinguish the proton and neutron states (see § 1.3); $f_{(p,n)}(\boldsymbol{x}, \boldsymbol{p})$ are the corresponding particle distribution functions.

If we substitute this expression in (4.30) and (4.33), and use (1.11) (without the Coulomb term), we obtain immediately the following expression for the Hamiltonian of the particles:

$$\hat{T} + \hat{W} = \hat{T} + \hat{W}_{(p)}\zeta_{(p)} + \hat{W}_{(n)}\zeta_{(n)}, \tag{7.3}$$

where

$$\hat{W} = \hat{B} - \hat{A}$$

$$\left.\begin{aligned} \hat{B}_{(p)} &= 2\int d^3x'\, d^3p'\, \{(V_{(c)} + \tfrac{1}{4}V_{(a)})\, f_{(p)} \\ &\qquad + (V_{(c)} + \tfrac{1}{2}V_{(a)})\, f_{(n)}\}, \\ \hat{B}_{(n)} &= 2\int d^3x'\, d^3p'\, \{(V_{(c)} + \tfrac{1}{2}V_{(a)})\, f_{(p)} \\ &\qquad + (V_{(c)} + \tfrac{1}{4}V_{(a)})\, f_{(n)}\}, \end{aligned}\right\} \tag{7.4}$$

and

$$\left.\begin{aligned} \hat{A}_{(p)} &= \int d^3p'\, \{(v_{(c)} - \tfrac{1}{2}v_{(a)})\, f_{(p)} - v_{(a)} f_{(n)}\}, \\ \hat{A}_{(n)} &= \int d^3p'\, \{-v_{(a)} f_{(p)} + (v_{(c)} - \tfrac{1}{2}v_{(a)})\, f_{(n)}\}. \end{aligned}\right\} \tag{7.5}$$

In (7.4) $\boldsymbol{x} - \boldsymbol{x}'$ is the argument of the potentials, $\boldsymbol{x}', \boldsymbol{p}'$ are the arguments of the distribution function. In (7.5) v is the Fourier transform of the corresponding potential and is a function of $\boldsymbol{p}'$; $\boldsymbol{x}, \boldsymbol{p}' + \hat{p}$ are the arguments of the distribution functions.

An expression for the distribution functions can be obtained from the general equation (4.18)†

$$f_{(i)}(\boldsymbol{x}, \boldsymbol{p}) = \langle \hat{\varrho}_{(i)} \rangle_{\boldsymbol{p}}, \tag{7.6}$$

where $\hat{\varrho}_{(i)} = \theta(\varepsilon_{F(i)} - \hat{T} - \hat{W}_{(i)})$, and $\varepsilon_{F(i)}$ represents the Fermi energies of the proton and neutron distributions, which can be found from the corresponding normalization conditions.

The energy of the system can be determined from the relations introduced in § 4.9. Introducing the kinetic energy density of the protons and neutrons

$$\mathscr{E}_{k(i)} = 2 \int d^3\boldsymbol{p}\, \overline{f_{(i)}(\boldsymbol{x}, \boldsymbol{p})\, p^2/2M},$$

we obtain equations similar to (4.38) (there is supposed to be no external field U):

$$\mathscr{E} = \mathscr{E}_{(p)} + \mathscr{E}_{(n)},$$

$$\mathscr{E}_{(i)} = \tfrac{1}{2} \left\{ \mathscr{E}_{k(i)} - \int_{-\infty}^{\varepsilon_{F(i)}} \bar{\varrho}_{(i)} d\varepsilon_{F(i)} + \varepsilon_{F(i)} \bar{\varrho}_{(i)} \right\}, \tag{7.7}$$

where the number density for the corresponding particles is

$$\varrho_{(i)} = 2 \int d^3\boldsymbol{p} f_{(i)}(\boldsymbol{x}, \boldsymbol{p}). \tag{7.8}$$

We still have to take into account the Coulomb interaction between the protons. The corresponding interaction parameter $(a_0 p_0)^{-1} \sim 1/30$ (see § 1.8). Hence, only the direct Coulomb interaction has to be considered, and this can be written as

$$B_{\text{Coul}} = 2e^2 \int \frac{d^3\boldsymbol{x}'\, d^3\boldsymbol{p}}{|\boldsymbol{x} - \boldsymbol{x}'|} f_{(p)}(\boldsymbol{x}', \boldsymbol{p}).$$

The contribution to the total energy is

$$\mathscr{E}_{\text{Coul}} = \tfrac{1}{2} \overline{\varrho_{(p)} B_{\text{Coul}}}$$

These relations are completely equivalent to the Hartree–Fock equations for multi-nucleon systems. The Hartree–Fock approximation is fairly widely applicable (in a much simplified form) in

† The suffix i signifies both p and n.

nuclear theory, and forms the basis of models which, in one form or another, proceed from the single-particle description of a system of nucleons [37, 73, 74]. The potential $\hat{B} - \hat{A}$ is not determined by some self-consistent procedure here, but is given by some semi-empirical method.

7.2. The Thomas–Fermi approximation may be used to describe a spatially uniform distribution of nucleons, or a sufficiently heavy nucleus. If we neglect the fact that the operators in (7.6) do not commute, we have

$$f_{(i)}(\boldsymbol{x}, \boldsymbol{p}) = \theta[\varepsilon_{F_{(i)}} - \varepsilon_{p(i)}(\boldsymbol{x})],$$

where

$$\varepsilon_{p(i)} = (\hat{T} + \hat{W}_{(i)})|_{\hat{p} \to p}.$$

The distribution function can be rewritten in a more convenient form:

$$f_{(i)}(\boldsymbol{x}, \boldsymbol{p}) = \theta[p^2_{0(i)}(x) - p^2], \tag{7.9}$$

where the limiting momenta of the proton and neutron distributions are determined by the equations

$$\left.\frac{p^2_{0(i)}}{2M} + \hat{B}_{(i)} - \hat{A}_{(i)}\right|_{p \to p_{0(i)}} = \varepsilon_{F_{(i)}}. \tag{7.10}$$

The quasi-classical condition requires that both the functions $p^2_{0(i)}$ be sufficiently smooth functions of position.

The number density of the particles, and the kinetic energy density, may be found easily from (7.7)

$$\varrho_{(i)} = p^3_{0(i)}/3\pi^2, \tag{7.11}$$

$$\mathscr{E}_{k(i)} = \overline{p^5_{0(i)}}/10\pi^2 M. \tag{7.12}$$

The total energy of the system is determined by

$$\mathscr{E} - \mathscr{E}_{k(p)} - \mathscr{E}_{k(n)} - \mathscr{E}_{\text{Coul}} =$$

$$\overline{\tfrac{1}{2}\varrho_{(p)}(B_{(c)(p)} + \tfrac{1}{4}B_{(a)(p)} + B_{(c)(n)} + \tfrac{1}{2}B_{(a)(n)})}$$
$$+ \overline{\tfrac{1}{2}\varrho_{(n)}(B_{(c)(n)} + \tfrac{1}{4}B_{(a)(n)} + B_{(c)(p)} + \tfrac{1}{2}B_{(a)(p)})}$$
$$+ \int d^3\boldsymbol{p}\, d^3\boldsymbol{p}' \,\{-\overline{f_{(p)}f_{(p)}(\nu_{(c)}} - \tfrac{1}{2}\nu_{(a)})$$
$$- \overline{f_{(n)}f_{(n)}}(\overline{\nu_{(c)}} - \tfrac{1}{2}\nu_{(a)}) + 2\overline{f_{(p)}f_{(n)}\nu_{(a)}}\}, \tag{7.13}$$

where

$$B_{(a,c)(i)} = \int d^3x' \varrho_{(i)}(x')\, V_{(a,c)}(x - x') \tag{7.14}$$

The arguments of the distribution functions in (7.13) are $\boldsymbol{p}$ and $\boldsymbol{p}'$ respectively; the argument of the functions ν is $\boldsymbol{p} - \boldsymbol{p}'$.

The relations which we have introduced in this section come directly from the Hartree-Fock equations for uniform systems. For atomic nuclei these relations also follow from the Hartree–Fock equations, but only so long as the mass number A is sufficiently large.

The density distribution of nucleons in heavy nuclei may be approximated in the following way [75]. The density is practically constant in the central region of the nucleus (radius $\sim A^{1/3}$ fermi), and is of the order 1/fermi^3. In a comparatively narrow surface layer, of width ~ 1 fm, the density falls to zero. The quantum parameter,

$$\xi^2 \sim \frac{1}{p_0^4}\left(\frac{\partial p_0}{\partial x}\right)^2 \sim \varrho^{-8/3}\left(\frac{\partial \varrho}{\partial x}\right)^2,$$

is therefore nearly zero inside the nucleus, and of the order of unity in the surface layer. It can be seen that the effective value of ξ^2 averaged over the nuclear volume is of the order $A^{-1/3}$. Hence, it follows that, as for an atom, the accuracy of the quasi-classical approximation increases with the number of particles.

Unlike the systems with Coulomb interactions considered in the previous section, the fact that the quantum effects are small here by no means guarantees that the exchange effects will also be small. Suppose we estimate the ratio of the terms in (7.5) to the kinetic energy ($\sim p_0^2/M$). The term corresponding to the potential $V_{(a)}$ gives an order of magnitude MV_0/p_0^2. So, as for Coulomb systems, the ratio is determined by the interaction parameter α (see § 1), and need not, therefore, be small. The analogous ratio for the potential $V_{(c)}$ is of the order cp_0; a term of this order gives the corresponding potential for the direct interaction $\hat{B}$. It follows from what we have said that we cannot ignore the exchange terms in the foregoing equations. This evidently complicates their solution considerably.

As the density increases, the relative importance of the exchange interaction compared with the direct interaction decreases (the ratio for $V_{(a)}$ is of the order $(ap_0)^{-3}$). This is because only the direct interaction term is affected by the increase in the number of particles which lie within the sphere of influence of any other particle.

7.3. It is difficult to use the equations we have obtained, because they contain the quantities $\int d^3\boldsymbol{x}\, V_{(c)}(\boldsymbol{x})$ and $v_{(c)}(\boldsymbol{p})$, which tend to infinity for a hard-core potential. It is thus impossible to apply the Hartree–Fock approximation for such potentials.

If there were no other interactions between particles besides $V_{(c)}$, then, with the requirement (7.1) satisfied,† this difficulty could be overcome easily enough. One would simply replace $V_{(c)}$ by the pseudopotential, defined in eqns. (1.17) to (1.19), and the Hartree–Fock approximation could be used immediately. In our case, however, there is an attractive force—described by the potential $V_{(a)}$—as well as a repulsive force (the Coulomb forces are unimportant from our present viewpoint). We must therefore consider the mutual influence, or interference, of the interactions $V_{(c)}$ and $V_{(a)}$. The approach discussed below will include the major part of this interference, whilst remaining within the framework of the Hartree–Fock approximation.

We will first consider this approximation in terms of $V_{(a)}$ only. This produces some average field represented by a potential well. The particles move in this well with a non-quadratic dispersion, $\varepsilon_p/p^2 \neq$ const. They behave as hard spheres, their repulsion tending to infinity upon contact. This physical picture is rather a good one, inasmuch as the transition to the Hartree–Fock approximation does not appear to produce any obvious errors in $V_{(a)}$ (see Chapter 4).

When we deduced an expression for the pseudopotential in § 1, we assumed that the law of dispersion for the particles was quadratic. The mass M, which characterizes this law, is contained explicitly in the pseudopotential, which is therefore no longer suitable

† The parameter $p_0 c \sim 0{\cdot}5$ for nuclear matter, i.e. not very small compared with unity. However, the convergence of the corresponding series is sufficiently rapid.

for our purposes. We will, instead, express the pseudopotential in the form:

$$\hat{V}_{\text{eff}} = \alpha\delta(\boldsymbol{r})\left(1 + r\frac{\partial}{\partial r} + \beta r\right), \tag{7.15}$$

where α and β are certain quantities as yet undetermined.

In the general case, when the dispersion ε_p is arbitrary, the Schrödinger equation in the centre-of-mass system is

$$\left[\varepsilon\left(\frac{\boldsymbol{P}}{2} - i\nabla\right) + \varepsilon\left(-\frac{\boldsymbol{P}}{2} - i\nabla\right) - \varepsilon\left(\frac{\boldsymbol{P}}{2} + \boldsymbol{k}\right) - \varepsilon\left(-\frac{\boldsymbol{P}}{2} - \boldsymbol{k}\right)\right]\psi(\boldsymbol{r}) = -V(\boldsymbol{r})\psi(\boldsymbol{r}),$$

where $\boldsymbol{P} = \boldsymbol{p}_1 + \boldsymbol{p}_2$ is the total momentum, and $\boldsymbol{k} = (\boldsymbol{p}_1 - \boldsymbol{p}_2)/2$ is the momentum transfer. The solution of this equation, which replaces (1.13), is therefore†

$$\psi(\boldsymbol{r}) = \exp i(\boldsymbol{k}\cdot\boldsymbol{r}) - [\Delta\varepsilon(-i\nabla) - i\delta]^{-1}\, V(\boldsymbol{r})\psi(\boldsymbol{r}),$$

where $\Delta\varepsilon(-i\nabla)$ corresponds to the expression in square brackets in the previous equation.

We now substitute the pseudopotential (7.15) for V, and denote by γ the value of $(1 + r\,\partial/\partial r + \beta r)\psi$ for $r = 0$. If we limit ourselves to a consideration of the s-state, we can write,‡

$$\psi(r) = \frac{\sin kr}{kr} - \alpha\gamma\int\frac{d^3\boldsymbol{p}\sin pr}{pr[\Delta\varepsilon(\boldsymbol{p}) - i\delta]}.$$

We will also split off from $[\Delta\varepsilon(\boldsymbol{p}) - i\delta^{-1}]$ the quantity $M(p^2 - \tilde{k}^2 - i\delta)^{-1}$ (the momentum $\tilde{k}$ will be defined below). Then the integral in the last equation can be written, for small r, as

$$\frac{M}{4\pi r}(1 + i\tilde{k}r) + I,$$

where

$$I = \int d^3\boldsymbol{p}\left(\frac{1}{\Delta\varepsilon(\boldsymbol{p}) - i\delta} - \frac{M}{p^2 - \tilde{k}^2 - i\delta}\right).$$

† Expressions of the type $(a \pm i\delta)^{-1}$ are investigated in Appendix C.

‡ This expression is obtained by expanding the δ-function (7.15) in a Fourier integral, and splitting the zero spherical harmonic into the exponents $\exp i(\boldsymbol{k}\cdot\boldsymbol{r})$, and $\exp i(\boldsymbol{p}\cdot\boldsymbol{r})$.

The convergence of I is ensured by the fact that, when $p \to \infty$, $\varepsilon_p \to p^2/2M$.

It is convenient to choose the parameter β equal to $(-4\pi/M)\,I$. Then, calculating $(1 + r\,\partial/\partial r + \beta r)\,\psi$ for $r = 0$, we obtain

$$\gamma = \left(1 + \alpha \frac{i\tilde{k}M}{4\pi}\right)^{-1}.$$

We have so far been considering this problem only in general terms: the specific characteristics of the initial potential V have not been used at all. We will now require that the function $\psi(r)$ goes to zero at $r = c$, corresponding to an infinite repulsion at this point. Then the following expression is obtained for α:

$$\alpha^{-1} = \frac{kc}{\sin kc} \int \frac{d^3\boldsymbol{p} \sin pc}{pc[\Delta\varepsilon(\boldsymbol{p}) - i\delta]} - \frac{Mi\tilde{k}}{4\pi}.$$

We will choose $\tilde{k}$ in the following way:

$$\frac{Mi\tilde{k}}{4\pi} = \frac{i\pi kc}{\sin kc} \int d^3\boldsymbol{p} \frac{\sin pc}{pc} \,\delta[\Delta\varepsilon(\boldsymbol{p})].$$

Using the general relation $(a - i\delta)^{-1} = \mathrm{P}a^{-1} + i\pi\delta(a)$, we find

$$\alpha^{-1} = \frac{kc}{\sin kc}\,\mathrm{P}\int \frac{d^3\boldsymbol{p} \sin pc}{pc\Delta\varepsilon(\boldsymbol{p})},$$

where the symbol $\mathrm{P}\int d^3\boldsymbol{p}$ denotes the principal value of the integral (see Appendix C).

We may summarize the results in the following way. For a non-quadratic dispersion law (which tends to the quadratic one for higher momenta, however) the pseudopotential is determined by ($kc \ll 1$):

$$\hat{V}_{\text{eff}} = \frac{4\pi c}{M_{\text{eff}}}\,\delta(\boldsymbol{r})\left(1 + r\frac{\partial}{\partial r} + \beta r\right), \tag{7.16}$$

where†

$$M_{\text{eff}} = 4\pi \mathrm{P} \int \frac{d^3\boldsymbol{p}}{p} \cdot \frac{\sin pc}{\Delta\varepsilon(\boldsymbol{p})},$$

† M_{eff} varies very little with $\boldsymbol{P}$ or $\boldsymbol{k}$, but depends very sensitively on p_0 [46].

$$\beta = -\frac{4\pi}{M}\int d^3p\left(\frac{1}{\Delta\varepsilon(p)} - \frac{M}{p^2 - k^2}\right).$$

It is evident, that, when $\varepsilon(p) = p^2/2M$, we regain the results obtained in § 1. In particular we have $\alpha = 4\pi c/M$, $\beta = 0$, $\tilde{k} = k$.

We note that, when $\varepsilon = p^2/2M$, one obtains the equality†

$$\delta(\boldsymbol{r})\left(1 + r\frac{\partial}{\partial r}\right)\mathrm{P}\int\frac{d^3\boldsymbol{p}}{pr}\cdot\frac{\sin pr}{\boldsymbol{p}^2 - k^2} = 0;$$

in the general case

$$\delta(\boldsymbol{r})\left(1 + r\frac{\partial}{\partial r} + \beta r\right)\mathrm{P}\int\frac{d^3\boldsymbol{p}}{pr}\cdot\frac{\sin pr}{\Delta\varepsilon(p)} = 0.$$

We will use this in § 18.

When $p \gg p_0$, $\varepsilon(p)$ tends to $p^2/2M$, the free particle energy. On the other hand, collisions correspond to small separations, $r \sim c$, and, consequently, large $p \sim 1/c \gg p_0$. The difference between M_{eff} and M is of a higher (first) order in p_0c. Consequently, the fact that the dispersion law for nucleons is non-quadratic can be ignored when considering p-scattering and other higher order scattering.

We will also omit terms of the order $(cp_0)^3$, which will be considered in § 18 together with correlation terms.

7.4. We will consider in this section several of the properties of uniform models of nuclear and neutron matter. By nuclear matter, we mean an infinite, uniform system of neutrons and protons, present in equal numbers. It is necessarily supposed in this case that there are no Coulomb forces (see § 5.3). This model therefore ignores the Coulomb interaction, the surface effects and the inequality in the number of protons and neutrons. It is, nevertheless, to some extent suitable for describing the material in the core of a heavy atomic nucleus. Despite its crudity, it plays a large part in microscopic theories of the nucleus.‡

† The integral, which figures here, is proportional to $\cos kr/kr$.
‡ A contemporary survey of the theory of nuclear matter is given in [76].

Putting $\varepsilon_{F(p)} = \varepsilon_{F(n)}$ and, correspondingly, $f_{(p)} = f_{(n)}$ in the foregoing equations, one obtains the dispersion relation, ε_p, and the energy of the nuclear matter. Using, first of all, eqns. (7.3) to (7.5), we have

$$\varepsilon_p = \frac{p^2}{2M} + \frac{p_0^3}{3\pi^2}\int d^3\xi\left(2V_{(c)} + \frac{3}{4}V_{(a)}\right) - \int d^3\boldsymbol{p}' f(\boldsymbol{x},\boldsymbol{p}')\left(v_{(c)} - \frac{3}{2}v_{(a)}\right), \tag{7.17}$$

where $\xi = \boldsymbol{x} - \boldsymbol{x}'$. This expression may be conveniently written in the form

$$\varepsilon_p = \int_0^p \frac{p\,dp}{M_0(p)} + V(p_0), \tag{7.18}$$

where $M_0(p) = [(1/p)\cdot(\partial\varepsilon_p/\partial p)]^{-1}$ is the effective mass.

A simple calculation shows that†

$$\left.\begin{aligned} \frac{M}{M_0(p)} &= 1 + \frac{3\pi}{4}\cdot\frac{a^2}{(a-c)^2}\varphi(p,p_0), \\ \varphi(p,p_0) &= \frac{1}{x}\frac{\partial}{\partial x}\left(\frac{\sin x}{x}\right)\sin y - \frac{1}{2x}\left[\frac{\sin(x+y)}{(x+y)} - \frac{\sin(x-y)}{(x-y)}\right], \end{aligned}\right\} \tag{7.19}$$

$$V(p_0) = -\frac{\pi}{12M(a-c)^2}[y^3 + 9\mathrm{Si}(y) - 9\sin y] + \frac{2cp_0^3}{\pi M_{\mathrm{eff}}}, \tag{7.20}$$

where $x = ap$; $y = ap_0$. We have introduced the integral sine $\mathrm{Si}(x) = \int_0^x (\sin t/t)\,dt$.

The first term in (7.18) is shown in Fig. 2, and may conveniently be approximated in the following way:

$$\int_0^p \frac{p\,dp}{M_0(p)} = \frac{p^2}{2M} + \begin{cases} \dfrac{p^2}{2M}\mu & p < \lambda p_0 \\ \dfrac{p_0^2}{2M}\lambda^2\mu & p > \lambda p_0 \end{cases} \tag{7.21}$$

† For $V_{(c)}$ we write (7.16) to lowest order in c. To the required approximation, we can replace the operator $1 + r\,\partial/\partial r + \beta r$ by unity.

where $\lambda(p_0)$ and $\mu(p_0)$ are obtained by a least squares fit [46]. For the physical value $p_0 = 1{\cdot}4\ \text{fm}^{-1}$, we have $\mu = 0{\cdot}75$, $\lambda = 1{\cdot}3$, $M_{\text{eff}}/M = 0{\cdot}7$. Thus, for nucleons lying below the Fermi surface, the effective mass is equal to $M_0 \approx 0{\cdot}57M$. Semi-empirical formulations usually give $M_0 = 0{\cdot}5M$.

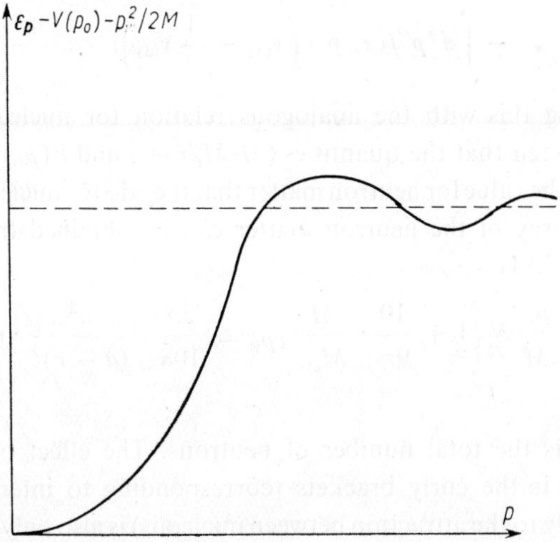

FIG. 2

The energy of the nuclear matter may be calculated from eqn. (7.12) to (7.14) (remembering that, in this case, $f_{(p)} = f_{(n)}$, $\varrho_{(n)} = \varrho_{(p)}$, $e^2 = 0$); it is proportional to the mass number A:†

$$E = \frac{3}{10}\frac{p_0^2}{M}A\left\{1 + \frac{10}{3\pi}\frac{M}{M_{\text{eff}}}cp_0 - \frac{5\pi}{36}\frac{a^2}{(a-c)^2}\varphi_1(y)\right\},$$

$$\varphi_1(y) = y + \frac{9}{y^2}\,\text{Si}(2y) - \frac{9}{2y^3}(3 - \cos 2y) + O(y^{-4}). \quad (7.22)$$

The pseudopotential was used in deducing this expression; it is therefore unsuitable for large values of cp_0.

There has of late developed an interest in the other model, namely, the neutron matter model, mentioned previously. This consists of an infinite, uniform aggregate of neutrons [48, 77].

† Numerically, when $p_0 = 1{\cdot}4\,\text{fm}^{-1}$, $y = 3{\cdot}22$ and $\varphi_1(y) = 4{\cdot}20$.

If we substitute in the equations of §§ 7.1 and 7.2, $\varepsilon_{F(p)} = -\infty$, $f_{(p)} = 0$, $\varrho_{(p)} = 0$, we find the following expression for the dispersion relation for neutrons

$$\varepsilon_p = \frac{p^2}{2M} + \frac{p_0^3}{3\pi^2} \int d^3\xi \left(V_{(c)} + \frac{1}{4} V_{(a)} \right) - \int d^3p' f(x, p') \left(\nu_{(c)} - \frac{1}{2} \nu_{(a)} \right).$$

Comparing this with the analogous relation for nuclear matter, it can be seen that the quantities $(M/M_0) - 1$ and $V(p_0)$ have only one-third the value for neutron matter that they do for nuclear matter.

The energy of the neutron matter can be obtained from eqns. (7.12) to (7.14)

$$E = \frac{3}{10} \frac{p_0^2}{M} N \left\{ 1 + \frac{10}{9\pi} \cdot \frac{M}{M_{\text{eff}}} c p_0 - \frac{5\pi}{108} \cdot \frac{a^2}{(a-c)^2} \cdot \varphi_1(y) \right\}, \tag{7.23}$$

where N is the total number of neutrons. The effect of the last two terms in the curly brackets (corresponding to interactions—particularly to the attraction between nucleons) is also only one-third as much as in (7.22). This raises the question of whether neutron matter can exist in a bound state.

7.5. The solution of the Thomas–Fermi equation for a real nucleus is complicated by the presence of the exchange terms [52]. It is expedient therefore, to apply a direct variational method. We must set up some trial function $\varrho_{(p,n)}(x)$ containing a number of variational parameters, substitute into the expression for the total energy of the system and then vary the parameters. Having calculated the values of the variational parameters for which the energy is minimum, both the particle distribution and the energy of the system can be found [78, 79].

We will take the following as the trial functions:

$$\varrho_{(p,n)}(x) = \varrho_{(p,n)}^{(0)} \begin{cases} 1 & x < R - d/2 \\ \dfrac{R + (d/2) - x}{d} & R - d/2 < x < R + d/2 \\ 0 & x > R + d/2. \end{cases} \tag{7.24}$$

A trapezoidal density distribution of this type is in close agreement with experiment.† It is supposed, for simplicity, that the average radius of the distribution R and the width of the surface layer d are equal for the proton and neutron distributions. The constants $\varrho^{0}_{(p,n)}$ are found directly from the normalization condition

$$\varrho^{(0)}_{(p)} = \frac{3Z}{4\pi R^3}\left[1 + O\left(\frac{d^2}{R^2}\right)\right]; \quad \varrho^{(0)}_{(n)} = \frac{3(A-Z)}{4\pi R^3}\left[1 + O\left(\frac{d^2}{R^2}\right)\right].$$

The selected trial function contains three variational parameters for a given A. The first is the nuclear charge Z, which differs comparatively little from $A/2$. We therefore introduce a small parameter $\sigma = (A - 2Z)/A$, which defines the relative excess of neutrons in the nucleus: we will expand energy terms up to σ^2 inclusive.

The second variational parameter is the width of the surface layer d. This is independent of A and can be considered small compared with $R \sim A^{1/3}$ (but not compared to the range of the nuclear forces). We will take into account terms of not higher than first order in d/R. We will also ignored cross-terms of the type $\sigma d/R$, $\sigma^2 d/R$, etc.

Finally, the third variational parameter is the radius of the distribution R (or, equivalently, the Fermi momentum p_0). The energy, as a function of R, does not have a minimum in the Hartree–Fock approximation (see § 7.7). We cannot, therefore, consider R as a variational parameter in this; we will substitute instead its empirical value $R \approx 1{\cdot}1 A^{1/3}$ fm (corresponding to $p_0 = 1{\cdot}4\ \text{fm}^{-1}$). If we substitute (7.24) into eqns. (7.12) to (7.14), and allow for the fact that, as discussed above, σ and d/R are small, we obtain the following general expression for the energy:

$$E = C_1 A + C_2 \frac{Z^2}{A^{1/3}} + C_3 A^{2/3} + C_4 \left(\frac{A-2Z}{A}\right)^2. \quad (7.25)$$

Weizsäcker's semi-empirical formula has an analogous form if one omits the even–odd anomaly. The first term is the volume-

† This choice naturally gives only a relative energy minimum. Reference [80], which we are following here, provides arguments in favour of this particular density distribution.

proportional part of the energy. It is given by (7.22) in the Hartree–Fock approximation. The empirical value of the coefficient $C_1 \approx -15$ MeV.

The second term is the Coulomb repulsion energy of the protons. If the small exchange and surface Coulomb energies are ignored, we find

$$C_2 = \frac{6}{5} \frac{e^2 p_0}{(9\pi)^{1/3}}.$$

It is convenient to introduce a quantity

$$C_2' = 4C_2 \frac{Z^2}{A^2} = C_2(1 - 2\sigma + \sigma^2). \tag{7.26}$$

Then the second term of (7.25) takes the form $C_2' A^{5/3}/4$. The asymmetry between the protons and neutrons is represented by the presence of term linear in σ in C_2'. The empirical value of $C_2 = 0{\cdot}7$ MeV; this is also given by the above equation.

The third term is the surface energy. A simple, but cumbersome, calculation gives

$$C_3 = \beta_1 \frac{a}{d} + \beta_2 \frac{d}{a}, \tag{7.27}$$

$$\beta_1 = \frac{\pi}{40(9\pi)^{1/3} M(a-c)^2} y^3,$$

$$\beta_2 = \frac{9}{40(9\pi)^{1/3} Ma^2} y^3 \left[-1 - \frac{40}{9\pi} \frac{M}{M_{\text{eff}}} c p_0 + \frac{5\pi}{27} \frac{a^2}{(a-c)^2} \varphi_2(y) \right],$$

$$\varphi_2(y) = y + \frac{81}{10y^3} + \frac{81}{4} \frac{\sin 2y}{y^4} + \frac{27}{2y^5} + 81 \frac{\cos 2y}{y^5} + O(y^{-6}),$$

where $y = ap_0$. The empirical value of $C_3 \approx 18$ MeV.

Finally, the last term of (7.25) corresponds to the so-called symmetry energy, which arises from the inequality in the number of protons and neutrons.† The coefficient C_4, which has an empirical

† The origin of this energy is related to Pauli's principle: if all the nucleons are replaced by protons (or neutrons), there is an increase in the energy of the system due to the increase in the exchange repulsion.

value ≈ 24 MeV, has the form

$$C_4 = \frac{y^2}{6Ma^2}\left[1 - \frac{2cp_0}{\pi}\cdot\frac{M}{M_{\text{eff}}} + \frac{\pi}{12}\cdot\frac{a^2}{(a-c)^2}\varphi_3(y)\right], \quad (7.28)$$

$$\varphi_3(y) = y + \frac{6}{y} + \frac{3}{2y^2}\sin 2y - \frac{9}{y^3}\sin^2 y.$$

Numerically, the above value of p_0 gives $\varphi_2(y) = 3{\cdot}70$; $\varphi_3(y) = 5{\cdot}09$; whence it follows that $C_4 \approx 30$ MeV.

7.6. We must now determine the equilibrium values of d and σ, and must therefore vary (7.25) with respect to these parameters.

We begin with the symmetry energy. The part of (7.25) which contains σ has, from (7.26), the form:

$$-\frac{C_2}{2}A^{5/3}\sigma + \left(C_4 A + \frac{C_2 A^{5/3}}{4}\right)\sigma^2.$$

The second term in the coefficient of σ^2 is sufficiently small to be ignored. If we minimize with respect to σ, we find

$$\sigma \approx \frac{C_2}{4C_4}A^{2/3} \approx 0{\cdot}006A^{2/3}. \quad (7.29)$$

This asymmetry obviously arises from the Coulomb forces. It increases with A, because the Coulomb forces, which similarly increase, need to be compensated. The form obtained above is, on the whole, in quite good agreement with experiment.

The coefficient C_3 is the only one dependent on d. Varying (7.27) with respect to this parameter, we have

$$d = \left(\frac{\beta_1}{\beta_2}\right)^{1/2} a \approx 2{\cdot}5 \text{ fm}. \quad (7.30)$$

This too is in quite good agreement with experiment. The observed value for the distance over which the density falls from 0·9 to 0·1 of its maximum value is about 2·4 fm. The calculated value for this distance, obtained from (7.24) and (7.30), is 2.0 fm.† Thus, the

† This is somewhat increased by the correlation effects (see § 18).

theory not only gives the qualitative variation of d with mass number, but also gives the magnitude of this parameter.

The constant C_3 may be determined by substituting (7.30) in (7.27). The value obtained is distinctly larger than the experimental value. This can be explained in terms of dynamical correlation effects which we have not yet considered. In principle, quantum effects could be important in the surface regions of the nucleus; however, they are numerically small. We will estimate, from (7.24), the quantity $(1/72M) \int d^3\boldsymbol{x}(\nabla\varrho)^2/\varrho$ [see (5.23)].† It can be seen that the quantum correction adds about 1 MeV to C_3; similarly, d is little changed.

We see, therefore, that several nuclear properties can be calculated correctly from the Hartree–Fock approximation. However, other properties depend in an important way on the dynamical correlation effects (see § 18).

7.7. We have seen that the radius of the nucleus R cannot be used as a variational parameter in the Hartree–Fock approximation. In fact, for small R (or large p_0), (7.22) becomes

$$E \sim -p_0^3 \sim -R^{-3}$$

and the absolute magnitude of this increases without limit as R decreases. Thus a system of nucleons in this approximation tends to collapse spontaneously. An ever larger number of nucleons therefore fall within the range of the nuclear forces: the interaction energy becomes proportional to A^2, and the nuclear forces lose their property of saturation.

The difficulty, then, is that the number of effectively interacting particles increases with the density. This tendency is counteracted by the presence of repulsion at small distances. However, the effect of the repulsion cannot be allowed for accurately in the Hartree–Fock approximation. One can, nevertheless, in this approximation find a sufficient criterion that the nucleus will collapse (it will be shown in Chapter 4 that this criterion is also necessary).

† The last term of (5.23) goes to zero from Gauss's theorem. Strictly speaking, we cannot use (5.23) here, since the exchange effects play an essential part. We may, nevertheless, suppose that this will not change our estimate to any great extent.

It is only necessary to remember that the energy in the Hartree–Fock approximation is always higher than its true value. This can be seen most directly from the derivation of the Hartree–Fock approximation by a variational method. If, therefore, we can find a criterion for the energy to go to $-\infty$ in the Hartree–Fock approximation, this will also hold for any more accurate approach. We may look for the criterion by investigating how the energy behaves for large p_0 in the Hartree–Fock approximation, and finding under what conditions it tends to infinity.

It is sufficient to consider a uniform distribution of nucleons with $f_{(p)} = f_{(n)}$. The exchange terms in the energy are of relatively little importance for large p_0, compared with the direct interaction between particles. The general expression for the energy may, therefore, be written as

$$\mathscr{E} = \mathrm{Tr}_{\sigma\tau} \int d^3\boldsymbol{p}\,\theta(p_0^2 - p^2)\left(\frac{p^2}{2M} + \frac{B}{2}\right),$$

$$B = \frac{p_0^3}{3\pi^2} \int d^3\xi\, \hat{V}(\xi).$$

Taking $\hat{V}$ in its most general form (1.7), we obtain the following equation for $\mathscr{E}$:

$$\mathscr{E} = \frac{1}{9\pi^4}\varkappa p_0^6 + \frac{p_0^5}{5\pi^2 M}, \tag{7.31}$$

where

$$\varkappa = \int d^3\xi\, \{V_W(\xi) + \tfrac{1}{2}[V_B(\xi) + V_H(\xi)] - \tfrac{1}{4}V_M(\xi)\}.$$

The first term predominates for large p_0, and the condition $\mathscr{E} \to -\infty$ requires that [81]

$$\varkappa < 0. \tag{7.32}$$

It is evident from this that both the Wigner forces, and the Serber forces, lead inevitably to a collapsed state of nuclear matter.

Introducing a "hard-core", $\varkappa \to +\infty$,† and so removes this difficulty—as is also obvious from a simple, physical picture.

Equation (7.32) provides a quantitative refinement of the well-known conclusions about the change in sign of the potential in the problem of saturation of nuclear forces [37].

† The pseudopotential cannot be introduced for large p_0, since (1.12) is then disobeyed.

CHAPTER 3

Perturbation Theory and Diagram Technique

8. Hole Formalism

8.1. We make the transition to the Hartree–Fock approximation by substituting the Hamiltonian $\hat{H}_0$ (4.29) for the exact Hamiltonian $\hat{H}$ (3.19). Consider the difference

$$\hat{H}' = \hat{H} - \hat{H}_0 = \tfrac{1}{2} \int dq\, dq' V(q, q') \{\hat{\psi}^+(q)\, \hat{\psi}^+(q')\, \hat{\psi}(q')\, \hat{\psi}(q) - 2R(q', q')\, \hat{\psi}^+(q)\, \hat{\psi}(q) + 2R(q, q')\, \hat{\psi}^+(q)\, \hat{\psi}(q') + R(q, q)\, R(q', q') - R(q, q')\, R(q', q)\}. \tag{8.1}$$

We suppose for simplicity that the potential V is a function of the coordinates only, and is not an operator.

The Hamiltonian $\hat{H}'$ represents the correlation interaction, and causes the breakdown of the individual particle description of the system. These effects are particularly reflected in the correlations between the instantaneous properties of different particles. For example, the probability of observing a particle at a given point in space depends on the distribution of the remaining particles at the same time.

Such correlations due to the interaction between particles are usually referred to as dynamical correlations.† The easiest way of including the effects of these correlations is by formally expanding the physical quantities as a series in $\hat{H}'$, i.e. by setting up the corresponding perturbation theory. We will proceed in this

† There is, besides the dynamical correlation, a statistical (exchange) correlation which is connected with Pauli's principle. It is included in the Hartree–Fock approximation by the expressions containing the exchange operator $\hat{\mathscr{P}}$.

chapter to work out the perturbation theory in a field form, as has been done (by analogy with quantum field theory) in [82–85].

The field form of the perturbation theory is simpler, more concise and more graphic than the usual Schrödinger perturbation theory. This difference is essentially due to diagram techniques and the formulation of the Feynman rules, which provide an easy way of setting up graphical constructions and analytical expressions for any term of the perturbation expansion.

Perturbation theory is not confined to problems where the correlation effects are small so that only the first few terms of the expansion need to be considered. It is also an essential ingredient in formulating the accurate theoretical equations, and, what is more important from the practical viewpoint, in constructing approximate expressions containing an infinite series of terms. An analysis of the different perturbation theory diagrams enables us to select the most important of these series and to sum them.

8.2. Before we start working out the perturbation theory, we must change slightly our interpretation of operators in the occupation-number representation. To this end, we introduce the concept of holes.

If we wish to construct the wave function for any state of a system, we must apply the particle creation operators to the wave function of the vacuum state. It is necessary to begin our calculations from this state; hence creation operators for all the particles in the system appear in the wave function.

However, this approach is, for various reasons, unsuitable. It is much more convenient to begin with the wave function of the system itself, Ψ_0, in the Hartree–Fock approximation. Then we can describe the state of the system in which we are interested by its deviation from Ψ_0. The difference is due to a redistribution of particles amongst the levels: some levels which were previously empty are now filled, and *vice versa*.

The advantages of this approach are as follows. First of all, the new starting point—the Ψ_0 state—has a much greater physical significance than the vacuum state; in particular it corresponds to the same number of particles as does the real state of the system.

Moreover, it frequently occurs that the real state of the system differs from Ψ_0 only by a comparatively minor rearrangement of the particles. In this case the present approach is neater and less complicated. There is the further advantage that several formal defects of the old approach now disappear, and many-body theory approximates more closely in structure to relativistic quantum field theory (see § 8.4).

8.3. We will reconsider the basic theory of the occupation-number representation. The dynamical variables are now the number of particles, $n_\nu^{(p)}$, in levels which were empty in the state Ψ_0, and the number of free places—holes—$n_\nu^{(h)}$, in levels which were occupied in the state Ψ_0. The state Ψ_0, itself, evidently corresponds to having the occupation numbers of particles and holes equal to zero.

We must now bring the expression for the field operator itself into agreement with this new description. We will rewrite the series (3.9) in the following form:

$$\hat{\psi}(q) = \sum_\nu [(1 - \sqrt{n_\nu})\,\hat{A}_\nu + \sqrt{(n_\nu)}\,\hat{A}_\nu]\,\chi_\nu(q), \tag{8.2}$$

where n_ν is the occupation number in the state Ψ_0. We will consider the ground state of the system. The first term in (8.2) differs from zero only for unoccupied levels ($\varepsilon_\nu > \varepsilon_F$), where $n_\nu = 0$, and can, as before, be interpreted as a particle-annihilation operator. The second term only differs from zero for occupied states ($n_\nu = 1$). It can be interpreted as a particle-annihilation operator for the occupied states, or, which is the same thing, as a hole-creation operator. In a similar way the operator $\hat{A}_\nu^+$, acting below the Fermi surface, leads to the creation of particles (or the annihilation of holes).

The series (3.9) may consequently be written in the form

$$\left.\begin{aligned} \hat{\psi}(q) &= \sum_\nu \hat{A}_\nu \chi_\nu(q), \\ \hat{\psi}^+(q) &= \sum_\nu \hat{A}_\nu^+ \chi_\nu^*(q), \end{aligned}\right\} \tag{8.3}$$

where

$$\left.\begin{aligned} \hat{A}_\nu &= \hat{a}_\nu\ (\varepsilon_\nu > \varepsilon_F), & \hat{A}_\nu &= \hat{b}_\nu^+(\varepsilon_\nu < \varepsilon_F), \\ \hat{A}_\nu^+ &= \hat{a}_\nu^+(\varepsilon_\nu > \varepsilon_F), & \hat{A}_\nu^+ &= \hat{b}_\nu(\varepsilon_\nu < \varepsilon_F). \end{aligned}\right\} \tag{8.4}$$

Here $\hat{a}$, $\hat{a}^+$ and $\hat{b}$, $\hat{b}^+$ are the annihilation and creation operators for particles and holes respectively.

Employing the definition of the hole operators and the general commutation rules (3.7), we can write

$$\left.\begin{aligned} [\hat{b}_\mu^+, \hat{b}_\nu]_+ = \delta_{\mu\nu}, \quad [\hat{a}_\mu^+, \hat{a}_\nu]_+ = \delta_{\mu\nu}, \\ [\hat{b}_\mu, \hat{b}_\nu]_+ = [\hat{b}_\mu^+, \hat{b}_\nu^+]_+ = [\hat{a}_\mu, \hat{a}_\nu]_+ = [\hat{a}_\mu^+, \hat{a}_\nu^+]_+ = 0. \end{aligned}\right\} \tag{8.5}$$

The particle and hole operators anti-commute with one another since the operators $\hat{a}$ and $\hat{b}$ always refer to different states.

The operators $\hat{a}$, $\hat{a}^+$ and $\hat{b}$, $\hat{b}^+$ act respectively on the occupation numbers for particles $n_\nu^{(p)}$ and holes $n_\nu^{(h)}$, in accordance with eqns. (3.5). Since the wave function Ψ_0 corresponds to occupation numbers $n_\nu^{(p)}$ and $n_\nu^{(h)}$ equal to zero, we have the equalities

$$\left.\begin{aligned} \hat{a}_\nu \Psi_0 = \hat{b}_\nu \Psi_0 = 0, \\ \Psi_0^* \hat{a}_\nu^+ = \Psi_0^* \hat{b}_\nu^+ = 0. \end{aligned}\right\} \tag{8.6}$$

The occupation-number operators for particles and holes can be evidently be written as

$$\hat{n}_\nu^{(p)} = \hat{a}_\nu^+ \hat{a}_\nu; \quad \hat{n}_\nu^{(h)} = \hat{b}_\nu^+ \hat{b}_\nu. \tag{8.7}$$

The transition to the hole description does not require any substantial change in the theory; it is connected rather with a reinterpretation of the quantities involved.

8.4. The transition to the hole formalism leads to a further analogy between many-body theory and relativistic quantum Fermi-field theory, which describes not only particles, but also anti-particles. These latter may be interpreted as holes in an unobservable background of negative-energy particles.

Holes in many-body theory and anti-particles in field theory have much in common. Many processes in many-body theory can be interpreted in the sort of language used in field theory. For example, in the simplest transition of a system from its ground state to an excited state a particle is transferred from a filled level to one with an energy greater than the Fermi energy. In hole language this corresponds to the creation of a particle–hole pair (the hole being the unoccupied level). The return of the excited system to its ground state corresponds to the mutual annihilation of particle and hole. These processes are analogous to the creation and annihilation of particles and anti-particles in quantum field theory. An interpretation of this sort, borrowed from relativistic quantum field theory, seems particularly helpful.

However, the analogy is not complete. There is, in relativistic field theory, a complete symmetry between particles and holes, which corresponds to a sign degeneracy of the particle energy levels. This symmetry (charge symmetry) has a whole series of consequences for example, Furry's theorem (see § 13).

In many-body theory there is no such symmetry between particles and holes. This can be seen from the fact that the regions corresponding to the particle and hole spectra differ considerably; one of them extends to infinite energies ($\infty > \varepsilon_\nu > \varepsilon_F$), whilst the other is limited to low-lying levels only ($0 < \varepsilon_\nu < \varepsilon_F$). Symmetry, therefore, is only observed in certain cases when the important particles and holes are very close to the Fermi surface.

8.5. The transition to the hole formalism results in the division of the field operators into a sum of a creation and an annihilation part, each of which contains only creation and annihilation operators (see § 8.3):

$$\hat{\psi}(q) = \hat{\psi}(q)_{(+)} + \hat{\psi}(q)_{(-)},$$

$$\hat{\psi}^+(q) = \hat{\psi}^+(q)_{(+)} + \hat{\psi}^+(q)_{(-)}.$$

Here the suffixes (+) and (−) refer respectively to the creation and annihilation parts. We note that, with respect to conjugation $(+) \rightleftarrows (-)$, in particular $\hat{\psi}^+(q)_{(+)} = [\hat{\psi}(q)_{(-)}]^+$.

It can be seen from the commutation rules (8.5), that the only anti-commutators of the operators $\hat{\psi}_{(\pm)}$ and $\overset{+}{\hat{\psi}}_{(\pm)}$ which differ from zero, have the following form:

$$\left.\begin{aligned} [\hat{\psi}(q)_{(+)}, \hat{\psi}^+(q')_{(-)}]_+ &= \sum_\nu n_\nu \chi_\nu^*(q')\, \chi_\nu(q), \\ [\hat{\psi}(q)_{(-)}, \hat{\psi}^+(q')_{(+)}]_+ &= \sum_\nu (1 - n_\nu)\, \chi_\nu^*(q')\, \chi_\nu(q). \end{aligned}\right\} \quad (8.8)$$

We have used here the relation $(1 - \sqrt{n_\nu})^2 = 1 - n_\nu$, which follows from the fact that $n_\nu = 0, 1$. The first anti-commutator is simply the density matrix $R(q, q')$ [see (4.7)]; the second, from the completeness requirement, is equal to $\delta(q - q') - R(q,q')$.

We now introduce the important concept of a normal product (an N-product) of the field operators. In order to go from an ordinary operator product to a normal product we must split each of the operators into a sum of creation and annihilation parts, and rearrange these parts so that all the creation operators are to the left of the annihilation operators. In this process each rearrangement of a pair of operators must be accompanied by a change of sign.

The following are very simple examples of N-products:

$$N(1) = 1, \quad N[\hat{\psi}_{(-)}\hat{\psi}^+_{(-)}] = \hat{\psi}_{(-)}\hat{\psi}^+_{(-)},$$

$$N(\hat{\psi}) = \hat{\psi}, \quad N[\hat{\psi}_{(+)}\hat{\psi}^+_{(+)}] = \hat{\psi}_{(+)}\hat{\psi}^+_{(+)},$$

$$N(\hat{\psi}^+) = \hat{\psi}^+, \quad N[\hat{\psi}_{(-)}\hat{\psi}^+_{(+)}] = -\hat{\psi}^+_{(+)}\hat{\psi}_{(-)}.$$

We can also take a more complicated example:

$$N[\hat{\psi}^+(q)\,\hat{\psi}(q')] = \hat{\psi}^+(q)_{(+)}\,\hat{\psi}(q')_{(+)} + \hat{\psi}^+(q)_{(+)}\,\hat{\psi}(q')_{(-)}$$
$$+ \hat{\psi}^+(q)_{(-)}\,\hat{\psi}(q')_{(-)} - \hat{\psi}(q')_{(+)}\,\hat{\psi}^+(q)_{(-)}.$$

This expression differs from a simple operator product in its last term, which would otherwise be $\hat{\psi}^+(q)_{(-)}\,\hat{\psi}(q')_{(+)}$. Hence the difference between the simple and the normal products for these particular operators lies in the appearance of the corresponding anti-commutator. We obtain from (8.8):

$$N[\hat{\psi}^+(q')\,\hat{\psi}(q)] = \hat{\psi}^+(q')\,\hat{\psi}(q) - R(q, q'). \tag{8.9}$$

In a similar, but more complicated way, we can obtain the corresponding relation for the product of four field operators:

$$N[\hat{\psi}^+(q)\,\hat{\psi}^+(q')\,\hat{\psi}(q')\,\hat{\psi}(q)] = \hat{\psi}^+(q)\,\hat{\psi}^+(q')\,\hat{\psi}(q')\,\hat{\psi}(q)$$
$$- R(q, q)\,\hat{\psi}^+(q')\,\hat{\psi}(q') - R(q', q')\,\hat{\psi}^+(q)\,\hat{\psi}(q)$$
$$+ R(q, q')\,\hat{\psi}^+(q)\,\hat{\psi}(q') + R(q', q)\,\hat{\psi}^+(q')\,\hat{\psi}(q)$$
$$+ R(q, q)\,R(q', q') - R(q, q')\,R(q', q). \tag{8.10}$$

Normal products of field operators have two important properties. In the first place any rearrangement of the operators may be made so long as the overall sign of the product is changed the appropriate number of times. If the rearrangement does not change the relative order of the creation and annihilation operators, then this assertion is obvious, since the anti-commutator of the exchanged operators is zero in this case. Suppose now that operators of different types are rearranged. A change of this sort will also produce no alteration in the result (so long as the sign condition is fulfilled), since the operators, in the final count, still conform to the definition of an N-product.

The other property of any N-product (except $N(1) = 1$) is that its average over the state Ψ_0 is equal to zero:

$$\langle \Psi_0 | N(\ldots) | \Psi_0 \rangle = 0. \tag{8.11}$$

This important equality is an immediate consequence of (8.6).

8.6. By introducting the concept of a normal product, we can considerably simplify the expressions for the Hamiltonians $\hat{H}_0$ and $\hat{H}'$.

We can use (8.9) to rewrite (4.1) in the form

$$\begin{aligned}\hat{H}_0 &= \int_{q' \to q} dq\,(\hat{T} + \hat{W})_q\,\hat{\psi}^+(q')\,\hat{\psi}(q) + C \\ &= \int dq\,(\hat{T} + \hat{W})_q\,\{N[\hat{\psi}^+(q')\,\hat{\psi}(q)] + R(q, q')\} + C.\end{aligned}$$

The last two terms in this expression can be put in the form [see Appendix A and eqn. (4.27)]

$$\mathrm{Tr}\,\hat{T}\hat{\varrho} + \frac{1}{2}\mathrm{Tr}\,(\hat{V}(q, q')\,(1 - \hat{\mathscr{P}}_{qq'})\,\hat{\varrho}_q\hat{\varrho}_{q'}) = \mathrm{Tr}\left[\left(\hat{T} + \frac{\hat{W}}{2}\right)\hat{\varrho}\right].$$

From (4.35), this quantity is equal to the energy in the Hartree–Fock approximation. Hence:

$$\hat{H}_0 = E_0 + \int dq\,N[\hat{\psi}^+(q)\,(\hat{T} + \hat{W})\,\hat{\psi}(q)]. \tag{8.12}$$

Thus if we reckon the energy from its zero-approximation value, the Hamiltonian $\hat{H}_0$ reduces to a normal product. In a similar way the particle-number operator $\hat{N}$ (see § 3) can be written as

$$\hat{N} = N + \int dq\,N[\hat{\psi}^+(q)\,\hat{\psi}(q)]. \tag{8.13}$$

When the expansion (8.3) is substituted in the operator on the right-hand side of this equation, it can be written, according to (8.7), in the form

$$\sum_{\nu} (\hat{n}_{\nu}^{(p)} - \hat{n}_{\nu}^{(h)}),$$

i.e. as the difference between the number of particles and holes. This operator, like $\hat{N}$, evidently commutes with the total Hamiltonian of the system.

We now consider how to simplify the interaction Hamiltonian. Comparing the expression in curly brackets in (8.1) with (8.10), we find

$$\hat{H}' = \tfrac{1}{2} \int dq\, dq' V(q, q')\, N[\hat{\psi}^{+}(q)\, \hat{\psi}^{+}(q')\, \hat{\psi}(q')\, \hat{\psi}(q)].$$

One can show easily enough that, for a potential $\hat{V}$ (which is an operator), this takes the form

$$H' = \tfrac{1}{2} \int dq\, dq' N[\hat{\psi}^{+}(q)\, \hat{\psi}^{+}(q')\, \hat{V}(q, q')\, \hat{\psi}(q')\, \hat{\psi}(q)]. \quad (8.14)$$

The possibility of representing the interaction Hamiltonian as a normal product is caused directly by the choice of the Hartree–Fock approximation as the zero-order one. This simplifies the following discussion considerably.

9. The Scattering Matrix

9.1. We will now consider how to set up a perturbation theory in which the physical quantities are expanded as a series in $\hat{H}'$.

We will take the field operators, $\hat{\psi}(q)$, $\hat{\psi}^{+}(q)$, in the interaction representation, signifying by (x) a combination of the q-coordinate and the time t. From the general equation (2.7)

$$\hat{\psi}_{\mathrm{Int}}(x) = \exp(i\hat{H}_0 t)\, \hat{\psi}(q) \exp(-i\hat{H}_0 t),$$

where $\hat{H}_0$ is chosen to be the Hamiltonian in the Hartree–Fock approximation (4.1). Similarly, the correlation Hamiltonian $\hat{H}'$ determines the time variation of the wave function of the system.

We will use the following general relation to calculate $\hat{\psi}_{\mathrm{Int}}(x)$ (see Appendix B):

$$\exp(\hat{a})\, \hat{b} \exp(-\hat{a}) = \sum_{n=0}^{\infty} \frac{1}{n!} \underbrace{[\hat{a}[\hat{a} \ldots [\hat{a}, \hat{b}]_{-} \ldots]_{-}]_{-}}_{n},$$

where $\hat{a}$ and $\hat{b}$ are any arbitrary operators. From the commutation rule (3.10), we have

$$[\hat{H}_0, \hat{\psi}(q)]_- = -(\hat{T} + \hat{W})\,\hat{\psi}(q),$$

whence

$$\overbrace{[\hat{H}_0 \ldots [\hat{H}_0}^{n}, \hat{\psi}(q)]_- \ldots]_- = (-1)^n\,(\hat{T} + \hat{W})^n\,\hat{\psi}(q)$$

and

$$\hat{\psi}_{\mathrm{Int}}(x) = \exp\,[-it(\hat{T} + \hat{W})]\,\hat{\psi}(q). \tag{9.1}$$

Differentiating this relation with respect to time, we find the equation of motion for the operator $\hat{\psi}_{\mathrm{Int}}(x)$:

$$i\,\frac{\partial \hat{\psi}_{\mathrm{Int}}(x)}{\partial t} = (\hat{T} + \hat{W})\,\hat{\psi}_{\mathrm{Int}}(x). \tag{9.2}$$

Substituting (8.3) into (9.1) and using (4.2), we arrive at a final expression for the field operators:

$$\left.\begin{aligned} \hat{\psi}_{\mathrm{Int}}(x) &= \sum_{\nu} \hat{A}_\nu \chi_\nu(q) \exp\,(-i\varepsilon_\nu t), \\ \hat{\psi}^+_{\mathrm{Int}}(x) &= \sum_{\nu} \hat{A}^+_\nu \chi^*_\nu(q) \exp\,(i\varepsilon_\nu t). \end{aligned}\right\} \tag{9.3}$$

It can be seen that the result of this transformation is to replace the time-independent part of the particle wave function $\chi_\nu(q)$ by the complete wave function $\chi_\nu(q)\exp\,(-i\varepsilon_\nu t)$.

We will now establish the commutation rules for the operators $\hat{\psi}_{\mathrm{Int}}(x)$, $\hat{\psi}^+_{\mathrm{Int}}(x)$. From eqns. (8.5) and (9.3), we have

$$[\hat{\psi}_{\mathrm{Int}}(x), \hat{\psi}^+_{\mathrm{Int}}(x')]_+ = \sum_{\nu} \chi^*_\nu(q')\,\chi_\nu(q) \exp\,[-i\varepsilon_\nu(t - t')]. \tag{9.4}$$

The remaining anti-commutators are equal to zero. When $t = t'$, the right-hand side of expression (9.4) becomes $\delta(q - q')$, which evidently agrees with eqn. (2.11).

We will also transform the Hamiltonian $\hat{H}'$ into the interaction representation. From (2.10), this can be written:

$$\hat{H}'_{\mathrm{Int}}(t) = \tfrac{1}{2} \int dq\,dq' N[\hat{\psi}^+_{\mathrm{Int}}(x)\,\hat{\psi}^+_{\mathrm{Int}}(x')\,\hat{V}\hat{\psi}_{\mathrm{Int}}(x')\,\hat{\psi}_{\mathrm{Int}}(x)], \tag{9.5}$$

where $x = (q, t)$; $x' = (q', t')$.

9.2. The time dependence of the wave function of a system in the interaction representation is determined by the general eqn. (2.6)

$$i\frac{\partial\Psi_{\mathrm{Int}}(t)}{\partial t} = \hat{H}'_{\mathrm{Int}}(t)\,\Psi_{\mathrm{Int}}(t).$$

We introduce the operator $\hat{S}(t, t_0)$ (the S-matrix or scattering matrix), which relates the values of the wave function at times t and t_0:

$$\Psi_{\mathrm{Int}}(t) = \hat{S}(t, t_0)\,\Psi_{\mathrm{Int}}(t_0). \tag{9.6}$$

This S-matrix has to satisfy certain conditions. In the first place it must be a unitary matrix

$$\hat{S}\hat{S}^{+} = \hat{S}^{+}\hat{S} = 1.$$

This property enables the norm of the wave function to be conserved. In the second place, the S-matrix must possess the following obvious properties:†

$$\left.\begin{aligned}\hat{S}(t_1, t_3) &= \hat{S}(t_1, t_2)\,\hat{S}(t_2, t_3),\\ \hat{S}(t_1, t_2) &= \hat{S}^{+}(t_2, t_1).\end{aligned}\right\} \tag{9.7}$$

The second of these relations stems from the unitary requirement, and from the condition

$$\hat{S}(t_0, t_0) = 1. \tag{9.8}$$

We now find the equation satisfied by the S-matrix. Substituting (9.6) into the Schrödinger equation, we find

$$i\frac{\partial\hat{S}(t, t_0)}{\partial t} = \hat{H}'_{\mathrm{Int}}(t)\,\hat{S}(t, t_0). \tag{9.9}$$

The initial conditions required to solve this equation are given by (9.8).

We next consider the procedure for expanding the S-matrix as a series in H'_{Int}:

$$\hat{S}(t, t_0) = 1 + \sum_{n=1}^{\infty}\hat{S}_n(t, t_0). \tag{9.10}$$

† The evolution of the system from t_3 to t_1 can be treated as the result of successive evolutions, firstly from t_3 to t_2 and then from t_2 to t_1.

Here $\hat{S}_n$ contains the product of n operators $\hat{H}'_{\text{Int}}$; the zero term of the expansion is chosen to be unity since, when H'_{Int} is absent, there is no change of Ψ_{Int} with time.

We will transform eqn. (9.9) into integral form, integrating both parts of it over t, and taking into account the requirement (9.8):

$$\hat{S}(t, t_0) = 1 - i \int_{t_0}^{t} d\tau\, \hat{H}'_{\text{Int}}(\tau)\, \hat{S}(\tau, t_0).$$

Substituting the series (9.10) into this equation, we obtain the recurrence relation:

$$\hat{S}_n(t, t_0) = -i \int_{t_0}^{t} d\tau\, \hat{H}'_{\text{Int}}(\tau)\, \hat{S}_{n-1}(\tau, t_0)$$

from which we can write

$$\hat{S}_n(t, t_0) = (-i)^n \int_{t_0}^{t} d\tau_1 \int_{t_0}^{\tau_1} d\tau_2 \ldots \int_{t_0}^{\tau_{n-1}} d\tau_n\, \hat{H}'_{\text{Int}}(\tau_1) \ldots \hat{H}'_{\text{Int}}(\tau_n). \tag{9.11}$$

The operators $\hat{H}'_{\text{Int}}$ in this expression are in chronological order $(\tau_1 > \tau_2 \ldots > \tau_n)$ i.e. their time arguments increase monotonically from right to left.

9.3. Owing to the complicated integration limits, eqn. (9.11) is inconvenient to use. This deficiency can be circumvented by introducing the concept of the *T*-product (the time-ordered chronological product) of the operators.

The *T*-product is obtained from the usual product of field operators by rearranging these latter so that their time arguments increase from right to left. The overall sign depends on whether an odd or even permutation of the operators is required. In particular the *T*-product of two field operators $\hat{F}_1(x_1)$, $\hat{F}_2(x_2)$ is equal to†

$$T[\hat{F}_1(x_1)\, \hat{F}_2(x_2)] = \begin{cases} \hat{F}_1(x_1)\, \hat{F}_2(x_2) & t_1 > t_2 \\ -\hat{F}_2(x_2)\, \hat{F}_1(x_1) & t_1 < t_2 \end{cases} \tag{9.12}$$

or

$$T[\hat{F}_1(x_1)\, \hat{F}_2(x_2)] = \theta(t_1 - t_2)\, \hat{F}_1\hat{F}_2 - \theta(t_2 - t_1)\, \hat{F}_2\hat{F}_1. \tag{9.12'}$$

† The order of the operators can be altered in the *T*-product (as in the *N*-product), so long as the overall sign is chosen properly. Again, as for the *N*-product, the operators will finally be in the necessary (in this case chronological) order.

This chronological ordering can be applied not only to the field operators, but also to combinations of them which contain a product of field operators referred to one specific moment of time. We can, in particular, have a T-product of the Hamiltonians $\hat{H}'_{\text{Int}}$:

$$T[\hat{H}'_{\text{Int}}(\tau_1) \dots \hat{H}'_{\text{Int}}(\tau_n)]$$

which supposes, as before, that the operators $\hat{H}'_{\text{Int}}$ are so distributed that τ increases from right to left. In this case, however, a rearrangement of the operators $\hat{H}'_{\text{Int}}$ does not produce a change in sign, since $\hat{H}'_{\text{Int}}$ contains the product of an even number of field operators. Hence

$$T[\hat{H}'_{\text{Int}}(\tau_1)\, \hat{H}'_{\text{Int}}(\tau_2)] = \theta(\tau_1 - \tau_2)\, \hat{H}'_{\text{Int}}(\tau_1)\, \hat{H}'_{\text{Int}}(\tau_2) + \theta(\tau_2 - \tau_1)\, \hat{H}'_{\text{Int}}(\tau_2)\, \hat{H}'_{\text{Int}}(\tau_1). \qquad (9.12'')$$

T-ordering does not affect the order of field operators in $\hat{H}'_{\text{Int}}$ referring to one particular time: the order of the operators remains unchanged. The statement that operators may be rearranged in a T-product refers only to the operators $\hat{H}'_{\text{Int}}$ as a whole, and not to their component parts.

We can now rewrite (9.11) through the substiution

$$\hat{H}'_{\text{Int}}(\tau_1) \dots \hat{H}'_{\text{Int}}(\tau_n) \to T[\hat{H}'_{\text{Int}}(\tau_1) \dots \hat{H}'_{\text{Int}}(\tau_n)].$$

If we expand the limits of integration to include the highest value of t, then the integral obtained can be written as a sum of $n!$ (corresponding to the number of permutations of the $\tau_1 \dots \tau_n$) integrals of the type (9.11) which are equal to one another. We can therefore write

$$\hat{S}_n(t, t_0) = \frac{(-i)^n}{n!} \int_{t_0}^{t} d\tau_1 \dots \int_{t_0}^{t} d\tau_n\, T[\hat{H}'_{\text{Int}}(\tau_1) \dots \hat{H}'_{\text{Int}}(\tau_n)].$$

Consider, for purposes of illustration, the term with $n = 2$:

$$\hat{S}_2 = -\frac{1}{2}\left\{\int_{t_0}^{t} d\tau_1 \int_{t_0}^{\tau_1} d\tau_2\, \hat{H}'_{\text{Int}}(\tau_1)\, \hat{H}'_{\text{Int}}(\tau_2) + \int_{t_0}^{t} d\tau_1 \int_{\tau_1}^{t} d\tau_2\, \hat{H}'_{\text{Int}}(\tau_2)\, \hat{H}'_{\text{Int}}(\tau_1)\right\}.$$

We have used here (9.12″). Changing the order of integration in the second integral $\int_{t_0}^{t} d\tau_1 \int_{\tau_1}^{t} d\tau_2 \to \int_{t_0}^{t} d\tau_2 \int_{t_0}^{\tau_2} d\tau_1$ and making the replacement $\tau_1 \rightleftarrows \tau_2$, we can show that the terms in S_2 are equal to one another, and we therefore revert to our former expression (9.11) with $n = 2$.

9.4. We shall later be interested in S-matrices which have one (or both) of the quantities t, t_0 equal to infinity. We consider first the S-matrix $\hat{S}\ (\infty, -\infty)$. If we substitute the expression for $\hat{S}_n$ obtained above in (9.10), we can write†

$$\hat{S} = \sum_{n=0}^{\infty} \frac{(-i)^n}{n!} \int_{-\infty}^{\infty} d\tau_1 \ldots \int_{-\infty}^{+\infty} d\tau_n\, T[\hat{H}'_{\text{Int}}(\tau_1) \ldots \hat{H}'_{\text{Int}}(\tau_n)]$$

$$\equiv T \exp\left[-i \int_{-\infty}^{\infty} d\tau\, \hat{H}'_{\text{Int}}(\tau)\right]. \tag{9.13}$$

$\hat{H}'_{\text{Int}}(\tau)$ as a function of τ has, in general, an oscillatory character. The characteristic integrals in (9.13) do not have a definite limit as $t \to \infty$ and $t_0 \to -\infty$. If these integrals are to be defined we must introduce a factor $\exp(-\delta|t|)$ in the exponent of the expression under the integral sign, with $\delta \to +0$. In other words we must make the substitution

$$\hat{H}'_{\text{Int}}(\tau) \to \hat{H}'_{\text{Int}}(\tau) \exp(-\delta|\tau|), \tag{9.14}$$

which will in the future, be implicitly understood. We must emphazise that no physically observable quantity is any way indeterminate in the limit $\delta \to 0$. Using (9.14), we may finally write:

$$\hat{S} = T \exp\left\{-\frac{i}{2} \int d^4x_1\, d^4x_2 \exp(-\delta|t_1|) \right.$$
$$\left. \times N[\hat{\psi}^{+}_{\text{Int}}(x_1)\, \hat{\psi}^{+}_{\text{Int}}(x_2)\, \hat{V}(x_1, x_2)\, \hat{\psi}_{\text{Int}}(x_2)\, \hat{\psi}_{\text{Int}}(x_1)]\right\}. \tag{9.15}$$

Here $d^4x = dq\, dt$; $\hat{V}(x_1, x_2) = \hat{V}(q_1, q_2)\, \delta(t_1 - t_2)$ (retardation is not taken into account). The integration is carried out over all four-dimensional space.

† The $\hat{S}$-matrix $S(\infty, -\infty)$ is denoted simply by $\hat{S}$.

Besides the operator $\hat{S}(\infty, -\infty)$, we also require the operator $\hat{S}(0, -\infty)$. The corresponding expression evidently contains a time integration between the limits $-\infty$ and 0.

9.5. We will now look for a relation which embodies the law of conservation of energy in S-matrix language. We first convince ourselves that $\hat{S}(t, -\infty)$ can be written in the form

$$\hat{S}(t, -\infty) = 1 + \int_{-\infty}^{t} d\tau \exp(i\hat{H}_0\tau)\,\hat{\sigma} \exp(-i\hat{H}_0\tau) \qquad (9.16)$$

where $\hat{\sigma}$ is an operator independent of τ. We can prove this by equating the quantities

$$i\frac{\partial \hat{S}}{\partial t} = i\exp(i\hat{H}_0 t)\,\hat{\sigma} \exp(-i\hat{H}_0 t)$$

and

$$\hat{H}'_{\text{Int}}\hat{S} = \exp(i\hat{H}_0 t)\,\hat{H}' \exp(-i\hat{H}_0 t) \times \left\{\int_{-\infty}^{t} d\tau \exp(i\hat{H}_0\tau)\,\hat{\sigma} \exp(i\hat{H}_0\tau) + 1\right\}.$$

Substituting $\tau \to t + x$ in this last expression, we obtain

$$\exp(i\hat{H}_0 t)\,\hat{H}' \left\{\int_{-\infty}^{0} dx \exp(i\hat{H}_0 x)\,\hat{\sigma} \exp(-i\hat{H}_0 x - \delta\,|t + x|) + 1\right\} \times \exp(-i\hat{H}_0 t).$$

We thus see that the operator $\hat{\sigma}$ is indeed independent of time, since the time disappears when the foregoing expressions are compared.

We now consider the matrix element $\langle \Psi_{0(m)} | \hat{S} - 1 | \Psi_{0(n)} \rangle$, where $\Psi_{0(n)}$ is an eigenfunction of the operator $\hat{H}_0$ with an eigenvalue $E_{0(n)}$. Using (9.16) with $t \to \infty$, and the relations

$$\exp(-i\hat{H}_0 t)\,\Psi_0 = \exp(-iE_0 t)\,\Psi_0,$$

$$\Psi_0^* \exp(i\hat{H}_0 t) = \Psi_0^* \exp(iE_0 t),$$

which may be easily demonstrated by a series expansion, we can see that the matrix element in which we are interested is proportional to the quantity

$$\int_{-\infty}^{\infty} \exp\,[i(E_{0(m)} - E_{0(n)})\,t]\,dt = 2\pi\delta(E_{0(m)} - E_{0(n)}). \qquad (9.16')$$

Thus

$$\langle \Psi_{0(m)}|\hat{S} - 1|\Psi_{0(n)}\rangle = 2\pi\delta(E_{0(m)} - E_{0(n)})\,\langle \Psi_{0(m)}|\hat{\sigma}|\Psi_{0(n)}\rangle. \quad (9.17)$$

The law of conservation of energy appears, therefore, in that the matrix element (9.17) is equal to zero for $E_{0(m)} \neq E_{0(n)}$. In other words a system evolving from the state $\Psi_{0(n)}$ can only transfer to a state with the same energy. This conclusion only holds for a process of total evolution of the system from $-\infty$ to ∞. Indeed if the limits of integration in (9.16′) are finite, we do not obtain the δ-function, but a function which differs from zero for any value of the difference $E_{0(m)} - E_{0(n)}$.

We now make the assumption that the state $\Psi_{0(n)}$ is non-degenerate, i.e. that there are no other states with the same energy. This is true in general for the ground state of a system (for excited states see §§ 21 and 22). For this to be true we must limit our discussion to systems with closed shells (see § 4.3), for which the occupation number depends only on the energy of the level. If we compare the states of such a system, which correspond to the same energy, then we find that they differ, at most, by a permutation of the particles. We are in fact dealing with one state only. Things are different for partly filled shells. Suppose, for example, that there is one occupied level above the closed shells. If we change the direction of the spin of the particle, we obtain a different state of the same energy (we suppose for simplicity that the forces are spin independent).

On the assumption that the ground state of the system is non-degenerate, we see that only the diagonal element $\langle \Psi_{0(n)}|\hat{S}|\Psi_{0(n)}\rangle$ of all the matrix elements (9.17) differs from zero. From the general relation

$$\hat{S}\Psi_{0(n)} = \sum_m \langle \Psi_{0(n)}|\hat{S}|\Psi_{0(m)}\rangle\,\Psi_{0(m)},$$

we obtain

$$\hat{S}\Psi_0 = \langle\Psi_0|\hat{S}|\Psi_0\rangle\, \Psi_0, \tag{9.18}$$

and from

$$\langle\Psi_0|\hat{S}^+\hat{S}|\Psi_0\rangle = 1 = |\langle\Psi_0|\hat{S}|\Psi_0\rangle|^2,$$

we find

$$(\hat{S}\Psi_0)^* = \Psi_0^*\hat{S}^+ = \langle\Psi_0|\hat{S}|\Psi_0\rangle^{-1}\, \Psi_0^*. \tag{9.18'}$$

These equations express the "stability" requirement for the ground state of a system: the function $\hat{S}\Psi_0$, which represents the result of the total evolution of the system, coincides (apart from an inessential factor) with the initial state Ψ_0. This state therefore undergoes no change during the evolution of the system.

9.6. We will find it necessary in the following to change from the interaction representation to the Heisenberg representation, and vice versa. The corresponding transformation operator $\hat{U}$ (see § 2)

$$\Psi_H = \hat{U}\Psi_{\text{Int}}; \quad \hat{\psi}_H(x) = \hat{U}\hat{\psi}_{\text{Int}}(x)\, \hat{U}^{-1}$$

is directly connected with the S-matrix:

$$\hat{U} = \hat{S}(0, t) = \hat{S}^+(t, 0). \tag{9.19}$$

For proof of this, we need only remember that, when $t = 0$, both of these representations coincide. Therefore

$$\Psi_H = \Psi_{\text{Int}}(0) = \hat{S}(0, t)\, \Psi_{\text{Int}}(t).$$

The relation (9.19) can also be deduced in another way, via (2.9):

$$\hat{U} = \exp(i\hat{H}t) \exp(-i\hat{H}_0 t).$$

Differentiating with respect to t, we may easily confirm that $\hat{U}$ satisfies the same equation as $\hat{S}(0, t)$, besides which, $\hat{U} = 1$ when $t = 0$, i.e. it coincides completely with $\hat{S}(0, t)$. In a similar way we can prove the general relation

$$\hat{S}(t, t_0) = \exp(i\hat{H}_0 t) \exp[-i\hat{H}(t - t_0)] \exp(-i\hat{H}_0 t_0). \tag{9.20}$$

It is, however, inconvenient to use owing to the presence in the exponential index of a sum of the non-commuting operators $\hat{H}_0$ and $\hat{H}'$.

We may, therefore, write

$$\Psi_H = \hat{S}(0, t)\, \Psi_{\text{Int}}(t); \quad \hat{\psi}_H(x) = \hat{S}(0, t)\, \hat{\psi}_{\text{Int}}(x)\, \hat{S}^{-1}(0, t). \quad (9.21)$$

In particular, when $t \to \infty$, we obtain

$$\Psi_H = \hat{S}(0, -\infty)\, \Psi_{\text{Int}}(-\infty).$$

Ψ_H coincides with the time-independent part of the exact Schrödinger wave function for the system. We may write, therefore,

$$(\hat{H}_0 + \hat{H}' - E)\, \hat{S}(0, -\infty)\, \Psi_{\text{Int}}(-\infty) = 0, \quad (9.22)$$

where $\hat{H}' = \hat{H}'_{\text{Int}}(0)$ is the interaction Hamiltonian in the Schrödinger representation; E is the true energy of the state.

If we want to find the precise wave function of the system Ψ_H, it is not enough to know only the S-matrix; we must also have some expression for the wave function $\Psi_{\text{Int}}(-\infty)$. We will devote the next few paragraphs to considering this function.

From (9.14), we may conclude that $\Psi_{\text{Int}}(-\infty)$ must coincide with the wave function in the zero approximation Ψ_0. In fact when $t \to -\infty$, the Hamiltonian (9.14) effectively goes to zero, and the correlation interaction between particles in the system disappears. The interaction increases with t at an infinitesimal rate; it is switched on adiabatically as $\delta \to 0$, and we then arrive at the real state of the system. When $t \to +\infty$ the interaction disappears slowly in the same way, so that the functions $\Psi_{\text{Int}}(+\infty)$ and Ψ_0 must also coincide. This is the so-called adiabatic hypothesis.

A fundamental question, which must be asked, is what particular state Ψ_0 corresponds to a given state, in particular to the ground state Ψ_H, when $t \to -\infty$. We can bring forward several arguments to show that Ψ_0 describes the ground state of the system with the correlation interaction excluded.

We note, to this end, that the factor $\exp(-\delta|t|)$ in the interaction destroys the stationary nature of the states of the system. However, when $\delta \to 0$, we can talk about "nearly stationary" states, whose characteristics are infinitely weakly dependent on time. In particular the energy of a state is likewise dependent on the time as parameter. This dependence is, of course, negligible for finite intervals of time, but becomes important when $t \to \pm\infty$.

There is a general quantum mechanical theorem concerning the absence of overlapping in energy levels which depend on a single parameter and have the same symmetry [31]. In other words if these conditions are fulfilled, a level which is lowest for one value of the parameter will also be the lowest for any other value of the parameter. This theorem obviously provides a basis for our previous assertion: the ground state for $t = 0$ is also the ground state for $t \to -\infty$.

So far as the symmetry of these states is concerned, we may appeal to the following circumstances. As a rule, the ground state of a system takes the maximum symmetry permitted by the imposed boundary conditions. Therefore this discussion of overlapping energy levels only applies when they have the same, that is the maximum, symmetry.

9.7. The account given in the previous section requires a certain amount of additional justification. Using the S-matrix series expansion in perturbation theory, we can attempt a more rigorous discussion. To prove that the functions $\Psi_{\text{Int}}(\pm\infty)$ and Ψ_0 coincide, it is necessary that the relation should be consistent with the infinite series expansion in $\hat{H}'$.†

We will show that the function $\Psi_{\text{Int}}(-\infty)$ satisfies the equation (see § 9.8 for details):

$$\left\{\frac{1}{2}(\hat{H}_0 + \hat{S}^{-1}\hat{H}_0\hat{S}) + \frac{i\delta}{2}\hat{S}^{-1}\frac{d\hat{S}}{d\lambda} - E\right\}\Psi_{\text{Int}}(-\infty) = 0. \quad (9.23)$$

The parameter λ in this equation, has the following significance: the Hamiltonian $\hat{H}'$ must be replaced by $\lambda\hat{H}'$, so that the S-matrix becomes a function of λ. When all the calculations have been made, λ is put equal to one; we assume here that δ tends to zero.

We must now make sure that the function Ψ_0 also satisfies eqn. (9.23). To this end, we use the stability condition for Ψ_0 (9.18), and the equation $\hat{H}_0\Psi_0 = E_0\Psi_0$:

$$\hat{S}^{-1}\hat{H}_0\hat{S}\Psi_0 = E_0\hat{S}^{-1}\hat{S}\Psi_0 = E_0\Psi_0.$$

† One must be careful when there is a bound state present, the characteristics of which depend on $\hat{H}'$ in a non-analytic way.

Moreover, since Ψ_0 is independent of λ, we can write

$$\hat{S}^{-1}\left(\frac{d\hat{S}}{d\lambda}\right)\Psi_0 = \hat{S}^{-1}\frac{\partial}{\partial\lambda}(\hat{S}\Psi_0) = \frac{\partial}{\partial\lambda}[\ln\langle\Psi_0|\hat{S}|\Psi_0\rangle]\,\Psi_0.$$

It is extremely important that the operator $\hat{S}^{-1}\,d\hat{S}/d\lambda$, when acting on Ψ_0, reduces to a *c*-number, i.e. Ψ_0 is an eigenfunction of this operator. We obtain, as a result, the condition which Ψ_0 must fulfill in order to satisfy eqn. (9.23):

$$E - E_0 = \frac{i\delta}{2}\cdot\frac{\partial}{\partial\lambda}\ln\langle\Psi_0|\hat{S}|\Psi_0\rangle. \qquad (9.24)$$

This relation can of itself serve to determine the correlation energy of the system [84]. It will be shown later that $\langle\Psi_0|\hat{S}|\Psi_0\rangle$ does, in fact, take the form $\exp(L_0/\delta)$, where $L_0(\lambda)$ is some quantity independent of δ. Thus the difference $E - E_0$ has a value well-defined non-vanishing in the limit $\delta \to 0$.

We are led to conclude that Ψ_0 and $\Psi_{\text{Int}}(-\infty)$ correspond to one another from the requirement that the energy levels be non-degenerate, i.e. they are the same except for an unimportant phase factor.

The expression for the wave function of a system may be written finally in the form

$$\Psi_H = \hat{S}(0, -\infty)\,\Psi_0, \qquad (9.25)$$

where Ψ_0 is the wave function of the ground state of the system in the Hartree–Fock approximation. We must emphasize once more the physical significance of this relation. When $t = -\infty$, the correlation interaction disappears; the system is in the state Ψ_0 with eigenvalue E_0. As t increases, the interaction is slowly switched on until, when $t = 0$, Ψ_0 becomes Ψ_H—the precise wave function of a system with eigenvalue E.

A similar relation holds for times $t = 0$ and $t = +\infty$:

$$\Psi_H = \hat{S}^{+}(\infty, 0)\,\Psi_0 = \hat{S}(0, \infty)\,\Psi_0. \qquad (9.26)$$

This can be demonstrated in the same way. Both (9.25) and (9.26) still contain different phase factors. These drop out of the final results, which are of direct physical significance.

9.8. We now turn to the proof of eqn. (9.23), using the following relations [86]:

$$\left.\begin{aligned}[\hat{H}_0, \hat{S}(0, -\infty)]_- &= -\hat{H}'\hat{S}(0, -\infty) + i\delta\frac{\partial\hat{S}(0, -\infty)}{\partial\lambda},\\ [\hat{H}_0, \hat{S}(\infty, 0)]_- &= \hat{S}(\infty, 0)\,\hat{H}' - i\delta\frac{\partial\hat{S}(\infty, 0)}{\partial\lambda}\end{aligned}\right\} \quad (9.27)$$

(we will derive these equations below). Multiplying (9.23) from the left by $\hat{S}$ (0, $-\infty$), and dividing the S-matrix into two factors, $\hat{S}(\infty, 0)$ and $\hat{S}(0, -\infty)$, in accordance with (9.7), we can easily reduce (9.23) to (9.22), using eqns. (9.27).

We will now verify the first relation of (9.27) (the second may be derived in a similar way). To this end, we consider the commutator $[\hat{H}_0, \hat{S}(0, -\infty)]_-$, substituting into it the S-matrix expansion

$$[\hat{H}_0, \hat{S}(0, -\infty)]_- = \sum_{n=1}^{\infty}\frac{(-i)^n}{n!}\int_{-\infty}^{0} d\tau_1 \ldots \int_{-\infty}^{0} d\tau_n \exp\left(\delta\sum_{1}^{n}\tau_i\right)\hat{K}_n,$$

where (9.28)

$$\hat{K}_n = T[\hat{H}_0, \hat{H}'_{\text{Int}}(\tau_1) \ldots \hat{H}'_{\text{Int}}(\tau_n)].$$

Since $\hat{H}_0$ is time independent, it can be put inside the T-product.

The Hamiltonian $\hat{H}_0$, when commuting with any operator in the interaction representation, results in a differentiation of the operator with respect to time. We may therefore write

$$\hat{K}_n = -in\frac{\partial}{\partial\tau_1}T[\hat{H}'_{\text{Int}}(\tau_1) \ldots \hat{H}'_{\text{Int}}(\tau_n)].$$

We have used here, first of all the symmetry of the expression under the integral sign in (9.28) relative to a permutation of the points $\tau_1 \ldots \tau_n$, and, secondly, the possibility of interchanging $\partial/\partial\tau$ and T when a product of similar factors occurs. This can be seen most simply for the case $n = 2$. If we differentiate (9.12″) and remember that $(\partial\theta(\pm x)/\partial x) = \pm\delta(x)$, we find that

$$\left(\frac{\partial}{\partial\tau_1}T - T\frac{\partial}{\partial\tau_1}\right)\hat{H}'_{\text{Int}}(\tau_1)\,\hat{H}'_{\text{Int}}(\tau_2)$$
$$= \delta(\tau_1 - \tau_2)\,[\hat{H}'_{\text{Int}}(\tau_1), \hat{H}'_{\text{Int}}(\tau_2)]_- = 0.$$

We have used here the fact that, owing to the presence of the δ-function, we are dealing with the commutator of two equal operators; the result of this is naturally zero.

Integrating (9.28) by parts in terms of τ_1, we can write the following expression for $[\hat{H}_0, \hat{S}(0, -\infty)]_-$:

$$\sum_{n=1}^{\infty} \frac{(-i)^{n+1}}{(n-1)!} \left\{ \int_{-\infty}^{0} d\tau_2 \ldots \int_{-\infty}^{0} d\tau_n \exp\left(\delta \sum_{2}^{n} \tau_i\right) \lambda^n \hat{H}'_{\text{Int}}(0) \right.$$
$$\times T[\hat{H}'_{\text{Int}}(\tau_2) \ldots \hat{H}'_{\text{Int}}(\tau_n)] - \delta \int_{-\infty}^{0} d\tau_1 \ldots \int_{-\infty}^{0} d\tau_n$$
$$\left. \times \exp\left(\delta \sum_{1}^{n} \tau_i\right) \lambda^n T[\hat{H}'_{\text{Int}}(\tau_1) \ldots \hat{H}'_{\text{Int}}(\tau_n)] \right\}.$$

The first term of this expression is the same as $-\lambda \hat{H}'_{\text{Int}}(0)\, \hat{S}(0, -\infty)$. The second term can be rewritten in the form $i\delta\lambda\, \partial \hat{S}(0, -\infty)/\partial\lambda$. The corresponding differentiation ensures that the correct numerical coefficient is obtained, with the factor n in the numerator.

10. The Contraction of Operators

10.1. Before we proceed to any further transformations of the expression for the S-matrix, we must consider the operator contraction in the interaction representation, defined by the relation

$$\overbrace{\hat{F}_1 \hat{F}_2} = T(\hat{F}_1 \hat{F}_2) - N(\hat{F}_1 \hat{F}_2), \tag{10.1}$$

where the $\hat{F}_{1,2}$ represent the field operators $\hat{\psi}$ or $\hat{\psi}^+$.† From (9.12) and the expression for the N-product of two operators, we can write

$$\overbrace{\hat{F}_1(x_1)\, \hat{F}_2(x_2)} = \begin{cases} [\hat{F}_1(x_1)_{(-)}, \hat{F}_2(x_2)_{(+)}]_+ & t_1 > t_2 \\ -[\hat{F}_1(x_1)_{(+)}, \hat{F}_2(x_2)_{(-)}]_+ & t_1 < t_2 \end{cases} \tag{10.2}$$

It can be seen that the contractions of the operators $\overbrace{\hat{\psi}\hat{\psi}}$ and $\overbrace{\hat{\psi}^+\hat{\psi}^+}$ are identically zero (see § 8.5), and the contraction of the operators

† Sections 10 to 14 will deal exclusively with the interaction representation; we will therefore omit the suffix "Int".

$\hat{\psi}$ and $\hat{\psi}^+$ is a c-number. From (8.11), this latter property can be written:

$$\overbrace{\hat{\psi}(x_1)\,\hat{\psi}^+}(x_2) = \langle \Psi_0 | T[\hat{\psi}(x_1)\,\hat{\psi}^+(x_2)] | \Psi_0 \rangle. \tag{10.3}$$

An interchange of contracted operators alters the sign:

$$\overbrace{\hat{\psi}\hat{\psi}^+} = -\overbrace{\hat{\psi}^+\hat{\psi}}. \tag{10.4}$$

We will now introduce an explicit expression for the contraction (10.3). Substituting (8.8) into (10.2) we obtain

$$\overbrace{\hat{\psi}(x_1)\,\hat{\psi}^+}(x_2) = \sum_{\nu} \chi_\nu^*(q_2)\,\chi_\nu(q_1) \exp\left[-i\varepsilon_\nu(t_1 - t_2)\right] \times \begin{cases} 1 - n_\nu, & t_1 > t_2 \\ -n_\nu, & t_1 < t_2 \end{cases} \tag{10.5}$$

When $t_1 - t_2 \to -0$ the contraction coincides with the single-particle density matrix of the system.

We next look for an equation governing the contraction. To this end we write (10.3) in the explicit form:

$$\overbrace{\hat{\psi}(x_1)\,\hat{\psi}^+}(x_2) = \theta(t_1 - t_2)\langle \Psi_0 | \hat{\psi}\hat{\psi}^+ | \Psi_0 \rangle - \theta(t_2 - t_1)\langle \Psi_0 | \hat{\psi}^+\hat{\psi} | \Psi_0 \rangle.$$

From eqn. (9.2) for $\hat{\psi}$, the relation $\partial\theta(\pm t)/\partial t = \pm\delta(t)$ and the commutation relations (9.4) for an identical time, we obtain

$$\left(i\frac{\partial}{\partial t_1} - \hat{T} - \hat{W}\right)\overbrace{\hat{\psi}(x_1)\,\hat{\psi}^+}(x_2) = i\delta^4(x_1 - x_2), \tag{10.6}$$

where $\delta^4(x) = \delta(t)\,\delta(\boldsymbol{x})$. The quantity $G_0(x_1, x_2) = -i\overbrace{\hat{\psi}(x_1)\,\hat{\psi}^+}(x_2)$ is thus the Green function of eqn. (9.2) for the operator $\hat{\psi}(x)$. It is called the free-particle Green function.

The Green function corresponding to (10.5) may be conveniently Fourier-transformed in terms of the difference $t_1 - t_2 = \tau$.† We thus obtain the quantity

$$G_0(q_1, q_2, \varepsilon) = \int_{-\infty}^{\infty} d\tau\, G_0(x_1, x_2) \exp(i\varepsilon\tau),$$

† If the external field is time independent, the Green function depends only on this difference, and not on t_1 and t_2 separately.

which may easily be calculated from the following relations, given in Appendix C:

$$\int_0^{\infty} \exp(ia\tau - \delta\tau)\, d\tau = \frac{i}{a + i\delta},$$

$$\int_{-\infty}^{0} \exp(ia\tau + \delta\tau)\, d\tau = \frac{-i}{a - i\delta}.$$

As a result, we have

$$G_0(q_1, q_2, \varepsilon) = \sum_{\nu} \chi_{\nu}^*(q_2)\, \chi_{\nu}(q_1) \left\{ \frac{1 - n_{\nu}}{\varepsilon - \varepsilon_{\nu} + i\delta} + \frac{n_{\nu}}{\varepsilon - \varepsilon_{\nu} - i\delta} \right\}.$$

We can write for the ground state of the system, where $n_{\nu} = \theta(\varepsilon_F - \varepsilon_{\nu})$†

$$G_0(q_1, q_2, \varepsilon) = \sum_{\nu} \frac{\chi_{\nu}^*(q_2)\, \chi_{\nu}(q_1)}{\varepsilon - \varepsilon_{\nu} + i\delta \operatorname{sign}(\varepsilon_{\nu} - \varepsilon_F)}. \tag{10.7}$$

For a homogeneous system, $G_0(q_1, q_2, \varepsilon)$ depends only on the difference in the spatial coordinates. Carrying out a Fourier transformation in terms of $\boldsymbol{x}_1 - \boldsymbol{x}_2 = \boldsymbol{\xi}$, we have

$$G_0(\boldsymbol{p}, \varepsilon)_{\sigma_1\sigma_2, \tau_1\tau_2} = \int d^3\xi\, G_0(q_1, q_2, \varepsilon) \exp -i(\boldsymbol{p} \cdot \boldsymbol{\xi}).$$

Substituting (4.20) in (10.7), we obtain

$$G_0(\boldsymbol{p}, \varepsilon) = \frac{\delta_{\sigma_1\sigma_2}\delta_{\tau_1\tau_2}}{\varepsilon - \varepsilon_{\boldsymbol{p}} + i\delta \operatorname{sign}(p^2 - p_0^2)}. \tag{10.8}$$

In all these expressions δ is a positive, infinitesimally small quantity.

10.2. The Green function also has a comparatively simple form for a self-consistent field which varies slowly in space, i.e. in the Thomas–Fermi approximation (see § 5).

As a preliminary step, we write (10.7) in the symbolic operator form. Putting

$$(\hat{T} + \hat{W})\, \chi_{\nu}(q) = \varepsilon_{\nu} \chi_{\nu}(q),$$

we have

$$G_0(q_1, q_2, \varepsilon) = \{\varepsilon - (\hat{T} + \hat{W})_{q_1} + i\delta \operatorname{sign}[(\hat{T} + \hat{W})_{q_1} - \varepsilon_F]\}^{-1} \times \delta(q_1 - q_2). \tag{10.9}$$

† Here, sign $x = 1$, $(x > 0)$; sign $x = -1$, $(x < 0)$.

This gives the free-particle Green function in the Hartree–Fock approximation; as for the density matrix, its apparent simplicity is illusory owing to the non-commutation of the operators in the denominator.

If the Thomas–Fermi approximation is applicable, this non-commutation can be ignored. Going over to the Fourier transform in terms of the difference $\boldsymbol{x}_1 - \boldsymbol{x}_2$, we find

$$G_0(\boldsymbol{x}_1, \boldsymbol{p}, \varepsilon) = \frac{\delta_{\sigma_1\sigma_2}\delta_{\tau_1\tau_2}}{\varepsilon - \varepsilon_p(x_1) + i\delta \operatorname{sign}[\varepsilon_p(x_1) - \varepsilon_F]}. \quad (10.10)$$

Here $\boldsymbol{x}_1$ acts as a parameter on which G_0 is only weakly dependent.

If we can neglect the exchange effects as well, then the relations obtained are even simpler. Introducing the Fermi momentum, $p_0^2(x)$, which is connected to $\varepsilon_p(x)$ by the relation

$$\varepsilon_p(x) = \frac{p^2 - p_0^2(x)}{2M} + \varepsilon_F,$$

we find

$$G_0(q_1, q_2, \varepsilon) = \int \frac{d^3p \exp[i(\boldsymbol{p}\cdot\boldsymbol{x}_1 - \boldsymbol{x}_2)]\, \delta_{\sigma_1\sigma_2}\delta_{\tau_1\tau_2}}{\varepsilon - \varepsilon_F - ([p^2 - p_0^2(x_1)]/2M) + i\delta \operatorname{sign}[p^2 - p_0^2 x_1)]}. \quad (10.11)$$

To the same approximation, we can make the replacement $p_0^2(x_1) \to p_0^2(x_2)$ in the expression under the integral sign in (10.11):

$$G_0(q_1, q_2, \varepsilon) = \int \frac{d^3p \exp[i(\boldsymbol{p}\cdot\boldsymbol{x}_1 - \boldsymbol{x}_2)]\, \delta_{\sigma_1\sigma_2}\delta_{\tau_1\tau_2}}{\varepsilon - \varepsilon_F - ([p^2 - p_0^2(x_2)]/2M) + i\delta \operatorname{sign}[p^2 - p_0^2(x_2)]}. \quad (10.12)$$

Indeed the effective value of the difference $\boldsymbol{x}_1 - \boldsymbol{x}_2 \sim 1/p \sim 1/p_0$. If the quasi-classical condition is fulfilled,

$$p_0^2(x_2) - p_0^2(x_1) \sim |\boldsymbol{x}_1 - \boldsymbol{x}_2|\, \nabla p_0^2(x_1) \sim \xi p_0^2(x_1) \ll p_0^2(x_1).$$

The expressions (10.11) and (10.12) may be conveniently employed to calculate the correlation effects in weakly inhomogeneous systems. However, it will be shown in § 15 that, for these expressions to be used in correlation calculations, not only must the normal quasi-classical condition be fulfilled, but also another condition which is frequently more stringent.

10.3. We now turn to an investigation of the physical significance of the Green function. We will write the expression for the Green function in the form:

$$G_0(x_1, x_2) = -i \begin{cases} \langle \Psi_0 | \hat{\psi}(x_1)\, \hat{\psi}^+(x_2) | \Psi_0 \rangle & t_1 > t_2 \\ -\langle \Psi_0 | \hat{\psi}^+(x_2)\, \hat{\psi}(x_1) | \Psi_0 \rangle & t_1 < t_2 \end{cases}$$

We will consider the upper line first. The operator $\hat{\psi}^+(x_2)$, acting on Ψ_0, leads to the creation of a particle at the point q_2 and time t_2 (the annihilation part of this operator does not contribute). At a subsequent moment t_1, this particle is annihilated at a point q_1 and the system returns to its initial state. According to the second line, a hole is created at point q_1 and time t_1, which is then annihilated at q_2 and t_2.

We may say, therefore, that the Green function describes the propagation processes: of particles from the point x_2 to x_1, and of holes in the reverse direction. The Green function is, for this reason, often referred to as the propagator. The propagation is always causal in character: creation always precedes an nihilation. If we were using the usual operator product instead of the T-product, this highly important property would be lost.

It can be seen that the Green function is equal to a transition matrix element to within some factor, and this corresponds to propagation in the sense defined above. In fact the wave function of a system with a particle at the point q_2 at time t_2 is obviously equal to $\Psi(x_2) = \hat{\psi}^+(x_2)\, \Psi_0$. The matrix element for the transition is

$$\langle \Psi^*(x_1)\, \Psi(x_2) \rangle \quad t_1 > t_2$$

and is also, of course, the Green function.

It is convenient to represent propagation graphically. The corresponding rules are a constituent part of diagram technique, which will be discussed in detail below. The space–time point x is represented as a point on a graph (Fig. 3a), where the horizontal axis represents time, going from left to right. The propagation of particles, represented by G_0, will be represented by the directed line joining the points x_2 and x_1 ($t_2 < t_1$). In a similar way the propagation of holes is described by a line going from x_1 to x_2

$(t_2 > t_1)$. Both of these lines are directed along the time axis in such a way as to correspond to the causal nature of the propagation.

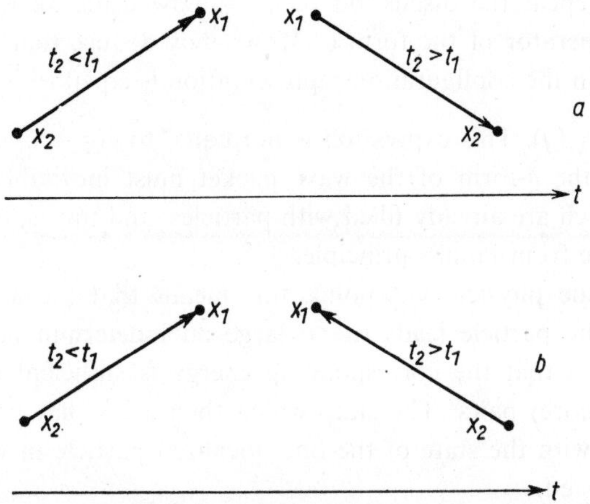

FIG. 3

The major inconvenience of this representation of the propagation function is the way in which the direction of the line (from x_2 to x_1, or the converse) depends on the relation between the times at x_1 and x_2. This can be overcome easily enough by supposing that holes propagate from their point of annihilation to their point of creation, and correspondingly directing the line for the holes (Fig. 3b) from x_2 to x_1. Thus one can choose a single direction for the line of the propagation function, for both $t_2 < t_1$ and $t_2 > t_1$, which coincides with the direction of propagation of the particle. Although the direction of propagation of the holes now seems to be opposite to the direction of the time axis, this does not contravene the causality principle, since this new condition has a purely formal significance. As we shall see later, this approach makes the graphical apparatus of the theory more compact.

10.4. We conclude by noting that the terms which were intronuced in the previous section, such as "creation at a point", dannihilation at a point", "propagation from point to point", must "ot be taken too literally.

The operator $\hat{\psi}^+(q_0)$, acting on the vacuum function Ψ_{vac}, does indeed give a state corresponding to a particle localized at the point q_0. This is not true, however, of the operator acting on Ψ_0—the wave function of the system.

If we repeat the discussion of § 3.4, now using an expression for the operator of the form (8.3), we may deduce that the wave function in the configuration representation is equal to $\sum_{\nu} (1 - n_\nu) \times \chi_\nu^*(q_0)\, \chi_\nu(q)$. This expression is not equal to $\delta(q - q_0)$. We can say that the δ-form of the wave packet must inevitably include states which are already filled with particles, and this is, of course, impossible from Pauli's principle.

From the physical viewpoint, this means that the attempt to localize the particle leads to so large an indeterminancy in the momentum that the corresponding energy is sufficient to create one (or more) pairs. The state which then arises has nothing in common with the state of the one, localized particle in which we are interested.

For this reason, a strict localization of particles is, generally speaking, impossible in many-body theory. This does not mean, however, that our previous discussion is useless. In constructing a field theory it is quite sufficient to treat the propagation process in the same conditional sense, and with the same reservations, as we have used above.

11. Graphical Representation of the Scattering Matrix Elements

11.1. We will now return to the S-matrix, and will consider the expression for it, obtained in § 9.3, as a sum of T-products. This expression is now unsuitable for the following reason.

Each of the T-products contains many creation and annihilation operators, which are arranged in a most varied order relative to each other. Hence, as the wave function of the system changes with time, the S-matrix, acting on $\Psi(t_0)$, not only annihilates particles in the state $\Psi(t_0)$ and creates particles which must be present in $\Psi(t)$, but it also creates, and subsequently annihilates, other particles which are present neither in $\Psi(t_0)$ nor in $\Psi(t)$. For example, the transition process from an initial state with two particles to a final state, also with two particles, corresponds not only to the S-matrix term containing

$$\hat{\Psi}^{+}_{(+)}\hat{\Psi}^{+}_{(+)}\hat{\Psi}_{(-)}\hat{\Psi}_{(-)},$$

but also to a term, shall we say, of the type

$$\hat{\Psi}^{+}_{(+)}\hat{\Psi}^{+}_{(+)}\hat{\Psi}_{(-)}\hat{\Psi}_{(-)}\hat{\Psi}_{(-)}\hat{\Psi}^{+}_{(+)}.$$

The last two operators here lead to the creation and subsequent annihilation of some "supernumerary" particle.

These supernumerary particles (they are called virtual particles, to distinguish them from the real particles which are present in $\Psi(t_0)$ or $\Psi(t)$) play an essential part in interpreting the higher terms of perturbation theory, and cannot be eliminated from the theoretical apparatus. The method of describing them in terms of T-products is, however, unsuitable. The point is that in this description the real and the virtual particles appear on equal terms. Hence, for example, the problem of discovering the part of the S-matrix which corresponds to a given transition process for real particles† is made much more complicated by the presence in the S-matrix expression of operators for both real and virtual particles.

It is possible to write the S-matrix in such a form that only the creation and annihilation operators for real particles appear. The creation and subsequent annihilation of virtual particles is then described by c-numbers (the contraction of operators introduced above). The simplification thus introduced is evident; in particular the choice of an S-matrix element, corresponding to a given process, can be made immediately from the form of the operators in the matrix element.

It is characteristic of virtual particles (or virtual holes) that their creation operator is necessarily to the right of the corresponding annihilation operator. Otherwise, the particle would necessarily be present in the initial and final states. The normal operator product (see § 8) is characterized by just the converse situation. The reorganization of the S-matrix, in which we are interested, is therefore connected with the possibility of representing a T-product as a sum of normal products of operators.

† The S-matrix describes all possible transition processes. The part of the S-matrix referring to a given process is called the element of the S-matrix for the process.

11.2. We will show that the reduction of a T-product of operators to a sum of normal products is always possible, and can be done by means of simple rules.

We already have a relation of the desired form for a pair of operators [see (10.1)]:

$$T(\hat{F}_1\hat{F}_2) = N(\hat{F}_1\hat{F}_2) + \overbrace{\hat{F}_1\hat{F}_2}N(1); \quad N(1) = 1. \tag{11.1}$$

The rules for constructing similar relationships in the general case are given by the following two theorems due to Wick.

1. A T-product of field operators can be represented as a sum of normal products, including as factors the contractions of all possible pairs of operators. The overall sign for each of these terms is determined by the number of rearrangements of the operators which is necessary to bring the contracted operators together:

$$\begin{aligned} T(\hat{F}_1\hat{F}_2 \ldots \hat{F}_n) &= N(\hat{F}_1\hat{F}_2 \ldots \hat{F}_n) + \overbrace{\hat{F}_1\hat{F}_2}N(\hat{F}_3 \ldots \hat{F}_n) \\ &- \overbrace{\hat{F}_1\hat{F}_3}N(\hat{F}_2 \ldots \hat{F}_n) + \cdots + \overbrace{\hat{F}_1\hat{F}_2}\overbrace{\hat{F}_3\hat{F}_4}N(\hat{F}_5 \ldots \hat{F}_n) + \cdots \end{aligned} \tag{11.2}$$

Here $\overbrace{\hat{F}_1\hat{F}_2}$ is the contraction of the operators $\hat{F}_1$ and $\hat{F}_2$.

We will consider an appropriate example:†

$$\begin{aligned} T[\hat{\psi}^+(1)\,\hat{\psi}^+(2)\,\hat{\psi}(3)\,\hat{\psi}(4)] &= N[\hat{\psi}^+(1)\,\hat{\psi}^+(2)\,\hat{\psi}(3)\,\hat{\psi}(4)] \\ &+ iG_0(3,1)\,N[\hat{\psi}^+(2)\,\hat{\psi}(4)] - iG_0(4,1)\,N[\hat{\psi}^+(2)\,\hat{\psi}(3)] \\ &- iG_0(3,2)\,N[\hat{\psi}^+(1)\,\hat{\psi}(4)] + iG_0(4,2)\,N[\hat{\psi}^+(1)\,\hat{\psi}(3)] \\ &+ G_0(3,1)\,G_0(4,2) - G_0(3,2)\,G_0(4,1). \end{aligned}$$

We have used here the definition of the Green function:

$$\overbrace{\hat{\psi}(1)\,\hat{\psi}^+(2)} = -\overbrace{\hat{\psi}^+(2)\,\hat{\psi}(1)} = iG_0(1,2). \tag{11.3}$$

The contractions of other pairs of operators are equal to zero.

Normal products can occur within the T-product from the beginning.

2. For such T-products which we shall call mixed T-products the foregoing theorem also holds; however, one must omit the contractions of those operators which have been within the same N-product from the start.

† The numerical argument is an abreviation for the corresponding space–time point: $(1) = x_1$, $(2) = x_2$, etc.

In particular

$$T\{N[\hat{\psi}^+(1)\,\hat{\psi}(2)]\,N[\hat{\psi}^+(3)\,\hat{\psi}(4)]\}$$
$$= N[\hat{\psi}^+(1)\,\hat{\psi}(2)\,\hat{\psi}^+(3)\,\hat{\psi}(4)] - iG_0(4,1)\,N[\hat{\psi}(2)\,\hat{\psi}^+(3)]$$
$$+ iG_0(2,3)\,N[\hat{\psi}^+(1)\,\hat{\psi}(4)] + G_0(4,1)\,G_0(2,3).$$

11.3. We now turn to the proof of Wick's theorems, using an induction method.

The first theorem is true for two operators. We will suppose that it is true for the product of n operators, and will consider $T(\hat{F}_1 \ldots \hat{F}_n \hat{F}_{n+1})$. Without loss of generality, we can suppose that the time for the operator $\hat{F}_{n+1}$ precedes the times for the remaining operators,† so that we can write the quantity in which we are interested, in the form $T(\hat{F}_1 \ldots \hat{F}_n)\,\hat{F}_{n+1}$.

Applying Wick's theorem to $T(\hat{F}_1 \ldots \hat{F}_n)$, we can reduce $T(\hat{F}_1 \ldots \hat{F}_n)\,\hat{F}_{n+1}$ to a sum of terms, each of which contains an operator product of the type $N(\ldots)\,\hat{F}_{n+1}$. We will consider a characteristic term of this type $N(\hat{F}_1 \ldots \hat{F}_k)\,\hat{F}_{n+1}$ $(k \leqq n)$, and we will show that the following equality holds

$$N(\hat{F}_1 \ldots \hat{F}_k)\,\hat{F}_{n+1} - N(\hat{F}_1 \ldots \hat{F}_k \hat{F}_{n+1}) = \Sigma, \tag{11.4}$$

where

$$\Sigma = \widehat{\hat{F}_k \hat{F}_{n+1}} N(\hat{F}_1 \ldots \hat{F}_{k-1}) - \widehat{\hat{F}_{k-1} \hat{F}_{n+1}} N(\hat{F}_1 \ldots \hat{F}_{k-2} \hat{F}_k) + \cdots$$
$$+ (-1)^k\, \widehat{\hat{F}_1 \hat{F}_{n+1}} N(\hat{F}_2 \ldots \hat{F}_k).$$

This statement is equivalent to Wick's theorem for $T(\hat{F}_1 \hat{F}_2 \ldots \hat{F}_{n+1})$.

We begin our proof by dividing $\hat{F}_{n+1}$ into a creation and an annihilation part, and put the left-hand side of (11.4) in the form

$$N(\hat{F}_1 \ldots \hat{F}_k)\,\hat{F}_{n+1(+)} - (-1)^k\, N(\hat{F}_{n+1(+)} \hat{F}_1 \ldots \hat{F}_k). \tag{11.5}$$

The annihilation part of the operator $\hat{F}_{n+1}$, as follows immediately from the definition of an N-product, disappears completely from

† We have already commented in § 9 on the possibility of rearranging the operators within a T-product, with a corresponding change in sign. It is therefore always possible to reduce a T-product to the form used in the text by a proper rearrangement and redesignation of the operators.

(11.5). We will next move the operator $\hat{F}_{n+1(+)}$ in the second term of (11.5) to the right, so that we reduce it to the same form as the first term. While doing this, the permutation rules must be kept in mind

$$[\hat{F}_{n+1(+)}, \hat{F}_{(i)}]_+ = [\hat{F}_{n+1(+)}, \hat{F}_{i(-)}]_+ = \overbrace{\hat{F}_i\hat{F}_{n+1}}.$$

These come from the expressions (10.2), (10.4), and from the condition $t_{n+1} < t_i$.

If we carry out the permutations the necessary number of times, we can compensate for the first term of (11.5); the remaining terms are equal to Σ. This completes the proof of the first of Wick's theorems.

We will consider, for purposes of illustration, the *T*-product of three operators $T(\hat{F}_1\hat{F}_2)\,\hat{F}_3$. Using (11.1), this is equal to

$$N(\hat{F}_1\hat{F}_2)\,\hat{F}_3 + \overbrace{\hat{F}_1\hat{F}_2}\hat{F}_3 = N(\hat{F}_1\hat{F}_2)\,\hat{F}_3 + \overbrace{\hat{F}_1\hat{F}_2}N(\hat{F}_3).$$

We have

$$N(\hat{F}_1\hat{F}_2)\,\hat{F}_3 - N(\hat{F}_1\hat{F}_2\hat{F}_3) = N(\hat{F}_1\hat{F}_2)\,\hat{F}_{3(+)} - N(\hat{F}_{3(+)}\hat{F}_1\hat{F}_2).$$

The last term on the right-hand side can be written in the form

$$-\overbrace{\hat{F}_1\hat{F}_3}N(\hat{F}_2) + \overbrace{\hat{F}_2\hat{F}_3}N(\hat{F}_1) - N(\hat{F}_1\hat{F}_2)\,\hat{F}_{3(+)}.$$

Thus, we may write finally that

$$T(\hat{F}_1\hat{F}_2\hat{F}_3) = \overbrace{\hat{F}_1\hat{F}_2}N(\hat{F}_3) - \overbrace{\hat{F}_1\hat{F}_3}N(\hat{F}_2) + \overbrace{\hat{F}_2\hat{F}_3}N(\hat{F}_1) + N(\hat{F}_1\hat{F}_2\hat{F}_3),$$

which evidently agrees with (11.2).

The second theorem is true for *T*-products of the type $T[N(\hat{F}_1)\,\hat{F}_2\,...\,\hat{F}_n]$. Let us suppose it to hold for the products $T[N(\hat{F}_1\,...\,\hat{F}_k)\,\hat{F}_{k+1}\,...\,\hat{F}_n]$, and let us consider the quantity $T[N(\hat{F}_1\,...\,\hat{F}_{k+1})\,\hat{F}_{k+2}\,...\,\hat{F}_n]$. Applying Wick's first theorem to the *N*-product here, i.e. writing it in the form $T(\hat{F}_1\,...\,\hat{F}_{k+1}) - \Sigma'$, we can transform this quantity to

$$T[T(\hat{F}_1\,...\,\hat{F}_{k+1})\,\hat{F}_{k+2}\,...\,\hat{F}_n] - T(\Sigma'\hat{F}_{k+2}\,...\,\hat{F}_n), \qquad (11.6)$$

where Σ' is the sum of all the possible products of the operator contractions $\hat{F}_1\,...\,\hat{F}_{k+1}$ with the corresponding *N*-products, where the number of factors in these latter does not exceed k. The theorem we have proved may therefore be applied to the second term of (11.6), enabling us to express it as a sum of the products of all possible operator contractions $\hat{F}_1 \ldots \hat{F}_{k+1}$ among themselves, of the contractions of these operators with the remaining operators, and of the N-products of the remaining operators. The term we are considering contains all the possible operator contractions $\hat{F}_1 \ldots \hat{F}_{k+1}$, whilst each of its components includes at least one such contraction.

The symbol T can be omitted from inside the T-product of the first term in (11.6), and, from Wick's first theorem, we arrive at a combination of terms which contain all possible contractions of all operators without exception. The second term in (11.6) cancels those terms of this combination, which contain at least one contraction of the operators $\hat{F}_1 \ldots \hat{F}_{k+1}$ with another. Thus the contractions of the operators which are under the N-product sign do indeed disappear from the considered expression.

We consider, as an illustration, the T-product $T[N(\hat{F}_1\hat{F}_2)\,\hat{F}_3]$. We will write this in the form

$$T(\hat{F}_1\hat{F}_2\hat{F}_3) - \overbrace{\hat{F}_1\hat{F}_2}\hat{F}_3$$

and apply Wick's theorem to the first term. The term with the contraction $\overbrace{\hat{F}_1\hat{F}_2}$ then disappears from the result.

11.4. If we apply Wick's theorem to the expansion of the S-matrix in terms of T-products,† we can express this latter as a sum of terms, each of which contains a known N-product of the field operators $\hat{\psi}$ and $\hat{\psi}^+$. Each of these operators is in turn the sum of

† The S-matrix contains T-products, not of the field operators, but of the interaction Hamiltonians $\hat{H}'_{\text{Int}}$. These latter contain a product of the field operators referring to one particular moment of time. The application of Wick's theorem to a product of this sort does not, in the general case, lead to a well-defined result, since the contraction of operators in $\hat{H}'_{\text{Int}}$ is not single-valued for a given time. These difficulties disappear, however, for the approach we are considering: on the one hand, the choice of $\hat{H}'$ in the form (8.1) makes it possible to write $\hat{H}'_{\text{Int}}$ as a normal product of the operators (see (9.5)); on the other hand, from Wick's second theorem, contractions of the operators within an N-product do not appear. This provides an additional argument for using the Hartree–Fock approximation as the zero approximation for the theory.

the corresponding creation and annihilation operators. We may therefore represent the S-matrix as a sum of terms containing N-products of the creation and annihilation operators $\hat{\psi}_{(\pm)}, \hat{\psi}^{+}_{(\pm)}$. These terms, which we have referred to above as the elements of the S-matrix, may be conveniently represented by graphs describing the corresponding transition process.

We will now turn to formulating rules for the construction of diagrams of this sort. An element of the S-matrix is made up of the following parts: the field operators referred to certain space–time points,† the contractions (or the free-particle Green functions) and the interaction potentials. These quantities may be represented in the following graphical form (Fig. 4).

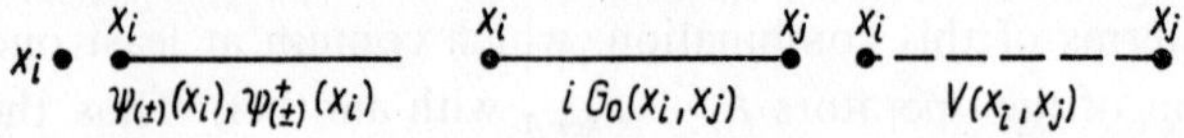

FIG. 4

Each space–time argument x_i, which enters into the operator fields, the contractions and the potentials, is represented by a dot on the graph. For each field operator $\hat{\psi}_{(\pm)}(x_i)$, $\hat{\psi}^{+}_{(\pm)}(x_i)$, there is a corresponding continuous line (the external line), attached to the point x_i at one end, and going to infinity at the other. Each contraction $iG_0(x_i, x_j)$ corresponds to a full-drawn line (a virtual line) connecting the points x_i and x_j. Finally, each interaction potential $V(x_i, x_j)$ corresponds to a dotted line (the interaction line), which connects the points x_i and x_j.

There are certain general remarks that should be made about the structure of diagrams. The fact that the total number of particles in a system does not change in the interaction process, so that, consequently, the difference between the number of particles and holes remains constant, necessitates that the full-drawn lines in the diagram should have no ends. These lines either go off to infinity, or form closed loops. The break-off of a full-drawn line at some node would signify the creation, or annihilation, of a particle (or

† These operators are called the operators of the external lines of the diagram.

hole) at this point, which would contradict the conservation law noted above.

If we consider the *n*th term of the expansion of the *S*-matrix in perturbation theory, it can be easily seen that the number of nodes in the corresponding diagrams is $2n$, and the number of interaction lines is n. The number of external lines F (necessarily even) may, like the number of virtual lines V be different for different processes. There is, however, a simple relation connecting F, V and n:

$$2V + F = 4n.$$

11.5. In the rules we have just formulated, four different external line operators, $\hat{\psi}_{(\pm)}(x)$ and $\hat{\psi}^{+}_{(\pm)}(x)$, corresponded to a single graphical representation—lines which connected the point x to an infinitely distant point. In order to make the correspondence between the transition process and the diagram unique, these rules must be supplemented so as to introduce the concept of direction and orientation of lines on the graph.

Using a convention similar to that in § 10.3, the propagation of a particle can be represented by a line directed from the point where the particle is created to the point where it is annihilated; this direction coincides with the direction of the time axis. The propagation of holes can be represented by a line directed from the point where the hole is annihilated to the point where it is created, i.e. in an opposite sense to the time axis.

We will choose the time-axis direction as always going from left to right and will not indicate it in the diagram. We will represent the direction of a full-drawn line by arrows. From the rules given

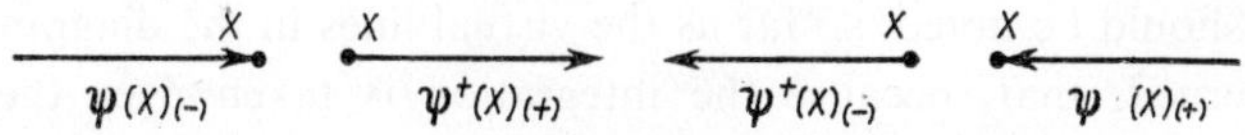

FIG. 5

in the previous paragraph for external lines, we can tell from the diagram whether we are dealing with particles or holes, with creation or annihilation (Fig. 5).

An external line corresponding to the annihilation of a particle (the operator $\hat{\psi}(x)_{(-)}$) must have the same direction as the time axis, and must finish at the annihilation point x. A line corresponding to the creation of a particle (the operator $\hat{\psi}^{+}(x)_{(+)}$), which has the same direction, must begin at a creation point x. A line corresponding to the annihilation of a hole (the operator $\hat{\psi}^{+}(x)_{(-)}$), must lie in the opposite direction to the time axis, and begin at an annihilation point x. Finally, a line which corresponds to the creation of a hole (the operator $\hat{\psi}(x)_{(+)}$), having the same direction, must finish at a creation point x. We can thus resolve the problem of obtaining a unique relation between the elements of the S-matrix and the diagram.

The direction of a virtual line is selected to correspond with the rules formulated in § 10.3. These rules lead to the components of a full-drawn line having only one permitted direction,

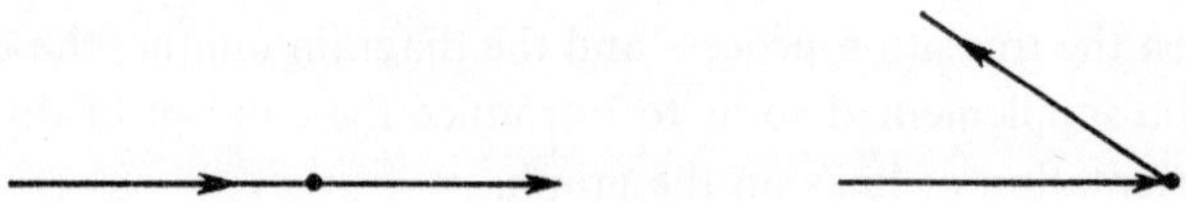

FIG. 6

i.e. no pair of neighbouring components can have opposite directions.† If we are dealing with a full-drawn line which is open, the unique direction is determined by the directions of the external line, i.e. by the type of process. If the full-drawn line is closed, then one can move along it in two opposite directions. It appears that the elements of the S-matrix contain, in all cases, a pair of terms corresponding to diagrams with two such ways of going round a closed curve. Thus the application of our rules does not lead to any ambiguity.

It should be noted, so far as the virtual lines in the diagram are concerned, that, because the integration is taken over the coordinates and time at each vertex, the relation between times in the propagation function can be arbitrary. In the same way the orien-

† From the requirement that the difference between the number of particles and the number of holes be conserved, the annihilation of a particle at a node is necessarily accompanied by the creation of another particle, or the annihilation of a hole, at the same node. In either case the direction along the full-drawn line remains the same (Fig. 6).

tation of a virtual line relative to the time axis can also vary† (Fig. 7). Our graph, however, is chosen to represent the simplest, unbroken configuration of lines. The subsequent integration over the coordinates and the time automatically accounts for all possibilities of this sort (see § 12 for details).

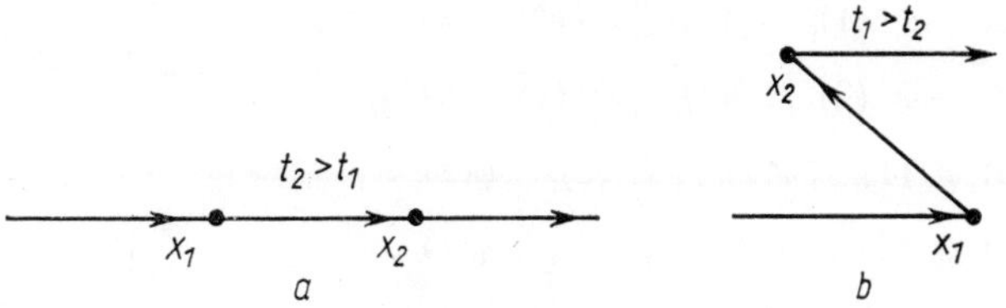

FIG. 7

We may note finally that the interaction line, the ends of which correspond to the same time, is naturally perpendicular to the time axis. We will devote the next section to considering illustrations of the rules so far introduced.

12. Lower-Order Processes

12.1. We will now make a systematic study of the low-order elements of the S-matrix. We will suppose for simplicity that V is not an operator but a c-number.

We will begin with first-order processes in $\hat{H}'$ ($n = 1$). From (9.15):

$$\left.\begin{aligned} \hat{S}_1 &= -\frac{i}{2}\int d1\, d2 \exp\left(-\delta|t_1|\right) V(1, 2)\, \hat{\tau}_1, \\ \hat{\tau}_1 &= TN[\hat{\psi}^+(1)\, \hat{\psi}^+(2)\, \hat{\psi}(2)\, \hat{\psi}(1)]. \end{aligned}\right\} \tag{12.1}$$

The symbol $d1$ here signifies $d^4x_1 = dq_1\, dt_1$. From Wick's second theorem, this T-product can be replaced by an N-product of the corresponding field operators. If we resolve the latter into a sum of terms, each of which is the product of creation and annihilation operators, we arrive at a total of $2^4 = 16$ terms. The corresponding diagrams may be conveniently divided into the following three classes.

† The process represented in Fig. 7a contains a virtual particle, that in Fig. 7b a virtual hole (a pair is created at x_2, the particle moves into the future, the hole annihilates with the original particle).

1. Scattering processes, which correspond to two creation operators and two annihilation operators:

(a) $\hat{\psi}^+(1)_{(+)}\,\hat{\psi}^+(2)_{(+)}\,\hat{\psi}(2)_{(-)}\,\hat{\psi}(1)_{(-)}$,

(b) $\hat{\psi}(1)_{(+)}\,\hat{\psi}(2)_{(+)}\,\hat{\psi}^+(2)_{(-)}\,\hat{\psi}^+(1)_{(-)}$,

(c) $-\hat{\psi}^+(1)_{(+)}\,\hat{\psi}(2)_{(+)}\,\hat{\psi}^+(2)_{(-)}\,\hat{\psi}(1)_{(-)}$,

(d) $-\hat{\psi}^+(2)_{(+)}\,\hat{\psi}(1)_{(+)}\,\hat{\psi}^+(1)_{(-)}\,\hat{\psi}(2)_{(-)}$,

(e) $\hat{\psi}^+(1)_{(+)}\,\hat{\psi}(1)_{(+)}\,\hat{\psi}^+(2)_{(-)}\,\hat{\psi}(2)_{(-)}$,

(f) $\hat{\psi}^+(2)_{(+)}\,\hat{\psi}(2)_{(+)}\,\hat{\psi}^+(1)_{(-)}\,\hat{\psi}(1)_{(-)}$.

The diagram in Fig. 8a describes the scattering process for a particle by a particle; Fig. 8b is the same for a hole by a hole. The four subsequent diagrams (Figs. 8c–8f) correspond to the scattering of a particle by a hole. Figures 8c and 8d are for simple scattering, when the particle and hole are annihilated and created each at its own vertex. Both elements of the *S*-matrix, corresponding to this diagram, are evidently equal, since they only differ by a change in the variables of integration. The diagrams in Figs. 8e and 8f correspond to what is called the annihilation scattering of a particle by a hole: the pair undergoing scattering is first annihilated at one

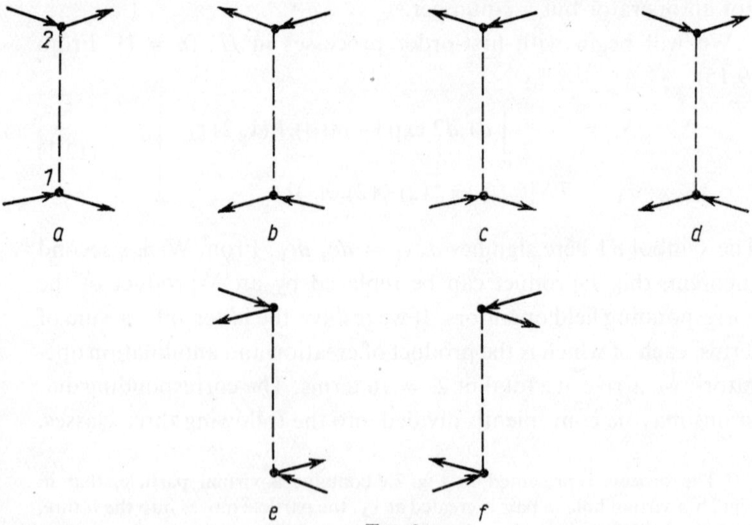

FIG. 8

node, and then recreated at the other. Both of these diagrams are likewise equal.

2. Processes in which a single pair participates: corresponding to three creation operators and one absorption (or vice versa). There are eight such diagrams; we will consider the typical ones (Fig. 9):

(a) $\hat{\psi}^+(1)_{(+)}\,\hat{\psi}^+(2)_{(+)}\,\hat{\psi}(2)_{(+)}\,\hat{\psi}(1)_{(-)}$,

(b) $\hat{\psi}^+(1)_{(+)}\,\hat{\psi}^+(2)_{(-)}\,\hat{\psi}(2)_{(-)}\,\hat{\psi}(1)_{(-)}$.

The diagram in Fig. 9a corresponds to the creation of a particle pair, i.e. excitation of the system due to interaction with a particle; Fig. 9b corresponds to the annihilation of a pair due to interaction with a particle. The remaining six diagrams of this class describe the analogous processes with holes, and also include duplicate diagrams with the replacement $1 \rightleftarrows 2$.

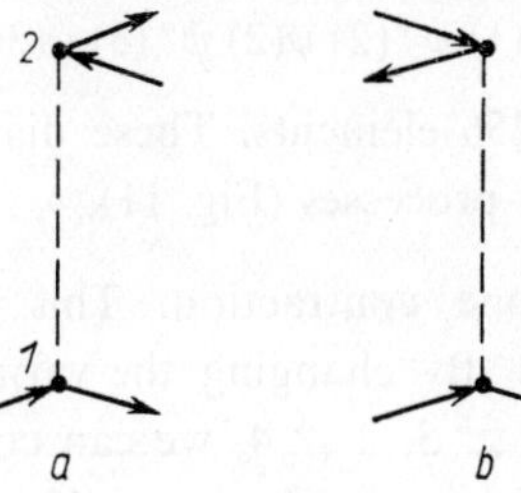

FIG. 9

3. Processes with the creation or annihilation of two pairs (four creation or annihilation operators):

(a) $\hat{\psi}^+(1)_{(+)}\,\hat{\psi}^+(2)_{(+)}\,\hat{\psi}(2)_{(+)}\,\hat{\psi}(1)_{(+)}$,

(b) $\hat{\psi}^+(1)_{(-)}\,\hat{\psi}^+(2)_{(-)}\,\hat{\psi}(2)_{(-)}\,\hat{\psi}(1)_{(-)}$.

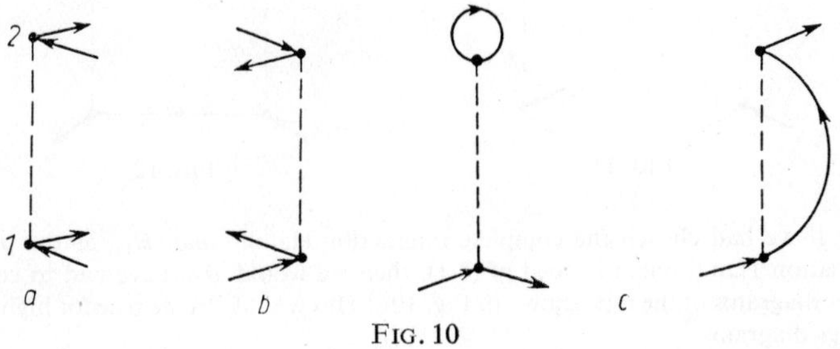

FIG. 10

The diagram in Fig. 10a corresponds to the self-excitation of a system with the creation of two pairs; Fig. 10b is the annihilation of two pairs.†

12.2. The S-matrix in second-order perturbation theory has the form:

$$\hat{S}_2 = -\tfrac{1}{8} \int d1\, d2\, d3\, d4 \exp\left[-\delta(|t_1| + |t_3|)\right] V(1, 2)\, V(3, 4)\, \hat{\tau}_2,$$
$$\hat{\tau}_2 = T\{N[\hat{\psi}^+(1)\, \hat{\psi}^+(2)\, \hat{\psi}(2)\, \hat{\psi}(1)]\, N[\hat{\psi}^+(3)\, \hat{\psi}^+(4)\, \hat{\psi}(4)\, \hat{\psi}\ (3)]\}. \tag{12.2}$$

Applying Wick's theorem to this T-product, we obtain a sum of terms which contain differing numbers of contractions (from 0 to 4). The overall number of elements of the matrix $\hat{S}_2$ is very large. We will only look at the most typical ones.

1. Diagrams without contractions. They arise from the term

$$\hat{\tau}_2 = N[\hat{\psi}^+(1)\, \hat{\psi}(1)\, \hat{\psi}^+(2)\, \hat{\psi}(2)\, \hat{\psi}^+(3)\, \hat{\psi}(3)\, \hat{\psi}^+(4)\, \hat{\psi}(4)],$$

which includes $2^8 = 256$ elements. These diagrams describe two independent first-order processes (Fig. 11).

2. Diagrams with one contraction. This contraction can be selected in eight ways. By changing the variables $1 \rightleftarrows 2$, $3 \rightleftarrows 4$ and, simultaneously, $1 \rightleftarrows 3$, $2 \rightleftarrows 4$, we can confirm that all these eight terms give the same contribution to $\hat{S}_2$. The corresponding part of the T-product leads to

$$\hat{\tau}_2 - 8iG_0(3, 1)\, N[\hat{\psi}^+(3)\, \hat{\psi}(1)\, \hat{\psi}^+(2)\, \hat{\psi}(2)\, \hat{\psi}^+(4)\, \hat{\psi}(4)]. \tag{12.3}$$

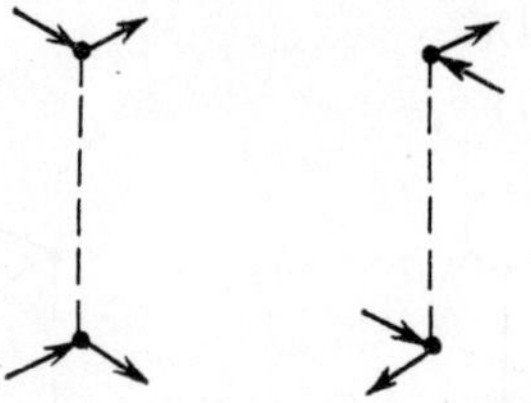

FIG. 11

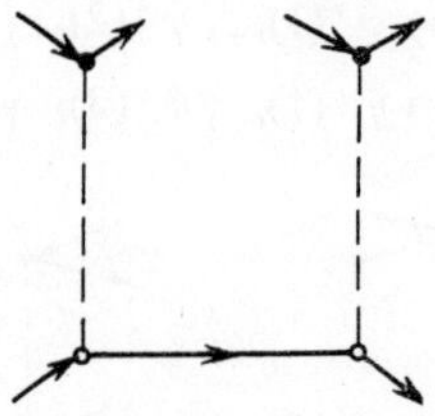

FIG. 12

† If we had chosen the complete interaction Hamiltonian $\hat{H}_{\text{Int}}$ as the perturbation Hamiltonian instead of (8.1), then we would also have had to consider diagrams of the type shown in Fig. 10c. This would also be true for higher-order diagrams.

This N-product can be split into $2^6 = 64$ separate terms. A typical one corresponds to the mutual scattering of three particles (Fig. 12). It also represents pair creation in two-particle scattering, the creation of two pairs of particles, etc.

3. Diagrams with two contractions. If we allow for all possible contractions, and change the variables as discussed above, we find

$$\begin{aligned}\hat{\tau}_2 = {} & 4iG_0(3, 1)\, iG_0(4, 2)\, N[\hat{\psi}^+(3)\, \hat{\psi}(1)\, \hat{\psi}^+(4)\, \hat{\psi}(2)] \\ & + 8iG_0(2, 3)\, iG_0(3, 1)\, N[\hat{\psi}^+(2)\, \hat{\psi}(1)\, \hat{\psi}^+(4)\, \hat{\psi}(4)] \\ & + 4iG_0(2, 4)\, iG_0(3, 1)\, N[\hat{\psi}^+(2)\, \hat{\psi}(4)\, \hat{\psi}^+(3)\, \hat{\psi}(1)] \\ & - 4iG_0(2, 3)\, iG_0(3, 2)\, N[\hat{\psi}^+(1)\, \hat{\psi}(1)\, \hat{\psi}^+(4)\, \hat{\psi}(4)]. \qquad (12.4)\end{aligned}$$

We have included here, besides other processes, the scattering of one particle by another in second-order perturbation theory. We will consider this in more detail (Fig. 13).

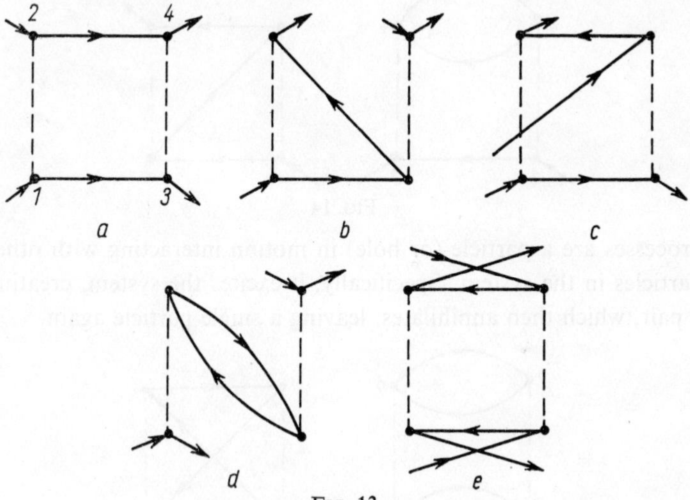

FIG. 13

The first term in (12.4) corresponds to direct scattering. Both of the scattered particles are either first absorbed, and then emitted (when $t_1 < t_3$, see Fig. 13a), or they are first emitted, and then absorbed (when $t_3 < t_1$, see Fig. 13e). In accordance with our previous stipulation our graph need only represent the simplest case, i.e. Fig. 13e can be discarded.

The processes corresponding to the second and third term of (12.4) (see Figs. 13b and 13c) have a more complicated interpretation. At $t_1 = t_2$ a pair is created; the particle moves off as an end-product of the reaction, but the hole is annihilated by one of the original particles. The other original particle is subject to simple scattering. Finally, the last term of (12.4) describes a process similar to first-order scattering, but containing one virtual pair (see Fig. 13d).

4. Diagrams with three contractions. A straightforward computation gives

$$\hat{\tau}_2 = -8iG_0(2,4)\, iG_0(4,2)\, iG_0(3,1)\, N[\hat{\psi}^+(3)\, \hat{\psi}(1)] + 8iG_0(2,3)\, iG_0(3,1)\, iG_0(4,2)\, N[\hat{\psi}^+(4)\, \hat{\psi}(1)]. \quad (12.5)$$

The corresponding diagrams are shown in Fig. 14. The equivalent

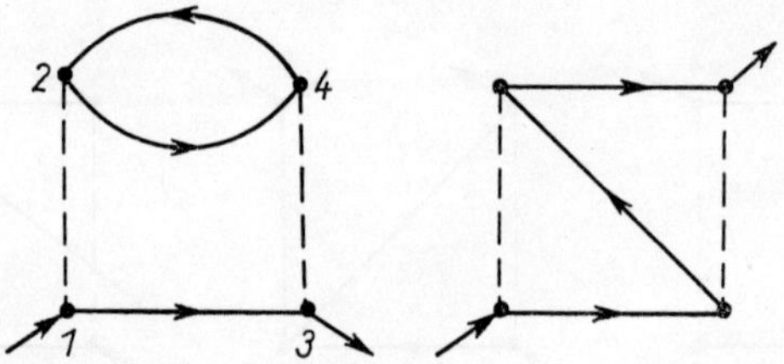

FIG. 14

processes are a particle (or hole) in motion interacting with other particles in the system. Specifically, it excites the system, creating a pair, which then annihilates, leaving a single particle again.

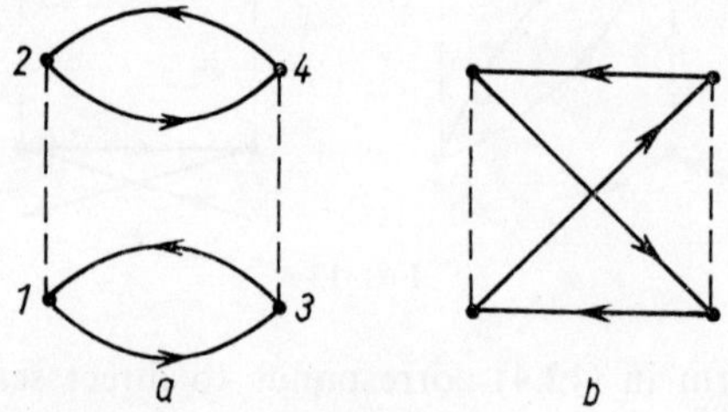

FIG. 15

5. A diagram with four contractions:

$$\hat{\tau}_2 = 2iG_0(1,3)\, iG_0(3,1)\, iG_0(2,4)\, iG_0(4,2) - 2iG_0(1,3)\, iG_0(2,4)\, iG_0(4,1)\, iG_0(3,2). \quad (12.6)$$

The diagrams which correspond to these expressions are given in Fig. 15. This is the so-called vacuum fluctuation: two pairs are created, which then annihilate.

12.3. The expressions which we have previously constructed for the elements of the S-matrix were written in the coordinate representation: the particles which took part in the corresponding process, were annihilated and then created at specific space–time points 1, 2, 3, 4. This form of the S-matrix is convenient for considerations of a general nature; however, it is expedient to go over to the energy representation for practical applications, since this corresponds to a description of the process with the particles in specific energy states (and with wave functions $\chi_\nu(q)$).

To do this, we take the expressions for the field operators and the Green function:

$$\hat{\psi}(x) = \sum_\nu \hat{A}_\nu \chi_\nu(q) \exp(-i\varepsilon_\nu t)$$

and

$$G_0(1, 2) = \int \frac{d\varepsilon}{2\pi} \exp[-i\varepsilon(t_1 - t_2)] \sum_\nu \chi_\nu^*(q_2)\, \chi_\nu(q_1)\, G_{0\nu}(\varepsilon),$$

where

$$\hat{A}_\nu = \begin{cases} \hat{a}_\nu(\varepsilon_\nu > \varepsilon_F) \\ \hat{b}_\nu^+(\varepsilon_\nu < \varepsilon_F) \end{cases} \qquad G_{0\nu}(\varepsilon) = \frac{1}{\varepsilon - \varepsilon_\nu + i\delta\, \mathrm{sign}(\varepsilon_\nu - \varepsilon_F)}. \quad (12.7)$$

We also need an expression for a matrix element of the interaction potential (see § 3.7):

$$\langle \sigma\lambda | \hat{V} | \nu\mu \rangle = \int dq\, dq' \chi_\nu^*(q)\, \chi_\mu^*(q')\, \hat{V}(q, q')\, \chi_\lambda(q')\, \chi_\sigma(q). \quad (12.8)$$

We will substitute this expression in the element of the S-matrix $\hat{S}_1$. Integrating in terms of t_1 and t_2, and using $\hat{V}(1, 2) = \hat{V}\delta(t_1 - t_2)$ gives the function $2\pi\delta(\varepsilon_{\nu_1} + \varepsilon_{\nu_2} - \varepsilon_{\nu_3} - \varepsilon_{\nu_4})$, which guarantees the conservation of energy in this process. Integration over the spatial coordinates leads to the matrix element (12.8). As a result, we obtain

$$\hat{S}_1 = -\frac{i}{2}(2\pi) \sum_{\nu_1\nu_2\nu_3\nu_4} \langle \nu_1\nu_2 | \hat{V} | \nu_3\nu_4 \rangle\, \delta(\varepsilon_{\nu_1} + \varepsilon_{\nu_2} - \varepsilon_{\nu_3} - \varepsilon_{\nu_4})$$

$$\times N(\hat{A}_{\nu_3}^+ \hat{A}_{\nu_1} \hat{A}_{\nu_4}^+ \hat{A}_{\nu_2}). \quad (12.9)$$

This expression describes a whole series of processes: the scattering of one particle by another, of a particle by a hole, etc., which can also be represented in diagram form (Fig. 16). Their sole difference from the other diagrams lies in the absence of space–time points; instead, the external lines acquire the indices of the corresponding states.

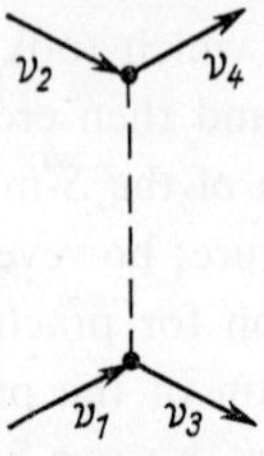

FIG. 16

We will now turn to second-order processes. If we repeat our previous calculations (in a more complicated, but completely analogous form), we obtain the following expression for an element of the S-matrix, for the processes considered in § 12.2:

$$\hat{S}_2 = -\tfrac{1}{8} \sum_{\nu_1\nu_2\nu_3\nu_4} \sum_{\mu_1\mu_2} \int d\varepsilon_1\, d\varepsilon_2 iG_{0\mu_1}(\varepsilon_1)\, iG_{0\mu_2}(\varepsilon_2) \times K_{\nu_1\nu_2\nu_3\nu_4} N(\hat{A}^+_{\nu_3}\, \hat{A}_{\nu_1} \hat{A}^+_{\nu_4}\, \hat{A}_{\nu_2}), \quad (12.10)$$

where

$$\begin{aligned} K = {} & 4\delta(\varepsilon_{\nu_1} + \varepsilon_{\nu_2} - \varepsilon_1 - \varepsilon_2)\, \delta(\varepsilon_1 + \varepsilon_2 - \varepsilon_{\nu_3} - \varepsilon_{\nu_4}) \\ & \times \langle \nu_1\nu_2|\hat{V}|\mu_1\mu_2\rangle \langle \mu_1\mu_2|\hat{V}|\nu_3\nu_4\rangle \\ & + 8\delta(\varepsilon_{\nu_1} + \varepsilon_2 - \varepsilon_{\nu_3} - \varepsilon_1)\, \delta(\varepsilon_1 + \varepsilon_{\nu_2} - \varepsilon_2 - \varepsilon_{\nu_4}) \\ & \times \langle \nu_1\mu_2|\hat{V}|\mu_1\nu_3\rangle \langle \mu_1\nu_2|\hat{V}|\mu_2\nu_4\rangle \\ & + 4\delta(\varepsilon_{\nu_1} + \varepsilon_2 - \varepsilon_{\nu_4} - \varepsilon_1)\, \delta(\varepsilon_1 + \varepsilon_{\nu_2} - \varepsilon_2 - \varepsilon_{\nu_3}) \\ & \times \langle \nu_1\mu_2|\hat{V}|\mu_1\nu_4\rangle \langle \nu_2\mu_1|\hat{V}|\mu_2\nu_3\rangle \\ & - 4\delta(\varepsilon_{\nu_1} + \varepsilon_1 - \varepsilon_{\nu_3} - \varepsilon_2)\, \delta(\varepsilon_2 + \varepsilon_{\nu_2} - \varepsilon_1 - \varepsilon_{\nu_4}) \\ & \times \langle \nu_1\mu_1|\hat{V}|\nu_3\mu_2\rangle \langle \mu_2\nu_2|\hat{V}|\mu_1\nu_4\rangle. \quad (12.11) \end{aligned}$$

The diagrams corresponding to these four processes (Fig. 17) are completely analogous to the diagrams in Fig. 13. We note that

$\mu_1\varepsilon_1$ and $\mu_2\varepsilon_2$ represent the indices of state and energy for virtual particles.

We must still consider the process of particle propagation (Fig. 18):

$$\hat{S}_2 = -\frac{1}{8(2\pi)} \sum_{\nu_1\nu_2} \sum_{\mu_1\mu_2\mu_3} \int d\varepsilon_1\, d\varepsilon_2\, d\varepsilon_3\, iG_{0\mu_1}(\varepsilon_1)\, iG_{0\mu_2}(\varepsilon_2)\, iG_{0\mu_3}(\varepsilon_3)$$

$$\times (-8)\, \delta(\varepsilon_{\nu_1} + \varepsilon_3 - \varepsilon_1 - \varepsilon_2)\, \delta(\varepsilon_1 + \varepsilon_2 - \varepsilon_3 - \varepsilon_{\nu_2})$$

$$\times \{\langle \nu_1\mu_3 | \hat{V} | \mu_1\mu_2\rangle \langle \mu_1\mu_2 | \hat{V} | \nu_2\mu_3\rangle$$

$$- \langle \nu_1\mu_3 | \hat{V} | \mu_1\mu_2\rangle \langle \mu_1\mu_2 | \hat{V} | \mu_3\nu_2\rangle\}\, N(\hat{A}^+_{\nu_2}\hat{A}_{\nu_1}). \qquad (12.12)$$

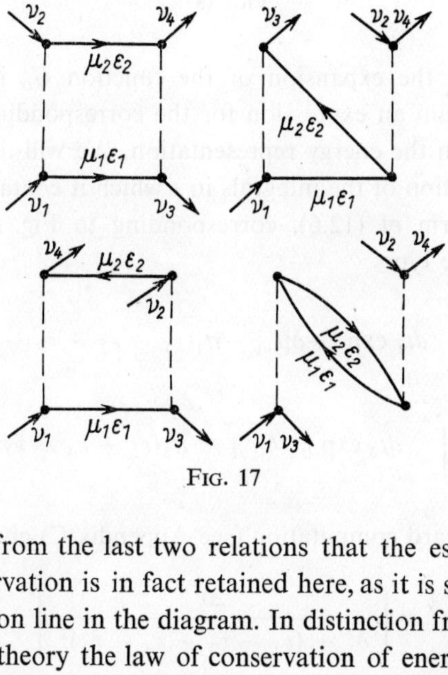

FIG. 17

It is evident from the last two relations that the essential law of energy conservation is in fact retained here, as it is satisfied along each interaction line in the diagram. In distinction from the "old" perturbation theory the law of conservation of energy holds here for virtual particles as well as real ones. This is brought about by the break down of the rigorous connection between the energy of a virtual particle ε, which enters into the Green function $G_{0\nu}(\varepsilon)$, and the quantity ε_ν (in the spatially uniform case $\varepsilon \neq p^2/2M$). This is one of the factors that leads to perturbation terms in field theory having a simple structure.

12.4. The results so far obtained do not depend on the parameter δ in the limit $\delta \to 0$; this is an inherent property of all diagrams with external lines. It is otherwise with vacuum fluctuation diagrams to which we must now turn.

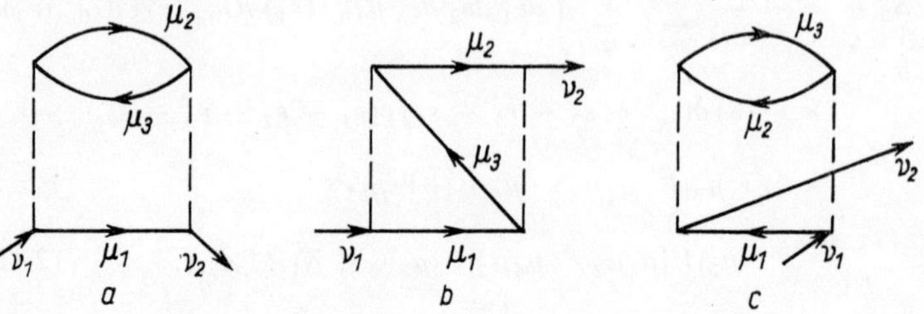

FIG. 18

Substituting the expansion of the function G_0 in (12.2) and (12.6), we obtain an expression for the corresponding element of the S-matrix in the energy representation. We will limit ourselves to a consideration of the integrals in t which it contains.

The first term of (12.6), corresponding to Fig. 15a, contains integrals of the type

$$I = \int_{-\infty}^{\infty} dt_1 \exp\left[-\delta|t_1| - it_1(\varepsilon_1 - \varepsilon_2 + \varepsilon_3 - \varepsilon_4)\right]$$

$$\times \int_{-\infty}^{\infty} dt_3 \exp\left[-\delta|t_3| - it_3(\varepsilon_2 + \varepsilon_4 - \varepsilon_1 - \varepsilon_3)\right].$$

A straightforward computation (see Appendix C) gives

$$I = \left[\frac{2\delta}{\delta^2 + (\varepsilon_1 - \varepsilon_2 + \varepsilon_3 - \varepsilon_4)^2}\right]^2.$$

In the limit $\delta \to 0$, the expression in the square brackets tends to $\delta(\varepsilon_1 - \varepsilon_2 + \varepsilon_3 - \varepsilon_4)$ apart from a numerical factor. Hence, as $\delta \to 0$, $I \to \infty$. More accurately, we have

$$I \sim \frac{1}{\delta}\delta(\varepsilon_1 + \varepsilon_3 - \varepsilon_2 - \varepsilon_4). \tag{12.13}$$

Similarly, the second term of (12.6) gives

$$I \sim \frac{1}{\delta}\delta(\varepsilon_1 - \varepsilon_2 + \varepsilon_3 - \varepsilon_4). \tag{12.14}$$

The remaining δ-function is eliminated in the subsequent integration over ε.

The results, thus obtained, are of a very general nature. Suppose we are considering an nth-order diagram with external lines. Then $(n-1)$ δ-functions in the corresponding element of the S-matrix (expressing the energy conservation law) disappear in the subsequent integration over ε. The remaining δ-function represents the energy conservation law for the external lines (for the diagram as a whole).

Any diagram which does not have external lines can be obtained from the corresponding diagram with external lines by closing these latter one by one, i.e. by identifying the energy of the outgoing and ingoing lines. In this case the law of conservation of energy for the external lines is identically true, and the corresponding δ-function has a zero argument.† A more rigorous consideration leads to a singularity $\sim 1/\delta$.

Thus any diagram of a vacuum fluctuation has a singularity of the type $1/\delta$. This fact is essential for the single-valuedness of the quantity (9.24) as $\delta \to 0$.

13. The Feynman Rules

13.1. The examples we have considered enable us to formulate general rules for the construction of an element of the S-matrix corresponding to a given process.

The nth-order element of the S-matrix contains the following constituent parts: a normal product of the field operators, n-fold products of the interaction potentials $\hat{V}$, a product of the contractions of the operators iG_0 and a numerical coefficient.

† For example, the diagram considered above with no external lines can be obtained from the diagram for second-order scattering by equating ε_{ν_1}, and ε_{ν_3}, ε_{ν_3} and ε_{ν_4} (see Fig. 17). It follows from (12.10) that we now have $\delta(\varepsilon_{\nu_1} + \varepsilon_{\nu_3} - \varepsilon_{\nu_3} - \varepsilon_{\nu_4}) = \delta(0)$.

We will take the normal product of the creation and annihilation operators with the same sign that they have in the expression:

$$N(\hat{\psi}_a^+ \hat{\psi}_a \hat{\psi}_b^+ \hat{\psi}_b \dots),$$

where the index a refers to one external line, b to another, and so on. The relative ordering of the groups $\hat{\psi}_a^+ \hat{\psi}_a$, $\hat{\psi}_b^+ \hat{\psi}_b$, is obviously unimportant.

It is essential to allow for the operator properties of the potential $\hat{V}$ in order to position it correctly in the expression for the S-matrix element. In all cases, which we will have to consider, we may suppose that the potential does not contain momentum operators. The only case in this book where it is necessary to deal with a momentum-dependent potential, is for the pseudopotential of the nuclear repulsive forces. However, the corresponding terms are sufficiently small for them to be considered in the Hartree–Fock approximation; they do not affect the S-matrix.

We will therefore suppose in the following that the potential $\hat{V}$ is a matrix in the discrete variables. This matrix "fits on" either to the external-line operators, $\hat{\psi}_\pm$, $\hat{\psi}_\pm^+$, or to the Green functions which depend on the corresponding arguments.

We will now find the magnitude of the numerical coefficient in the S-matrix element. From the general expression (9.13), there is a factor $(-i)^n/2^n n!$ in the nth-order term. There is also, as can be seen from the results of § 12, another integral coefficient k which arises from the combination of terms which differ only in the variables of integration. We can determine the magnitude of this coefficient [84].

We consider any diagram of the general type, and determine the number of ways of denoting its vertices. We can first transpose the arguments representing each of the n interaction lines. This gives 2^n possibilities. We can simultaneously change the places of the arguments of different interaction lines, which can be done in $n!$ ways. We therefore have altogether $2^n n!$ different ways of specifying the arguments of the S-matrix elements.

Some changes of the variables, however, may produce no change in the expression under the integral sign. This can happen when there is a definite symmetry to the diagram. If we denote the number of equivalent ways by $\varkappa$, and take into account that, when the S-matrix elements are actually computed, terms with the same expression under the integral sign do not appear, then we find that $k = 2^n n!/\varkappa$.

The quantity $\varkappa$ can be determined directly from an inspection of the diagram. We will now turn once more to the concrete processes considered in § 12. When $n = 1, k = 2/\varkappa$. The diagrams in Figs. 8a, 8b and 10a, 10b transform into each other when the replacement $1 \rightleftarrows 2$ is made. Hence $\varkappa = 2$ and $k = 1$ for them. The form of the remaining first-order diagrams changes with the substitution $1 \rightleftarrows 2$: Fig. 8c goes to 8d, 8e to 8f. Hence $\varkappa = 1$ and $k = 2$ for these cases.

We now turn to the more complicated case of second-order processes, where $k = 8/\varkappa$. No transformation exists for the diagrams of Fig. 12 which leaves them unchanged. Hence $\varkappa = 1$ and $k = 8$. Such a change does exist for Fig. 13a; this is the simultaneous replacement $1 \rightleftarrows 2$, $3 \rightleftarrows 4$, whence $\varkappa = 2$ and $k = 4$. For Fig. 13b, $\varkappa = 1$ and $k = 8$. The replacement $1 \rightleftarrows 4$, $2 \rightleftarrows 3$ leaves Fig. 13c invariant, whence $\varkappa = 2$ and $k = 4$. Finally, if we put $2 \rightleftarrows 3$, Fig. 13d is unchanged, and $\varkappa = 2$ in this case. This is all, of course, in complete agreement with (12.4).

The diagrams in Fig. 14 have $\varkappa = 1$ and $k = 8$. There are two changes in Fig. 15, which leave the diagrams invariant. These are $1 \rightleftarrows 3$, $2 \rightleftarrows 4$ and $1 \rightleftarrows 2$, $3 \rightleftarrows 4$. Hence $\varkappa = 4$ and $k = 2$.

We must finally establish the sign of the numerical coefficient. As can be seen from the results in § 12, an additional minus sign appears in $\hat{S}$ when there is an odd number of closed loops. This is hardly surprising, since, when dealing with closed loops, we have contractions with opposite directions; in the simplest case, for example, $\overbrace{\hat{\psi}(1)\,\hat{\psi}^+(2)}$ and $\overbrace{\hat{\psi}^+(2)\,\hat{\psi}(1)}$, and their product is equal to $-(iG_0)\,(iG_0)$.

Thus our final expression for the numerical coefficient of the S-matrix element has the form

$$\frac{(-i)^n}{\varkappa}\,(-1)^m,$$

where m is the number of closed loops in the diagram.

13.2. We can now formulate the rules required to form an expression for the S-matrix element. They were first established by Feynman for quantum electrodynamics.

The essentials are:

1. To make a graphical representation of all possible, topologically different diagrams of the process under consideration.†

The expression for the S-matrix element has the form of a sum of terms, each of which corresponds to one of these diagrams. We will be considering in future the individual diagrams.

2. To determine the number of changes in the arguments $\varkappa$, which do not change the form of the diagram. The numerical coefficient of the desired expression has the form

$$\frac{(-i)^n}{\varkappa}(-1)^m,$$

where m is the number of closed loops in the diagram.

3. The external lines of the diagram must be made to correspond with the operators $\hat{\psi}(x)_{(-)}$ (annihilation of a particle at point x), $\hat{\psi}^+(x)_{(+)}$ (creation of a particle at point x), $\hat{\psi}(x)_{(+)}$ (creation of a hole at point x), $\hat{\psi}^+(x)_{(-)}$ (annihilation of a hole at point x).

The required expression will contain the normal product.

$$N(\hat{\psi}^+_{(\pm)a}\hat{\psi}_{(\pm)a}\hat{\psi}_{(\pm)b}\hat{\psi}_{(\pm)b}\ldots),$$

where the common indices a, b, etc., denote operators referring to the same full-drawn line.

4. For each virtual line on a diagram, going from x to x', we must take the appropriate contraction $iG_0(x', x)$.‡

5. For each interaction line connecting the points x and x', we must have the appropriate operator $\hat{V}(x, x')$, which acts on the

† We mean here essentially different diagrams which cannot be transformed into one another by a change in notation, by a continuous deformation of the graph, etc. Thus the diagrams in Figs. 13a and 13d are topologically different, whereas Figs. 13a and 13e are the same. Figs. 8c and 8d, 8e and 8f, and Figs. 18a and 18c are also topologically the same.

‡ The sequence of the arguments of the functions G_0 is opposite to the direction of the virtual line.

quantities

$$\hat{\psi}(x),\ \hat{\psi}(x'),\ G_0(x, x_i),\ G_0(x', x_i)$$

in the particular expression.

6. Each interaction vertex x corresponds to an integration $\int d^4x = \int dq\, dt$.

In relativistic field theory there is one further rule (Furry's theorem): diagrams with closed loops, containing an odd number of components, need not be considered. In our case, however, this theorem does not hold, since charge symmetry (which is the basic requirement) is lacking (see § 8).

The rules we have formulated are very convenient for constructing an expression for an S-matrix element of any order.

13.3. We have introduced the Feynman rules in terms of the coordinate representation. We will now consider the corresponding rules in the energy representation.

1. Remains the same. It is merely necessary to arrange the indices of state and energy of the external lines ν, ε_ν, and of the virtual lines μ, ε on the diagram. The direction of the external lines is chosen to correspond with the type of process (absorption or emission) and the type of object (particle or hole). The direction of the virtual line must be such as to guarantee a unique direction along the continuous line.

2. Remains the same.

3. Remains the same, but the field operators must be replaced by the creation and annihilation operators, with a particle operator for $\varepsilon_\nu > \varepsilon_F$, and a hole operator for $\varepsilon_\nu < \varepsilon_F$.

4. For each virtual line with indices μ and ε, we have a factor

$$\frac{i}{2\pi} G_{0\mu}(\varepsilon) = \frac{i}{2\pi[\varepsilon - \varepsilon_\mu + i\delta \operatorname{sign}(\varepsilon_\mu - \varepsilon_F)]},$$

which is summed over μ, and integrated over ε.

5. Each interaction line which is entered by lines with indices of state a and b, and which is left by lines c and d corresponds to a matrix element

$$\langle ab|\hat{V}|cd\rangle = \int dq\, dq' \chi_c^*(q)\, \chi_d^*(q')\, \hat{V}\chi_b(q')\, \chi_a(q).$$

The indices a, c, and b, d, here correspond in pairs to the same full-drawn line. For each interaction line there is a factor

$$2\pi\delta(\varepsilon_a + \varepsilon_b - \varepsilon_c - \varepsilon_d).$$

13.4. We should pay particular attention to the case of a spatially uniform system, where the indices μ and ν refer to the momentum $\boldsymbol{p}$ and to the discrete indices denoted by ϱ. The energy ε_p depends only on momentum.

Rule 4 of the previous section can be reformulated in the following way: each virtual line corresponds to

$$i\sum_{\varrho}\int d^4p\,\frac{1}{\varepsilon - \varepsilon_p + i\delta\,\mathrm{sign}\,(\varepsilon_p - \varepsilon_F)}.$$

Here we have gone from a summation over the momenta to an integration, which leads to the appearance of an extra factor $(2\pi)^{-3}$.

The wave functions now have the form of plane waves. Substituting them into the matrix element $\hat{V}$, we get

$$\langle ab|\hat{V}|cd\rangle = (2\pi)^3\,\delta(\boldsymbol{p}_a + \boldsymbol{p}_b - \boldsymbol{p}_c - \boldsymbol{p}_d)\,\langle\varrho_a\varrho_b|\nu(\boldsymbol{p}_c - \boldsymbol{p}_a)|\varrho_c\varrho_d\rangle,$$

where $\nu(\boldsymbol{p})$ is the Fourier transform of the potential (see § 4); the matrix element is in the discrete indices. Hence, the overall contribution of the interaction line to the matrix element is given by

$$(2\pi)^4\,\delta^4(p_a + p_b - p_c - p_d)\,\langle\varrho_a\varrho_b|\nu(\boldsymbol{p}_c - \boldsymbol{p}_a)|\varrho_c\varrho_d\rangle,$$

where

$$\delta^4(p) = \delta(\boldsymbol{p})\,\delta(\varepsilon).$$

The remaining rules formulated in the previous section are unchanged.

If the laws of conservation of energy and momentum are obeyed at each vertex of a diagram, a more useful representation is possible using these laws explicitly. For this, we must introduce the momentum transferred along an interaction line $\boldsymbol{k} = \boldsymbol{p}_c - \boldsymbol{p}_a$ (the energy transerred is zero since retardation is not allowed for and v does not depend on the energy). The conservation laws will now be obeyed at each vertex of the diagram (Fig. 19).

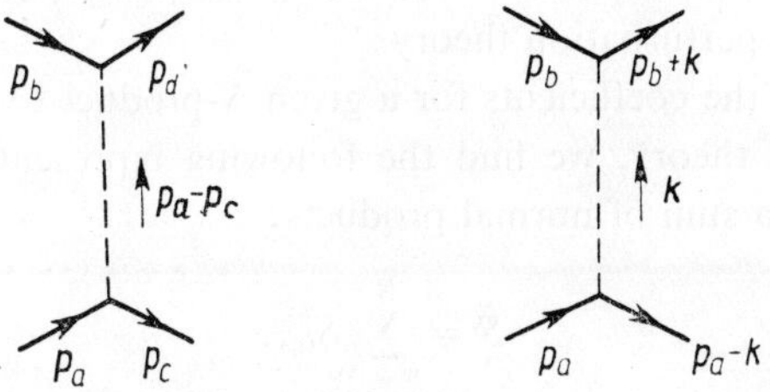

FIG. 19

There is often no need, particularly in the course of a general analysis, impossible to consider separately the S-matrix elements corresponding to the absorption of a particle and the emission of a hole (or vice versa). An expression for an S-matrix element containing all possibilities of this sort can be obtained from the Feynman rules with only one difference: in the third of these rules, the expression for the S-matrix element must now include the N-product $N(\hat{\psi}_a{}^+\hat{\psi}_a\hat{\psi}_b{}^+\hat{\psi}_b \ldots)$, where $\hat{\psi}^+$, $\hat{\psi}$ are the complete field operators. Similarly, in the energy representation, the quantity $N(A_\mu^+ A_\nu \ldots)$ appears without limitation on ε_ν, $\varepsilon_\mu \ldots \varepsilon_F$. The corresponding diagrams (which we will call unoriented) need only indicate one of the possible orientations of each external line.

The next paragraph will provide examples of the use of the Feynman rules.

14. The General Structure of the Scattering Matrix

14.1. The Feynman rules may be used to analyse the structure of the S-matrix as a whole, without any connection with lower-order perturbation theory.†

If we expand the S-matrix first of all in terms of T-products and then in terms of normal products, we see that each normal

† The number of particles in the system N is supposed to be large. Possible violations of the relationships used concern higher-order terms of perturbation theory, and provide a relative contribution of the order $1/N$.

product can appear in any order of perturbation theory; in the higher orders it will be accompanied by a correspondingly larger number of contractions of operators. In other words the contribution to the S-matrix element for a given process gives terms of all orders of perturbation theory.

If we sum the coefficients for a given N-product to all orders of perturbation theory, we find the following representation of the S-matrix as a sum of normal products:

$$\hat{S} = \sum_{n=0}^{\infty} \hat{S}_{(n)}, \tag{14.1}$$

where

$$\hat{S}_{(n)} = \int d1 \ldots dn\, d1' \ldots dn' K_n(1 \ldots n, 1' \ldots n') \times N[\hat{\psi}^+(1) \ldots \hat{\psi}^+(n)\, \hat{\psi}(n') \ldots \hat{\psi}(1')].$$

We will look at some of the first terms of this expansion. As will become evident, the corresponding coefficients, the functions K, play an important part in the theory.

14.2. We will begin with the zero-order term of (14.1)

$$\hat{S}_{(0)} = K_0 = \langle \Psi_0 | \hat{S} | \Psi_0 \rangle. \tag{14.2}$$

This is not an operator, and can be obtained by averaging (14.1) over the state Ψ_0 [see (8.6)]. It describes internal (vacuum) fluctuations of the system with no external particles or holes participating. Examples of diagrams for such transitions were given in Fig.15.

Transitions of this sort do not correspond to real processes in a system, so they may therefore be excluded from our consideration.

We will analyse the structure of K_0 in more detail. As a preliminary, we introduce the concept of connected diagrams. A diagram is called connected if it cannot be divided by a line without intersecting any line in the diagram.

K_0 includes the contributions of both connected (Figs. 20a and 20b) and unconnected (Figs. 20c and 20d) diagrams. We will denote by L the sum of all connected diagrams that contribute to K_0. Then K_0 can be put in the form

$$\langle \Psi_0 | \hat{S} | \Psi_0 \rangle = 1 + L + \frac{L^2}{2!} + \frac{L^3}{3!} + \cdots = \exp(L). \tag{14.3}$$

We have allowed here for the contributions of all (connected and unconnected) diagrams. The first term of this expansion is related to the zero-order term of the S-matrix. The second describes all connected diagrams. The third contains the sum of the contributions (or simply the sum) of all unconnected diagrams, each of which is composed of two connected ones. The fourth corresponds to unconnected diagrams composed of three connected ones, etc.

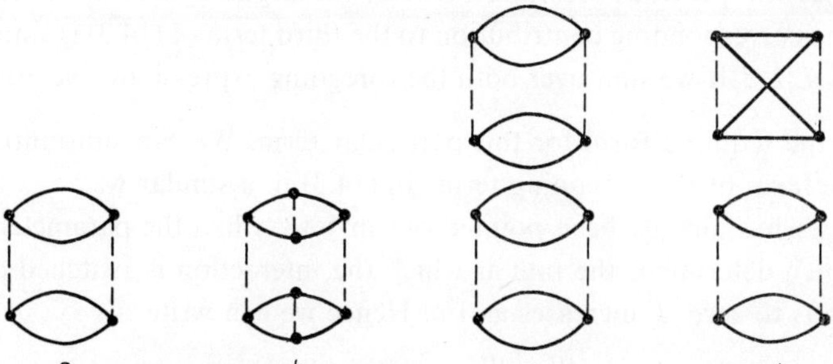

FIG. 20

We will clarify the structure of these terms on the basis of the Feynman rules. In the first place the S-matrix element of an unconnected diagram can be split into a product of independent elements which correspond to the connected diagrams. Thus we see that the appearance of terms in L^2, L^3, etc., is completely natural.

The appearance of the factors 1/2! 1/3!, etc., in the numerical coefficients can be explained in the following way. Suppose we consider for simplicity the third term in (14.3) and put $L = \Sigma_i L_i$ as the sum of the diagrams of different orders of perturbation theory. We will next construct all possible unconnected diagrams composed of pairs of connected ones. We must distinguish between two cases: identical pairs (see Fig. 20c) and distinguishable pairs (see Fig. 20d) of connected diagrams. For the first, Feynman rule 2 leads to the numerical coefficient $(-i)^{2n}(-1)^{2m}/\varkappa_{\text{sum}}$, where n and m are the order and the number of the loops in each of the connected diagrams and $\varkappa_{\text{sum}}$ —the number of ways of designating the arguments which do not change the expressions for the S-matrix elements—is equal to $2\varkappa^2$, where $\varkappa$ is the analogous number for each of the connected diagrams. The extra two reflects the fact that the number of rearrangements for a given unconnected diagram contains the interchange of coordinates of both the (identical) connected parts. Thus that part of the third term in (14.3) which is concerned with a pair of identical diagrams, has the form $\frac{1}{2}\sum_i L_i^2$.

For a pair of distinguishable diagrams, the factor 2 is absent, since the previous rearrangement of the arguments is impossible. The corresponding contribution to the third term of (14.3) is simply $\sum_{i \neq j} L_i L_j$. If we sum over both the foregoing expressions, we arrive at the required form for this particular term. We can substantiate the form of the remaining terms in (14.3) in a similar way.

As has already been pointed out in § 12, when the parameter δ, which determines the rate at which the interaction is switched on, tends to zero, L increases as $1/\delta$. Hence we can write

$$\langle \Psi_0 | \hat{S} | \Psi_0 \rangle = \exp (L_0/\delta),$$

where $L = L_0/\delta$.

Vacuum fluctuations always accompany a real process; in other words when we consider an element of the S-matrix, we must also allow for the unconnected diagrams which contain vacuum fluctuations.

Reasoning similar to that given above shows that the vacuum fluctuations in this case appear as the factor

$$\hat{S} = \langle \Psi_0 | \hat{S} | \Psi_0 \rangle \, \hat{S}_{\text{real}}, \tag{14.4}$$

where $\hat{S}_{\text{real}}$ is the S-matrix element of the process in which we are interested.

Thus we observe a complete independence of the vacuum and the real processes, which makes it possible to ignore completely the vacuum fluctuations.

14.3. We now consider the next term of the expansion (14.1):

$$\hat{S}_{(1)} = \int d1\, d2\, K_1(1, 2)\, N[\hat{\psi}^+(1)\, \hat{\psi}(2)]. \tag{14.5}$$

Taking separately the term in K_1 corresponding to vacuum fluctuations, we obtain

$$K_1(1, 2) = i \langle \Psi_0 | \hat{S} | \Psi_0 \rangle\, \Sigma(1, 2),$$

where Σ is referred to as the self-energy part. This name is taken from quantum field theory, where an analogous quantity describes the interaction of a particle with its own field.

In many-body theory $\hat{S}_{(1)}$ describes the transition processes for a particle between point 1 and point 2, or a hole in the opposite direction, i.e. the way in which they propagate allowing for their interaction with the other particles in the system. Figure 14 provides examples of such transitions.

The self-energy can also be included the in graphical representation (Fig. 21a). In diagram language Σ is the sum of the diagrams which have two external lines. Appropriate examples are given in Figs. 21b–21e.

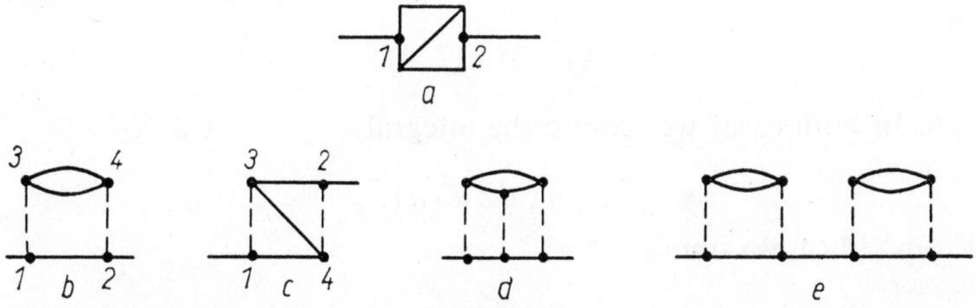

FIG. 21

As will be explained in the next section, Σ may be used to express important physical characteristics of a system of particles.

We will calculate $\Sigma(1, 2)$, using the Feynman rules in a low (second)-order perturbation theory (see Figs. 21b and 21c). Eliminating the factor $\langle\Psi_0|\hat{S}|\Psi_0\rangle$, which is equal to unity in this approximation, we can then construct the corresponding element of the S-matrix, which has the form

$$\hat{S} = i \int d1\, d2\, \Sigma(1, 2)\, N[\hat{\psi}^+(1)\, \hat{\psi}(2)]. \tag{14.6}$$

The operations described below will then be denoted by the same numbers as the corresponding Feynman rules of § 13.2.

1. The only two topologically different diagrams of the process are shown in Figs. 21b and 21c.

2. The factor $\varkappa$ was calculated in § 13.1, where we found that $\varkappa = 1$ for both diagrams. For the first diagram, $m = 1$; for the

second, $m = 0$. Hence the numerical factor for the S-matrix element is equal respectively to $+1$ and -1 ($n = 2$).

3. When we consider simultaneously all the possible propagation processes which correspond to these diagrams, we should take the N-product in the form $N[\hat{\psi}^+(1)\,\hat{\psi}(2)]$.

4. The virtual lines for the first diagram are described by the combination

$$iG_0(2, 1)\,iG_0(3, 4)\,iG_0(4, 3);$$

for the second by

$$iG_0(4, 1)\,iG_0(3, 4)\,iG_0(2, 3).$$

5. In both cases the interaction lines correspond to the factor

$$V(1, 3)\,V(2, 4).$$

6. In both cases we require the integral

$$\int d1\,d2\,d3\,d4.$$

From (14.6), we obtain

$$\Sigma(1, 2) = -\int d3\,d4\,V(1, 3)\,V(2, 4)\,\{G_0(2, 1)\,G_0(3, 4)\,G_0(4, 3) - G_0(4, 1)\,G_0(3, 4)\,G_0(2, 3)\}.$$

This can also be obtained from a comparison of (12.5) and (12.2) with (14.5).

14.4. We will now work out the self-energy in the energy representation, considering the transition $\mu \to \nu$ and rewriting (14.6) in the form

$$\hat{S}_{\mu\nu} = 2\pi i \int d\varepsilon\,\Sigma_{\mu\nu}(\varepsilon)\,N(\hat{A}_\nu^+\hat{A}_\mu)\,\delta(\varepsilon - \varepsilon_\mu)\,\delta(\varepsilon - \varepsilon_\nu).$$

The quantity $\Sigma_{\mu\nu}(\varepsilon)$ thus defined, satisfies the relation

$$\sum(1, 2) = \int \frac{d\varepsilon}{2\pi} \exp\left[-i\varepsilon(t_1 - t_2)\right] \sum_{\mu,\nu} \chi_\mu^*(q_2)\,\chi_\nu(q_1)\,\Sigma_{\mu\nu}(\varepsilon), \tag{14.8}$$

i.e. it is the coefficient of the expansion in terms of the χ_ν, and the Fourier transform with respect to $t_1 - t_2$ of the function $\Sigma(1, 2)$.

We now use the Feynman rules of § 13.3:

1. The topologically different energy diagrams are given in Figs. 18a and 18b.

2. This is the same as in § 14.3.

3. We have $N(\hat{A}_\nu^+ \hat{A}_\mu)$.

4. We have the combination

$$\left(\frac{i}{2\pi}\right)^3 \sum_{\mu_1\mu_2\mu_3} d\varepsilon_1\, d\varepsilon_2\, d\varepsilon_3 G_{0\mu_1}(\varepsilon_1)\, G_{0\mu_2}(\varepsilon_2)\, G_{0\mu_3}(\varepsilon_3).$$

5. The matrix elements in the first diagram have the form

$$\langle\mu, \mu_3|\hat{V}|\mu_1, \mu_2\rangle\, \langle\mu_1, \mu_2|\hat{V}|\nu, \mu_3\rangle;$$

in the second

$$\begin{aligned}&\langle\mu, \mu_3|\hat{V}|\mu_1, \mu_2\rangle\, \langle\mu_1, \mu_2|\hat{V}|\mu_3, \nu\rangle \\ &\quad = \langle\mu, \mu_3|\hat{V}|\mu_1, \mu_2\rangle\, \langle\mu_1, \mu_2|\hat{V}\hat{\mathscr{P}}|\nu, \mu_3\rangle,\end{aligned}$$

where $\hat{\mathscr{P}}$ is the exchange operator for the coordinates (or, which is the same thing, the exchange operator for the state indices).†

We also have the δ-function

$$(2\pi)^2\, \delta(\varepsilon_\mu + \varepsilon_3 - \varepsilon_1 - \varepsilon_2)\, \delta(\varepsilon_1 + \varepsilon_2 - \varepsilon_3 - \varepsilon_\nu).$$

We finally obtain

$$\begin{aligned}\Sigma_{\mu\nu}(\varepsilon) = &-\frac{1}{(2\pi)^2} \sum_{\mu_1\mu_2\mu_3} \int d\varepsilon_1\, d\varepsilon_2\, d\varepsilon_3\, \langle\mu, \mu_3|\hat{V}|\mu_1\mu_2\rangle \\ &\times \langle\mu_1\mu_2|\hat{V}(1 - \hat{\mathscr{P}})\,|\nu\mu_3\rangle\, G_{0\mu_1}(\varepsilon_1)\, G_{0\mu_2}(\varepsilon_2)\, G_{0\mu_3}(\varepsilon_3) \\ &\times \delta(\varepsilon + \varepsilon_3 - \varepsilon_1 - \varepsilon_2).\end{aligned}$$

We can integrate over the energies, using the results of Appendix C. This gives

$$\begin{aligned}\Sigma_{\mu\nu}(\varepsilon) = &\sum_{\mu_1\mu_2\mu_3} \langle\mu, \mu_3|\hat{V}|\mu_1\mu_2\rangle\, \langle\mu_1\mu_2|\hat{V}(1 - \hat{\mathscr{P}})|\nu\mu_3\rangle \\ &\times \left\{\frac{(1 - n_{\mu_1})(1 - n_{\mu_2})\, n_{\mu_3}}{\varepsilon + \varepsilon_{\mu_3} - \varepsilon_{\mu_1} - \varepsilon_{\mu_2} + i\delta} + \frac{n_{\mu_1} n_{\mu_2}(1 - n_{\mu_3})}{\varepsilon + \varepsilon_{\mu_3} - \varepsilon_{\mu_1} - \varepsilon_{\mu_2} - i\delta}\right\}. \quad (14.9)\end{aligned}$$

† The diagrams corresponding to this exchange operator $\hat{\mathscr{P}}$ are called exchange correlation diagrams.

The first term in the curly brackets corresponds to the diagrams represented in Figs. 18a and 18b: the states μ_1 and μ_2 are inherently particles and contribute to (14.9) only outside the Fermi sphere, the state μ_3 is a hole and is effective only within the Fermi sphere. The second term in the curly brackets corresponds to diagrams which are topologically equivalent to those considered previously (see Fig. 18c). Here the states μ_1, μ_2 correspond to holes, μ_3 to a particle. It is as if the diagram in Fig. 18a were turned inside out.

We next consider the self-energy in the momentum representation for a spatially uniform system. Starting from the Fourier expansion,

$$\Sigma(x) = \int d^4p\, \Sigma(\boldsymbol{p}, \varepsilon) \exp i[(\boldsymbol{p}\cdot\boldsymbol{x}) - \varepsilon t]$$

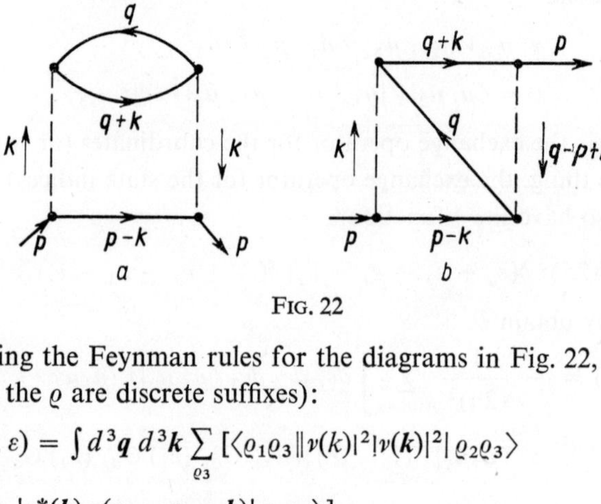

FIG. 22

and using the Feynman rules for the diagrams in Fig. 22, we find (where the ϱ are discrete suffixes):

$$\sum_{\varrho_1\varrho_2} (\boldsymbol{p}, \varepsilon) = \int d^3\boldsymbol{q}\, d^3\boldsymbol{k} \sum_{\varrho_3} [\langle \varrho_1\varrho_3 \| v(\boldsymbol{k})|^2 |v(\boldsymbol{k})|^2 | \varrho_2\varrho_3\rangle$$

$$- \langle \varrho_1\varrho_3 | v^*(\boldsymbol{k})\, v(\boldsymbol{p}-\boldsymbol{q}-\boldsymbol{k}) | \varrho_3\varrho_2\rangle]$$

$$\times \left\{ \frac{(1-n_{p-k})\,(1-n_{q+k})\, n_q}{\varepsilon + \varepsilon_q - \varepsilon_{p-k} - \varepsilon_{q+k} + i\delta} + \frac{n_{p-k}\, n_{q+k} (1-n_q)}{\varepsilon + \varepsilon_q - \varepsilon_{p-k} - \varepsilon_{q+k} - i\delta} \right\}, \tag{14.10}$$

where

$$n_p = \theta(p_0^2 - p^2).$$

For a potential which is simply a function of the coordinates, the expression in square brackets takes the form

$$\delta_{\varrho_1\varrho_2}\{g|v(\boldsymbol{k})|^2 - v^*(\boldsymbol{k})\, v(\boldsymbol{p}-\boldsymbol{q}-\boldsymbol{k})\},$$

where g is the statistical weight factor (see § 4).

14.5. The third term of the expansion (14.1) has the form

$$\hat{S}_{(2)} = \int d1\, d2\, d3\, d4 K_2(1, 2, 3, 4)\, N[\hat{\psi}^+(1)\, \hat{\psi}^+(2)\, \hat{\psi}(4)\, \hat{\psi}(3)]. \quad (14.11)$$

Separating out vacuum-type processes, we have

$$K_2 = \langle \Psi_0 | \hat{S} | \Psi_0 \rangle\, \Gamma(1, 2, 3, 4),$$

where Γ is the effective interaction potential, or the vertex potential.

The element $\hat{S}_{(2)}$ describes all the possible processes for which the diagrams have four external lines. Thus we are concerned with scattering processes of one particle by another, of one hole by another and of a particle by a hole, allowing for the correlation interaction with the particles in the system. Examples of such diagrams, which we have also considered previously, are given in Fig. 23. It can be seen that the vertex part, which is represented graphically by the symbol shown in Fig. 23a, describes a modified interaction law, and indicates how the interaction line is changed when correlation interactions are included.

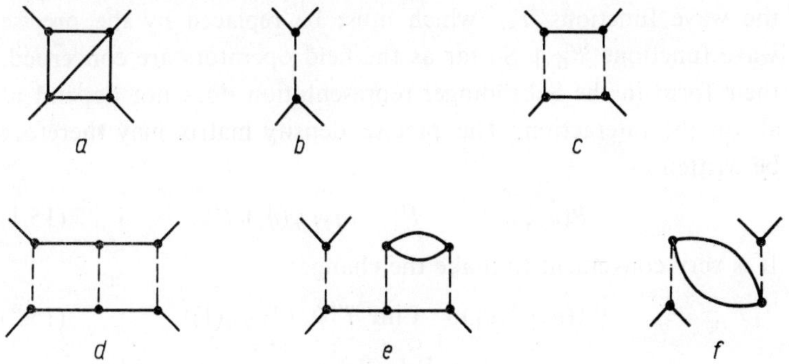

FIG. 23

As is evident from a comparison of (14.11) and (12.1), lowest order perturbation theory gives†

$$\Gamma(1, 2, 3, 4) = -\frac{i}{2} V(1, 2)\, \delta(1-3)\, \delta(2-4). \quad (14.12)$$

† If we allow for the antisymmetry of Γ for the permutation $1 \rightleftarrows 2$ or $3 \rightleftarrows 4$, we ought to change (14.12) to

$$\delta(1-3)\, \delta(2-4) \to \tfrac{1}{2}[\delta(1-3)\, \delta(2-4) - \delta(1-4)\, \delta(2-3)]$$

The self-energy and the vertex parts, which describe the influence of correlation effects on the particle line and the interaction line, are very important structural elements, which can be used to construct a diagram for any process (see §§ 20 and 25 for details).

15. The Scattering Matrix and Physical Quantities

15.1. Now that we have obtained an expression for the self-energy part of the S-matrix, we can determine a considerable amount of physical information about the many-body system.

We first of all find an expression for the single-particle density matrix, including the correlation interaction between particles. This quantity provides the same information on the properties of the system as the density matrix in the Hartree–Fock approximation, which we will here denote by R_0:

$$R_0(q_1, q_2) = \langle \Psi_0 | \hat{\psi}^+(q_2)\, \hat{\psi}(q_1) | \Psi_0 \rangle .$$

The inclusion of the correlation interaction obviously only affects the wave functions Ψ_0, which must be replaced by the precise wave functions Ψ_H.† So far as the field operators are concerned, their form in the Schrödinger representation does not depend at all on the interaction. The precise density matrix may therefore be written as

$$R(q_1, q_2) = \langle \Psi | \hat{\psi}^+(q_2)\, \hat{\psi}(q_1) | \Psi \rangle . \tag{15.1}$$

It is very convenient to make the change:

$$\hat{\psi}^+(q_2)\, \hat{\psi}(q_1) = \lim_{\substack{t_2 > t_1 \\ t_1, t_2 \to 0}} T[\hat{\psi}_H^+(2)\, \hat{\psi}_H(1)] . \tag{15.2}$$

Here, both t_2 and t_1 tend to zero, but t_2 remains larger than t_1. This first of all provides the required order of the operators, and, secondly, each operator coincides with the operator in the Schrödinger representation. We then obtain

$$R(q_1, q_2) = \lim_{t_2 - t_1 \to +0} \langle \Psi | T[\hat{\psi}_H^+(2)\, \hat{\psi}_H(1)] | \Psi \rangle . \tag{15.3}$$

The average value depends only on the time difference $t_1 - t_2$.

† For simplicity, we omit the suffix H from Ψ_H.

We now transform this expression from the Heisenberg representation to the interaction representation, using (9.21) and (9.26):

$$\hat{\psi}_H(1) = \hat{S}(0, t_1)\, \hat{\psi}_{\text{Int}}(1)\, \hat{S}(t_1, 0),$$

$$\hat{\psi}_H^+(1) = \hat{S}(0, t_1)\, \hat{\psi}_{\text{Int}}^+(1)\, \hat{S}(t_1, 0),$$

$$\Psi = \hat{S}(0, -\infty)\, \Psi_0,$$

$$\Psi^* = \Psi_0^* \hat{S}(-\infty, 0).$$

The average value in (15.3) has the following form for $t_2 > t_1$:

$$\langle \Psi_0 | \hat{S}(-\infty, t_2)\, \hat{\psi}_{\text{Int}}^+(2)\, \hat{S}(t_2, t_1)\, \hat{\psi}_{\text{Int}}(1)\, \hat{S}(t_1, -\infty) | \Psi_0 \rangle.$$

We have used here the group property of the S-matrix (9.7). Writing $\hat{S}(-\infty, t_2)$ as $\hat{S}^{-1}(\infty, -\infty)\, S(\infty, t_2)$, we can make use of the stability of the ground state (9.18′). The required average value can therefore be rewritten as

$$\frac{\langle \Psi_0 | \hat{S}(\infty, t_2)\, \hat{\psi}_{\text{Int}}^+(2)\, \hat{S}(t_2, t_1)\, \hat{\psi}_{\text{Int}}(1)\, \hat{S}(t_1, -\infty) | \Psi_0 \rangle}{\langle \Psi_0 | \hat{S} | \Psi_0 \rangle}.$$

It is evident from this expression that the operators in the expectation value are in true chronological order:† there is first the part of the S-matrix for times greater than t_2; then the operator at t_2; then the part of the S-matrix for times greater than t_1, but less than t_2; etc. Combining this expression with the analogous expression for $t_1 > t_2$, we may finally write

$$\langle \Psi | T[\hat{\psi}_H^+(2)\, \hat{\psi}_H(1)] | \Psi \rangle = \frac{\langle \Psi_0 | T[\hat{\psi}_{\text{Int}}^+(2)\, \hat{\psi}_{\text{Int}}(1)\, \hat{S}] | \Psi_0 \rangle}{\langle \Psi_0 | \hat{S} | \Psi_0 \rangle}. \quad (15.4)$$

This is a most important relation, of which we will make considerable use in the following.

Equation (15.3) can be directly expressed in terms of the self-energy part of the S-matrix. If we substitute the S-matrix expansion

† The representation of the S-matrix as a chronological product has the following graphic interpretation. We can split up $\hat{S}(\infty, -\infty)$ into an arbitrary number of factors, using the S-matrix property (9.7): $\hat{S}(\infty, -\infty) = \hat{S}(\infty, t_1) \times \hat{S}(t_1, t_2) \ldots \hat{S}(t_n, -\infty)$, where $t_1 > t_2 \ldots > t_n$. Putting in chronological order represents this explicity: each factor contains a time which is later than its right-hand neighbour, but earlier than its left-hand neighbour.

(14.1) in the numerator of (15.4), we see that we can limit our consideration to the first two terms of this expansion:

$$\hat{S} = \langle \Psi_0 | \hat{S} | \Psi_0 \rangle \{1 + i \int d1\, d2\, \Sigma(1, 2)\, N[\hat{\psi}^+_{\text{Int}}(1)\, \hat{\psi}_{\text{Int}}(2)] + \cdots\}.$$

Indeed, if we apply Wick's theorem, we can confirm that terms of higher order lead to the appearance of normal products, differing from $N(1)$, in the expectation value. Averaging them obviously leads to a zero result.

The first term of the $\hat{S}$ expansion gives simply

$$\langle \Psi_0 | T[\hat{\psi}^+_{\text{Int}}(2)\, \hat{\psi}_{\text{Int}}(1)] | \Psi_0 \rangle .$$

So far as the second term is concerned, we have to calculate

$$\langle \Psi_0 | T\{\hat{\psi}^+_{\text{Int}}(2)\, \hat{\psi}_{\text{Int}}(1)\, N[\hat{\psi}^+_{\text{Int}}(3)\, \hat{\psi}_{\text{Int}}(4)]\} | \Psi_0 \rangle .$$

Applying Wick's theorem for mixed T-products, it is easy to see that the only term differing from zero is

$$iG_0(1, 3)\,(-i)\, G_0(4, 2),$$

where G_0 is the free-particle Green function (see § 10). Thus

$$\langle \Psi | T[\hat{\psi}^+_H(2)\, \hat{\psi}_H(1)] | \Psi \rangle = \langle \Psi_0 | T[\hat{\psi}^+_{\text{Int}}(2)\, \hat{\psi}_{\text{Int}}(1)] | \Psi_0 \rangle + i \int d3\, d4 G_0(1, 3)\, \Sigma(3, 4)\, G_0(4, 2). \tag{15.5}$$

Turning to the density matrix (15.1), we have

$$R(q_1, q_2) = R_0(q_1, q_2) + \lim_{t_2 - t_1 \to +0} i \int d3\, d4\, G_0(1, 3)\, \Sigma(3, 4)\, G_0(4, 2). \tag{15.6}$$

The first term of this expression corresponds to the Hartree–Fock approximation, the second describes the correlation corrections to the density matrix.

15.2. It is easy to find an expression for the number density of particles in the system from (15.6):

$$\varrho(\boldsymbol{x}_1) = \text{Tr}_{\sigma\tau}\, R(\boldsymbol{x}_1, \sigma_1, \tau_1; \boldsymbol{x}_1, \sigma_2, \tau_2), \tag{15.7}$$

whence

$$\varrho(\boldsymbol{x}_1) - \varrho_0(\boldsymbol{x}_1) = \text{Tr}_{\sigma\tau}\, i \int_{\substack{2 \to 1 \\ t_2 > t_1}} d3\, d4 G_0(1, 3)\, \Sigma(3, 4)\, G_0(4, 2)$$

(ϱ_0 is the density in the Hartree–Fock approximation). This relation describes the redistribution of particles in space when subject to the correlation interaction. There is no such redistribution for a spatially uniform system.

It is of interest, however, to enquire how the particles in a spatially uniform system may be redistributed in terms of momenta. If we transform from (15.6) to the momentum representation (see § 4), we find:

$$\varrho(\boldsymbol{p}) - \varrho_0(\boldsymbol{p}) = \lim_{t_2 - t_1 \to +0} i \int_{-\infty}^{\infty} \frac{d\varepsilon}{2\pi} \exp[i\varepsilon(t_2 - t_1)] \times \frac{g\Sigma(\boldsymbol{p}, \varepsilon)}{[\varepsilon - \varepsilon_p + i\delta \operatorname{sign}(\varepsilon_p - \varepsilon_F)]^2}.$$

It is useful to express this in another form. We will continue the expression under the integral sign into the complex ε-plane. If we take into account that this expression decreases rapidly enough for large ε, and that the ε in the exponent, $i\varepsilon(t_2 - t_1)$, has a negative real part in the upper half-plane (i.e. the exponent decreases as $\operatorname{Im}\varepsilon$ increases), then, from Cauchy's theorem, we can replace the ε-integration along the real axis by an integration round the closed contour C, which is the boundary of a semi-circle in the upper half-plane (see Fig. 24). The integral along this semi-circle introduces a negligible contribution.

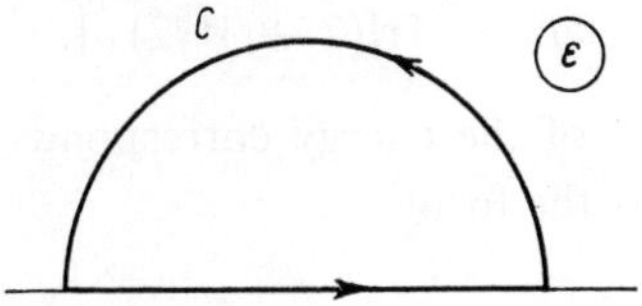

FIG. 24

We can now go to the limit $t_2 - t_1 \to 0$:

$$\varrho(\boldsymbol{p}) - \varrho_0(\boldsymbol{p}) = i \int_C \frac{d\varepsilon}{2\pi} \cdot \frac{g\Sigma(\boldsymbol{p}, \varepsilon)}{[(\varepsilon - \varepsilon_p) + i\delta \operatorname{sign}(\varepsilon_p - \varepsilon_F)]^2}. \quad (15.8)$$

15.3. We turn now to expressing the most important property of the ground state of a system—its energy E—in terms of Σ. We start off with the usual expression for E

$$E = \langle \Psi | \hat{H} | \Psi \rangle, \tag{15.9}$$

where we will write the Hamiltonian in the Schrödinger representation in its original form (3.19):

$$\hat{H} = \int dq\, \hat{\psi}^+(q)\, \hat{T}\hat{\psi}(q) + \tfrac{1}{2} \int dq\, dq'\, \hat{\psi}^+(q)\, \hat{\psi}^+(q')\, \hat{V}\hat{\psi}(q')\, \hat{\psi}(q).$$

We will next express this in terms of the Heisenberg operators $\hat{\psi}_H$ for $t = 0$. From the equation of motion for these operators (see § 3.9) we can write

$$\hat{H} = \operatorname*{Lim}_{t \to 0} \frac{1}{2} \int dq_1\, \hat{\psi}_H^+(1) \left(i \frac{\partial}{\partial t_1} + \hat{T} \right) \hat{\psi}_H(1)$$

$$= \frac{1}{2} \int dq_1 \operatorname*{Lim}_{\substack{2 \to 1 \\ t_1 \to 0}} \left(i \frac{\partial}{\partial t_1} + \hat{T} \right) \hat{\psi}_H^+(2)\, \hat{\psi}_H(1).$$

Putting $t_2 > t_1$, and substituting in (15.9), we can write

$$E = \frac{1}{2} \int dq_1 \operatorname*{Lim}_{\substack{2 \to 1 \\ t_2 - t_1 \to +0}} \left(i \frac{\partial}{\partial t_1} + \hat{T} \right) \langle \Psi | T[\hat{\psi}_H^+(2)\, \hat{\psi}_H(1)] | \Psi \rangle.$$

We can now use (15.5). The first term on the right-hand side of this equation corresponds to the Hartree–Fock approximation, and when substituted in E gives

$$E_0 = \mathrm{Tr}[(\hat{T} + \hat{W}/2)\, \hat{\varrho}].$$

The remaining part of the energy corresponds to the correlation interaction, and has the form

$$E - E_0 = \frac{i}{2} \int dq_1 \operatorname*{Lim}_{\substack{2 \to 1 \\ t_2 - t_1 \to +0}} \left(i \frac{\partial}{\partial t_1} + \hat{T} \right) \int d3\, d4\, G_0(1, 3)$$

$$\times\, \Sigma(3, 4)\, G_0(4, 2). \tag{15.10}$$

We will write this in a slightly different way, in order to clarify its physical significance. From the equation

$$\left(i \frac{\partial}{\partial t_1} - \hat{T} - \hat{W} \right) G_0(1, 2) = \delta(1 - 2),$$

we have

$$E - E_0 = \int dq_1 \lim_{q_2 \to q_1} (\hat{T} + \hat{W}/2)\,[R(q_1, q_2) - R_0(q_1, q_2)]$$

$$+ \frac{i}{2} \int dq_1 \lim_{\substack{2 \to 1 \\ t_2 - t_1 \to +0}} \int d3\, \Sigma(1, 3)\, G_0(3, 2). \qquad (15.11)$$

We have used (15.6) here. The first term of (15.11) describes the change in kinetic energy of the particles and of their self-consistent interaction as a result of their redistribution among the states due to correlation effects. The second term corresponds to a change in the character of the interaction between the particles, i.e. it allows for the self-correlation effects.

15.4. The expression for the energy of the system can be put in another form, if we use the method already applied in § 9. We replace the correlation interaction Hamiltonian $\hat{H}'$ by the quantity $\lambda\hat{H}'$, where λ is some parameter which will be put equal to unity when the calculations are completed. All the functions in the calculations, $\hat{\psi}_H$, $\hat{\psi}_H^+$, $\hat{S}$, Σ, etc., are functions of λ.

We will make use of the general quantum mechanical theorem [31]

$$\partial E/\partial\lambda = \langle \Psi | \partial \hat{H}/\partial\lambda | \Psi \rangle, \qquad (15.12)$$

which can be proved by differentiating the equation $(\hat{H} - E)\,\Psi = 0$ with respect to λ, multiplying the result from the left by Ψ^*, and using the Hermitian character of $\hat{H}$. In our case $\partial\hat{H}/\partial\lambda = \hat{H}'$, and we arrive at the equation

$$E - E_0 = \int_0^1 d\lambda\, \langle \Psi | \hat{H}' | \Psi \rangle. \qquad (15.13)$$

We use the fact that, when $\lambda = 0$, we get back to the Hartree–Fock approximation.

The operator $\hat{H}'$ may be expressed in the form

$$\hat{H}' = \frac{1}{\lambda}(\hat{H} - \hat{H}_0),$$

where the operator $\hat{H}$ can be written

$$\hat{H} = \int dq_1\, \hat{\psi}^+(q_1)\, [\hat{T} + (1-\lambda)\, \hat{W}]\, \hat{\psi}(q_1)$$

$$+ \frac{\lambda}{2} \int dq_1\, dq_2\, \hat{\psi}^+(q_1)\hat{\psi}^+(q_2)\, \hat{V}\hat{\psi}(q_2)\, \hat{\psi}(q_1) + (1-\lambda)\, C,$$

where $C = -\frac{1}{2} \int dq\, \langle \Psi_0 | \hat{\psi}^+(q)\, \hat{W}\hat{\psi}(q) | \Psi_0 \rangle$. It is easy to show that this expression for $\hat{H}$ is correct by collecting the terms proportional to λ, which give an N-product of the required type. If we repeat the arguments of the preceding section, we find

$$\hat{H} = \frac{1}{2} \int_{t_1 \to 0} dq_1\, \hat{\psi}_H^+(1) \left[i \frac{\partial}{\partial t_1} + \hat{T} + (1-\lambda)\, \hat{W} \right] \hat{\psi}_H(1) + (1-\lambda)\, C.$$

Subtracting

$$\hat{H}_0 = \int_{t_1 \to 0} dq_1\, \hat{\psi}_H^+(1)\, (\hat{T} + \hat{W})\, \hat{\psi}_H(1) + C,$$

we obtain

$$\hat{H}' = \frac{1}{2\lambda} \int dq_1\, \hat{\psi}_H^+(1) \left[i \frac{\partial}{\partial t_1} - \hat{T} - (1+\lambda)\, \hat{W} \right] \hat{\psi}_H(1) - C.$$

Finally, we may write from the equation $(i\partial/\partial t_1 - \hat{T} - \hat{W})\, \hat{\psi}_{\text{Int}}(1) = 0$, that

$$\langle \Psi | \hat{H}' | \Psi \rangle = \frac{1}{2\lambda} \int dq_1 \lim_{\substack{2 \to 1 \\ t_2 - t_1 \to +0}} \left[i \frac{\partial}{\partial t_1} - \hat{T} - (1+\lambda)\, \hat{W} \right]$$

$$\times \{ \langle \Psi | T[\hat{\psi}_H^+(2)\, \hat{\psi}_H(1)] | \Psi \rangle - \langle \Psi_0 | T[\hat{\psi}_{\text{Int}}^+(2)\, \hat{\psi}_{\text{Int}}(1)] | \Psi_0 \rangle \}.$$

Substituting this expression in (15.13), and using (15.5) and (10.6), we find that

$$E - E_0 = \frac{i}{2} \int_0^1 \frac{d\lambda}{\lambda} \int dq_1 \lim_{\substack{2 \to 1 \\ t_2 - t_1 \to +0}} \int d3 \Big\{ \Sigma(1,3)\, G_0(3,2)$$

$$- \lambda \hat{W} \int d4\, G_0(1,3)\, \Sigma(3,4)\, G_0(4,2) \Big\}. \tag{15.14}$$

For a spatially uniform system

$$E - E_0 = \frac{\Omega}{2} \operatorname{Tr}_{\sigma\tau} \int_0^1 \frac{d\lambda}{\lambda} \int d^4p$$

$$\times \{ \Sigma(p)\, G_0(p) - \lambda(\varepsilon_p - p^2/2M)\, G_0^2(p)\, \Sigma(p) \}. \tag{15.14'}$$

Σ depends on λ, and is the self-energy of the S-matrix corresponding to the Hamiltonian $\lambda \hat{H}'$.

The expansion of this expression in perturbation theory as a series in $\hat{H}'$ is done most easily by expanding the expression under the integral sign as a series in λ. In particular the lowest (second)-order perturbation theory gives

$$E - E_0 = \frac{i}{2}\int_0^1 \frac{d\lambda}{\lambda}\int dq_1 \int d3 \lim_{\substack{2\to 1\\ t_2 - t_1 \to +0}} \Sigma_0(1, 3)\, G_0(3, 2), \tag{15.15}$$

where Σ_0 is the self-energy part in the lowest order (see § 14); this quantity is proportional to λ^2.

If we now go over to the energy representation and introduce the quantity $\Sigma_{\nu\mu}(\varepsilon)$, we obtain

$$E - E_0 = \frac{i}{4}\int_C \frac{d\varepsilon}{2\pi}\sum_\nu \frac{\Sigma_{0\nu\nu}(\varepsilon)}{\varepsilon - \varepsilon_\nu + i\delta \operatorname{sign}(\varepsilon_\nu - \varepsilon_F)}.$$

Substituting (14.9) and carrying out the straightforward integration over ε (see Appendix C), we have†

$$E - E_0 = \frac{1}{2}\sum_{\mu_1\mu_2\mu_3\mu_4} n_{\mu_1} n_{\mu_2} (1 - n_{\mu_3})(1 - n_{\mu_4}) \times \frac{|\langle \mu_1\mu_2 | \hat{V}(1 - \hat{\mathscr{P}}) | \mu_3\mu_4\rangle|^2}{\varepsilon_{\mu_1} + \varepsilon_{\mu_2} - \varepsilon_{\mu_3} - \varepsilon_{\mu_4}}. \tag{15.16}$$

This expression is very reminiscent of the lowest term in the usual series expansion for Schrödinger perturbation theory, corresponding to the perturbation Hamilton $\hat{H}_I$ (see § 1). The difference is that, in (15.16), both the intermediate states μ_3 and μ_4 must separately differ from the initial states, whereas, in the usual perturbation theory only the simultaneous coincidence of the states μ_3, μ_4 with μ_1, μ_2 is forbidden (see § 17 for more details).

15.5. Let us consider the correlation energy of a system whose density distribution is nearly uniform. To be accurate, we

† The equality

$$|\langle \mu_1\mu_2 | \hat{V}(1 - \hat{\mathscr{P}}) | \mu_3\mu_4\rangle|^2 = 2\langle \mu_1\mu_2 | \hat{V} | \mu_3\mu_4\rangle \langle \mu_3\mu_4 | \hat{V}(1 - \hat{\mathscr{P}}) | \mu_1\mu_2\rangle$$

may be easily verified.

will suppose that particle motion in the self-consistent field is quasi-classical with

$$\xi = d\lambda/dx \sim d/x_0 \ll 1. \tag{15.17}$$

Here x_0 represents the scale-length of the inhomogeneity, i.e. the distance within which the properties of the self-consistent field change appreciably; $d \sim 1/p_0$ is the average distance between the particles. In the Hartree–Fock approximation d is a unique characteristic parameter with the dimensions of length. The self-consistent field, in practice, only comes into the quantity $p_0^2(x) \sim d(x)^{-2}$, where $d(x)$ is the average distance between particles near the specified point. Hence, if (15.17) is fulfilled, we may consider that the system is uniform in each volume element and make our computations of local properties by using a system of equations which hold for the uniform case, but which refer to the appropriate value of the density. If we go to another volume element, the density changes but the same equations still hold. The system is said to be quasi-uniform, the terms describing local properties depend on the coordinate x as a parameter. Integral quantities, which had the form $\Omega f(\varrho)$ for a uniform system, now change to $\int d^3x\, f[\varrho(x)]$. The accuracy of this change is determined by the quantum effects which are small when (15.17) is fulfilled.

If we go outside the limits of the Hartree–Fock approximation, the requirement (15.17) is insufficient for a system to be considered quasi-uniform. This is because the terms describing the correlation effects necessarily contain, besides d, another scale length—the range of the forces R.

Hence a system may be considered quasi-uniform only if the condition

$$R/x_0 \ll 1 \tag{15.18}$$

is fulfilled, as well as (15.17). In a dilute system with $R \ll d$, (15.18) becomes superfluous. In a dense system, however, with $R \gg d$, (15.18) becomes basic, and, if it is not satisfied, the corresponding expressions have a complicated non-local dependence on the coordinates.

We may take, as a characteristic example, a heavy atom for which (15.17) is obeyed ($\xi \sim Z^{-1/3}$). However, (15.18) is not obeyed, since $R \sim (a_0/p_0)^{1/2} \sim a_0 Z^{-1/3}$, and $x_0 \sim a_0 Z_0^{-1/3}$. This complicates considerably the investigation of correlation effects in inhomogeneous systems.

15.6. A natural way of constructing the S-matrix elements for a nearly uniform system (for which both (15.17) and (15.18) are obeyed) is to write down the corresponding quantities in the coordinate representation, and to use approximate expressions for the Green function (10.11).

We are now, however, interested in the lowest-order correlation correction to the energy, and we will therefore make a more general approach which will help us to demonstrate the assertions of the preceding section. We will use (15.16) for $E - E_0$ in the energy representation, and will rewrite it as

$$E - E_0 = \int dq_1\, dq_2\, dq_3\, dq_4\, V(q_1, q_2)\, V(q_3, q_4)$$

$$\times \sum_{\mu_1 \mu_2 \mu_3 \mu_4} \frac{n_{\mu_1} n_{\mu_2} (1 - n_{\mu_3})(1 - n_{\mu_4})}{\varepsilon_{\mu_1} + \varepsilon_{\mu_2} - \varepsilon_{\mu_3} - \varepsilon_{\mu_4}} \chi^*_{\mu_3}(q_1)\, \chi^*_{\mu_4}(q_2)\, \chi_{\mu_2}(q_2)$$

$$\times \chi_{\mu_1}(q_1) \chi_{\mu_3}(q_3)\, \chi_{\mu_4}(q_4)\, (1 - \hat{\mathscr{P}}_{q_3 q_4})\, \chi^*_{\mu_1}(q_3)\, \chi^*_{\mu_2}(q_4).$$

We have written the expressions for the matrix elements explicitly, and will suppose for simplicity that the potential V is a function of $\boldsymbol{x}_1 - \boldsymbol{x}_2$ only.

It is convenient to put the energy denominator in the form (see Appendix C)†

$$\mathrm{P}(\varepsilon_{\mu_1} + \varepsilon_{\mu_2} - \varepsilon_{\mu_3} - \varepsilon_{\mu_4})^{-1} = \frac{1}{2i} \int_{-\infty}^{\infty} dt\, \mathrm{sign}\,(t)$$

$$\times \exp\,[it(\varepsilon_{\mu_4} + \varepsilon_{\mu_2} - \varepsilon_{\mu_3} - \varepsilon_{\mu_4})].$$

We must obviously deal with quantities of the type:

$$\sigma(q_1, q_2, t) = \sum_{\mu} \chi^*_{\mu}(q_2)\, \chi_{\mu}(q_1)\, n_{\mu} \exp\,(it\varepsilon_{\mu}),$$

$$\sigma'(q_1, q_2, t) = \sum_{\mu} \chi^*_{\mu}(q_2)\, \chi_{\mu}(q_1)\, (1 - n_{\mu}) \exp\,(-it\varepsilon_{\mu}),$$

† Since the factors n_{μ_1}, $(1 - n_{\mu_2})$, etc., occur in the expression for $E - E_0$, the corresponding denominator does not go tozero and the integral must therefore be thought of as a principal value.

in terms of which the quantity we require may be written as:

$$E - E_0 = \frac{1}{2i}\int_{-\infty}^{\infty} dt \operatorname{sign}(t) \int dq_1\, dq_2\, dq_3\, dq_4\, V(q_1, q_2)\, V(q_3, q_4)$$
$$\times\, \sigma'(q_3, q_1, t)\, \sigma'(q_4, q_2, t)\, (1 - \hat{\mathscr{P}}_{q_3 q_4})\, \sigma(q_1, q_3, t)\, \sigma(q_2, q_4, t). \tag{15.19}$$

We will now change to an explicit operator formulation of the problem [81]. We will therefore replace the quantities, ε_μ, n_μ, respectively by

$$\hat{T} + \hat{W}, \quad \theta[\varepsilon_F - (\hat{T} + \hat{W})].$$

The quantities σ and σ' can then be put in the following operator form:

$$\sigma(q_1, q_2, t) = \theta[\varepsilon_F - (\hat{T} + \hat{W})_{q_1}] \exp[it(\hat{T} + \hat{W})_{q_1}] \times \delta(q_1 - q_2),$$

$$\sigma'(q_1, q_2, t) = \theta[(\hat{T} + \hat{W})_{q_1} - \varepsilon_F] \exp[-it(\hat{T} + \hat{W})_{q_1}] \times \delta(q_1 - q_2).$$

Neglecting the exchange effects and the fact that the operators $\hat{T}$ and $\hat{W}$ do not commute, we find

$$\left.\begin{aligned}
\sigma(q_1, q_2, t) &= \int d^3\boldsymbol{p}\, \theta[p_0^2(x_1) - p^2] \\
&\times \exp\left[it\left(\frac{p^2 - p_0^2(x_1)}{2M} - \varepsilon_F\right) + i(\boldsymbol{p}\cdot \boldsymbol{x}_1 - \boldsymbol{x}_2)\right]\delta_{\varrho_1\varrho_2}, \\
\sigma'(q_1, q_2, t) &= \int d^3\boldsymbol{p}\, \theta[p^2 - p_0^2(x_1)] \\
&\times \exp\left[-it\left(\frac{p^2 - p_0^2(x_1)}{2M} - \varepsilon_F\right) + i(\boldsymbol{p}\cdot \boldsymbol{x}_1 - \boldsymbol{x}_2)\right]\delta_{\varrho_1\varrho_2}.
\end{aligned}\right\} \tag{15.20}$$

This representation obviously corresponds to selecting the Green functions in the form of (10.11), (10.12).

If the expressions thus obtained are substituted in (15.19), p_0^2 appears as the function of two different arguments $\boldsymbol{x}_1$ and $\boldsymbol{x}_2$.

The distance between the points $\boldsymbol{x}_1$ and $\boldsymbol{x}_2$ does not exceed the range of the forces R as is indicated by the presence of the factor $V(q_1, q_2)$.† Hence if (15.18) is fulfilled, we can neglect the difference in value of these arguments in p_0^2. If we then carry out a straightforward but tedious integration, we arrive at the relationship

$$E - E_0 = -2M(2\pi)^3 \int d^3\boldsymbol{x} \int d^3\boldsymbol{p}_1\, d^3\boldsymbol{p}_2\, d^3\boldsymbol{k}\, \theta[p_0^2(x) - p_1^2]$$

$$\times \frac{\theta[p_0^2(x) - p_2^2]\,\theta[(\boldsymbol{p}_1 + \boldsymbol{k})^2 - p_0^2(x)]\,\theta[(\boldsymbol{p}_2 - \boldsymbol{k})^2 - p_0^2(x)]}{(\boldsymbol{p}_1 + \boldsymbol{k})^2 + (\boldsymbol{p}_2 - \boldsymbol{k})^2 - p_1^2 - p_2^2}$$

$$\times [g^2|\nu(\boldsymbol{k})|^2 - g\nu^*(\boldsymbol{k})\,\nu(\boldsymbol{p}_2 - \boldsymbol{p}_1 - \boldsymbol{k})], \tag{15.21}$$

where g is the statistical factor; $\nu(\boldsymbol{k})$ is the Fourier transform of the potential.

This expression is indeed quasi-uniform in character. If we wish to go over to the purely uniform case, all that is necessary is to replace $p_0^2(x)$ by p_0^2 and $\int d^3\boldsymbol{x}$ by Ω (we can also use (15.21) here for an arbitrary dispersion relation ε^p):

$$E - E_0 = -2M\,\Omega(2\pi)^3$$

$$\times \int \frac{d^3\boldsymbol{p}_1\, d^3\boldsymbol{p}_2\, d^3\boldsymbol{k}\, n_{p_1} n_{p_2} (1 - n_{p_1+k})(1 - n_{p_2-k})}{(\boldsymbol{p}_1 + \boldsymbol{k})^2 + (\boldsymbol{p}_2 - \boldsymbol{k})^2 - p_1^2 - p_2^2}$$

$$\times [g^2|\nu(\boldsymbol{k})|^2 - g\nu^*(\boldsymbol{k})\,\nu(\boldsymbol{p}_2 - \boldsymbol{p}_1 - \boldsymbol{k})]. \tag{15.22}$$

15.7. If the requirement (15.18) is not fulfilled, two Fermi momenta $p_0^2(x_1)$, $p_0^2(x_2)$, appear in (15.21), and this expression loses its local character. However, a much more fundamental question concerns the validity of the quasi-classical expressions themselves, such as (10.11) and (15.20), when (15.18) is not obeyed. What we must do is to clarify the part played by the quantum effects in the corresponding expressions.

We will use the operator expressions for σ and σ' which were given in the previous section. Expanding the δ-functions in a

† For a Coulomb system, diagrams of higher order must be included; this leads to Debye screening (see § 16).

Fourier integral, we may write the following expression for the characteristic combination in (15.19):†

$$\sigma(q_1, q_3, t)\, \sigma'(q_3, q_1, t) = \delta_{\varrho_1 \varrho_3} \int d^3\boldsymbol{p}_1\, d^3\boldsymbol{p}_2 \exp\,[i(\boldsymbol{p}_1 - \boldsymbol{p}_2) \cdot (\boldsymbol{x}_1 - \boldsymbol{x}_3)]\, \langle \theta(\varepsilon_F - \hat{T} - \hat{W}) \exp\,[it(\hat{T} + \hat{W})] \rangle_{\hat{p}_1} \times \langle \theta(\hat{T} + \hat{W} - \varepsilon_F) \exp\,[-it(\hat{T} + \hat{W})] \rangle_{\hat{p}_2} \qquad (15.23)$$

(the notation $\langle \ldots \rangle_{\hat{p}}$ was introduced in § 4).

We will neglect the exchange effects, for simplicity, since they do not alter the situation qualitively; we can then replace $\hat{T} + \hat{W}$ by $\{[\hat{p}^2 - p_0^2(x)]/2M\} + \varepsilon_F$. The expansion of the functions θ in terms of the commutators of the operators $\hat{p}^2$ and $p_0^2(x)$ might unnecessarily complicate the problem without qualitatively changing the conclusion. To simplify the calculations, we will therefore replace these functions by their quasi-classical expressions, and neglect the difference $\boldsymbol{x}_1 - \boldsymbol{x}_3$ in their arguments (see § 10). This gives

$$\sigma\sigma' = \delta_{\varrho_1 \varrho_3} \int d^3\boldsymbol{p}_1\, d^3\boldsymbol{p}_2 \exp\,[i(\boldsymbol{p}_1 - \boldsymbol{p}_2) \cdot (\boldsymbol{x}_1 - \boldsymbol{x}_3)]\, \theta[p_0^2(x_1) - p_1^2] \times \theta[p_2^2 - p_0^2(x_1)]\, E,$$

where

$$E = \left\langle \exp\left\{\frac{it}{2M}[\hat{\boldsymbol{p}}^2 - p_0^2(x_1)]\right\}\right\rangle_{p_1} \times \left\langle \exp\left\{-\frac{it}{2M}[\hat{\boldsymbol{p}}^2 - p_0^2(x_3)]\right\}\right\rangle_{p_2}.$$

If we expand the exponent as a series in the commutators (see Appendix B), and limit ourselves to first-order terms, we find

$$E = \exp\left\{\frac{it}{2M}[p_1^2 - p_2^2 + p_0^2(x_3) - p_0^2(x_1)]\right\} \times \left\{1 - \frac{t^2}{4M^2}[\nabla^2 p_0^2 + 2i(\boldsymbol{p}_1 + \boldsymbol{p}_2 \cdot \nabla p_0^2)] + \cdots\right\}.$$

† The second term in square brackets of (15.19), which corresponds to the exchange correlation effects, makes little contribution for dense systems.

Expanding this latter exponent as a series in $\boldsymbol{x}_1 - \boldsymbol{x}_3$, and substituting $\boldsymbol{p}_2 \to \boldsymbol{p}_2 - t\nabla p_0^2/2M$, we obtain, after expanding in terms of ∇p_0^2:

$$\sigma\sigma' = \delta_{\varrho_1\varrho_3} \int d^3\boldsymbol{p}_1\, d^3\boldsymbol{p}_2 \exp\left[i(\boldsymbol{p}_1 - \boldsymbol{p}_2)\cdot(\boldsymbol{x}_1 - \boldsymbol{x}_3)\right]$$

$$\times\, \theta[p_0^2(x_1) - p_1^2]\,\theta[p_2^2 - p_0^2(x_1)] \exp\left[\frac{it(p_1^2 - p_2^2)}{2M}\right]$$

$$\times \left\{1 - \frac{t^2}{4M^2}\left[\nabla^2 p_0^2 + 2i(\boldsymbol{p}_1 - \boldsymbol{p}_2\cdot\nabla p_0^2)\right] + \cdots\right\}.$$

We will estimate the contribution of the second (quantum) term in the curly brackets. The difference $\boldsymbol{p}_1 - \boldsymbol{p}_2$ has the same order of magnitude as $|\boldsymbol{x}_1 - \boldsymbol{x}_3|^{-1} \sim 1/R$; the quantities p_1, p_2, themselves, are of the order $p_0 \gg 1/R$. Hence $p_1^2 - p_2^2 \sim p_0/R$, and the quantity $t \sim M/|p_1^2 - p_2^2| \sim MR/p_0$. Hence, with $\nabla^2 p_0^2 \sim p_0^2/x_0^2$, $\nabla p_0^2 \sim p_0^2/x_0$, we obtain

$$\frac{t^2\Delta^2 p_0^2}{M^2} \sim \frac{R^2}{x_0^2}; \qquad \frac{t^2(\boldsymbol{p}_1 - \boldsymbol{p}_2\cdot\nabla p_0^2)}{M} \sim \frac{R}{x_0}.$$

Thus the contribution of the quantum effects is determined by terms of order $(R/x_0)^2$ and R/x_0 respectively. We may conclude that the quasi-classical Green functions (10.11), (10.12), can only be employed so long as the conditions (15.17) and (15.18) are obeyed. The system is then quasi-uniform; we may consider it as homogeneous and replace $\Omega f(\varrho)$ by $\int d^3\boldsymbol{x}\, f[\varrho(x)]$.

This situation may be interpreted physically in the following way. In dense systems the region in momentum space immediately adjacent to the Fermi surface is of fundamental importance: the effective width of this zone $p_1^2 - p_2^2$ $(p_2^2 > p_0^2 > p_1^2)$ is of the order $p_0/R \sim dp_0^2/R$. However, it is specifically in this region near the Fermi surface that the inhomogeneities of the system make an important contribution [57].

This is formally dependent on the fact that a measure of the inhomogeneity is given by the ratio of the average value of the commutators, such as $[\hat{p}^2, p_0^2]_- \sim \nabla^2 p_0^2$, to the corresponding power of $p^2 - p_0^2$, in this particular case $(p^2 - p_0^2)^2$. It is obvious that this ratio is large near the Fermi surface, even for small ξ. For this reason, the part played by inhomogeneity is more substantial in dense systems than in dilute ones.

We can say, in summary, that, when we are considering correlation effects in weakly inhomogeneous, dense systems, the fact that the usual, quasi-classical condition (15.17) is fulfilled does not substantially simplify the problem. Apart from the fact that integral-type expressions (particularly for the energy) are no longer local, there is also the need to include quantum effects.† The problem is only simplified if the requirement (15.18) is satisfied.

16. Selection of the Most Important Diagrams

16.1. As has already been emphasized several times, the exact solution of many-body problems, is in the overwhelming majority of cases, impossible. Success in applying many-body theory to the description of a given object is therefore very greatly dependent on finding the appropriate small parameters, and using them to simplify the problem.

It is from this standpoint that field methods appear most suitable, since they make it possible to go from the Hamiltonian containing the small parameters to the final physical results by the shortest and simplest route. Moreover, it is not simply a matter of expanding in a series of perturbation terms, and keeping those of lowest order. It is also frequently necessary to take into account an infinite number of terms in the perturbation series, for which field methods are far more suitable and justified.

We are therefore faced with the problem of finding the set of diagrams (finite or infinite), which play the most important part for given small parameters.

We must consider dimensionless parameters formed from the quantities which characterize the system of particles. We include here, in the first place, the interaction parameter α which has the same order-of-magnitude value as the ratio of the average inter-

† The original deduction of this result [81] was quantitatively incorrect. Nevertheless, the general qualitative conclusion concerning the fundamental role played by inhomogeneity, when R/x_0 is not small compared with one, is still completely true (see §§ 27, 28).

action energy of a pair of particles to their kinetic energy, and the condensation parameter η which is determined by the ratio of the effective range of the forces to the average distance between the particles. It is also necessary in several cases to include the effective number of particles, parameters characterising the external fields, the relative concentration of the particles (for systems containing different sorts of particle), etc.

To solve a given problem, we must select some quantity characteristic of the system under consideration, and analyse the relative contributions of the different diagrams to it. It is convenient to take the energy of the system as an example, since it is a very important property: the principal diagrams from the point of view of the energy also make the essential contribution to the other physical properties of the system. In this connection it is necessary to note the following. If we consider the expression for the energy in the momentum representation, it is easy to confirm that the main contribution to this expression is due to values of the energy and momentum of the order of the Fermi energy and Fermi momentum, respectively. Hence, our projected programme will not be successful in those (relatively rare) instances when we are especially interested in the values of the energy ε' and the momentum p' far from ε_F and p_0. A special analysis is required in this case, which takes into account the additional parameters $\varepsilon'/\varepsilon_F$ and p'/p_0.

We use (15.22) to investigate the lowest-order correlation energy. An analysis of the higher-order diagrams can be made using the self-energy $\Sigma(\boldsymbol{p}, \varepsilon)$: the most important set of diagrams from the energy standpoint is that which includes the most important contribution to the quantity $\Sigma(p_0, \varepsilon_F)$.

It is more convenient to replace the energy itself by the dimensionless quantity

$$\beta = ME/Np_0^2 \sim ME/\Omega p_0^5. \tag{16.1}$$

Here, the parameter $\beta \sim M\Sigma(p_0, \varepsilon_F)/p_0^2$ for the correlation part of the energy, as may be easily deduced from (15.14′).

16.2. It is expedient to begin by making an estimate of the parameter β for the self-consistent interaction. To some extent, this reproduces the results obtained in §§ 5 to 7.

We will choose β to be the ratio $\beta_{dir} \sim MB/p_0^2$ for direct interactions, and $\beta_{ex} \sim MA/p_0^2$ for exchange interactions. We have from (4.30″), $\beta_{dir} \sim M\varrho \int d^3\xi\, V/p_0^2$. We can therefore write for short-range forces in a dense system

$$\beta_{dir} \sim \alpha\eta^3. \tag{16.2}$$

This can be appreciable for dense systems, even if α is small. The reason for this has already been discussed in §§ 1 and 5. For a dilute system we have simply

$$\beta_{dir} \sim \alpha. \tag{16.2'}$$

A uniform system with Coulomb interactions has

$$\beta_{dir} = 0 \tag{16.3}$$

due to the complete cancellation between the charges of the system and the background (see § 5). If the system is inhomogeneous, then $\int d^3\xi\, V$ "cuts off" at a distance of the order x_1 and

$$\beta_{dir} \sim \frac{x_1^2}{da_0} \sim \alpha(x_1/d)^2.$$

The ratio x_1/d represents the number of particles which are present in a characteristic scale-length of the inhomogeneity in the system. This ratio is usually of the order $Z^{1/3}$, so that

$$\beta_{dir} \sim \alpha Z^{2/3}. \tag{16.4}$$

And in this case β_{dir} can be considerably larger than α.

We will now consider the role of the self-consistent exchange interaction. Starting with (4.33′), we have $A \sim \int d^3\mathbf{p}\, fv \sim \varrho \int d^3\xi\, V \times \exp\,[i(p_0\xi)]$. In dilute systems $p_0\xi \ll 1$, and we come to the same estimate as for β_{dir}:

$$\beta_{ex} \sim \beta_{dir}. \tag{16.5}$$

In dense systems, the exponent $\exp\,(ip_0\xi)$ oscillates violently, and the result is determined by the behaviour of the potential at small ξ. If $V(\xi) \sim V_0(\xi/R)^n$ $(n > -3)$, then $A \sim V_0(d/R)^n$ and

$$\beta_{ex} \sim \alpha/\eta^n. \tag{16.6}$$

In particular, for a square well, where $n = 0$,

$$\beta_{ex} \sim \alpha. \tag{16.7}$$

For a Coloumb system, where $n = -1$, $A \sim e^2/d$, and we have similarly

$$\beta_{ex} \sim \alpha. \tag{16.8}$$

16.3. We will consider the contribution from the terms in perturbation theory, beginning with the lowest-order term. From (15.22) and (16.1), we have

$$\beta \sim \frac{M^2}{p_0^5}\int d^3\boldsymbol{p}_1\, d^3\boldsymbol{p}_2\, d^3\boldsymbol{k}$$

$$\times \frac{n_{p_1}n_{p_2}(1 - n_{p_1+k})(1 - n_{p_2-k})}{(\boldsymbol{p}_1 + \boldsymbol{k})^2 + (\boldsymbol{p}_2 - \boldsymbol{k})^2 - p_1^2 - p_2^2}|v(\boldsymbol{k})|^2. \tag{16.9}$$

We suppose, for simplicity, that V is a function of the coordinates only and will discard the exchange correlation term containing the operator $\hat{\mathscr{P}}$. We also suppose that the dispersion law for the particles is quadratic. These simplifications do not change the order of magnitude of the results.

We will consider first dilute systems with short-range forces. The requirement for a system to be dilute means that the characteristic momentum transfer $k \sim 1/R \gg p_0$, i.e. $k/p_0 \sim \eta^{-1} \gg 1$.

Hence, if we neglect $p_{1,2}$ in comparison with k, we find that

$$\beta \sim M^2 p_0 \int dk\, |v(\boldsymbol{k})|^2 \sim \frac{M^2 p_0^2 V_0^2 R^6}{\eta}.$$

We have assumed here that $v \sim \int d^3\xi\, V \sim V_0 R^3$. Finally, from the definition of α, we obtain

$$\beta \sim \alpha^2/\eta. \tag{16.10}$$

16.4. We will now consider dense systems. We will first analyse the region of integration in (16.9), remembering that, from the requirement for a system to be condensed, $k \ll p_0$. The conditions $p_{1,2} < p_0$ and $|\boldsymbol{p}_{1,2} \pm \boldsymbol{k}| > p_0$ can obviously only be obeyed simultaneously for values of $p_{1,2}$ which are close to p_0. Putting

$p_{1,2} = p_0(1 - \alpha_{1,2})$, we can write (where x is the cosine of the angle between $\boldsymbol{p}$ and $\boldsymbol{k}$)

$$\left.\begin{aligned} kx_1/p_0 > \alpha_1 > 0 \qquad x_1 > 0 \\ -kx_2/p_0 > \alpha_2 > 0 \qquad x_2 < 0. \end{aligned}\right\} \tag{16.11}$$

A reduction in the effective region of integration leads to a diminished contribution from the correlation effects, which affects not only the lowest-order term, but all the remaining perturbation terms as well. Hence, the denser the system, the more accurate is the Hartree–Fock approximation.

This "suppression" of the correlation is a direct result of Pauli's principle [81, 88]. Each pair of interacting particles with momenta which initially lie within the Fermi sphere, must, as a result of interaction, go outside the limits of the sphere, since all the states within it are filled. It is also evident that, since particles can transfer only a small amount of momentum to each other in dense systems, the only particles free to interact are those with momenta immediately adjacent to the Fermi sphere.

We will now make an estimate of the integral (16.9) for a dense system. Taking into account (16.11), we can write

$$\int d^3\boldsymbol{p}_1\, d^3\boldsymbol{p}_2\, [(\boldsymbol{p}_1 + \boldsymbol{k})^2 + (\boldsymbol{p}_2 - \boldsymbol{k})^2 - p_1^2 - p_2^2]^{-1}$$

$$\sim p_0^3 \int_0^1 dx_1 \int_0^{kx_1/p_0} d\alpha_1 \int_{-1}^0 dx_2 \int_{-kx_2/p_0}^0 d\alpha_2\, [k(x_1 - x_2)]^{-1} \sim kp_0^4,$$

whence

$$\beta = \frac{M^2}{p_0^2} \int_0^{p_0} dk\, k^3 |v(\boldsymbol{k})|^2. \tag{16.12}$$

For values of $k \gg 1/R$, $|v(k)|^2$ is usually proportional to $1/k^4$. This is true for the most interesting cases: the square well and the Yukawa potential. To be more accurate, we must write $|v(k)|^2 \sim V_0^2 R^2/k^4$. Hence, if we remember that values of $k \sim p_0 \gg 1/R$ are the important ones in (16.12), we obtain

$$\beta \sim \alpha^2 \eta^2 \ln \eta. \tag{16.13}$$

The presence of the logarithmic term in this expression is characteristic. This is particularly obvious in dense systems with Coulomb interactions. If we substitute $\nu(k) \sim e^2/k^2$ in (16.12), we find

$$\beta \sim \alpha^2 \int_0^{p_0} \frac{dk}{k}, \tag{16.14}$$

i.e. it goes logarithmically to infinity. This difficulty is only important for the lowest-order perturbation theory (the precise solutions of the Schrödinger equation cannot contain any divergences); it will become evident later that, due to the redistribution of particles under the influence of the correlation effects, the Coulomb potential is in fact screened to within the Debye length, and $\nu(k)$ loses its singularity at the point $k = 0$, i.e. the forces lose their long-range character.

It is of interest to enquire how an inhomogeneity in the particle distribution affects the way in which the integral in (16.14) converges. It is obvious that, in general, this difficulty will be smoothed out, since particles, scattered by inhomogeneities, will exchange momenta with them so that, roughly speaking, the particles will depart from the region where momentum transfer is small. This conclusion is indeed supported by the calculations [81]. If the distance within which the particle distribution changes appreciably is less than the Debye length, then the divergence of (16.14) disappears. This is even clearer in a spatially limited system. In this case the momenta have a lower bound, and the integral (16.14) "cuts off" automatically when $k \sim 1/r$, where r is the radius of the system.

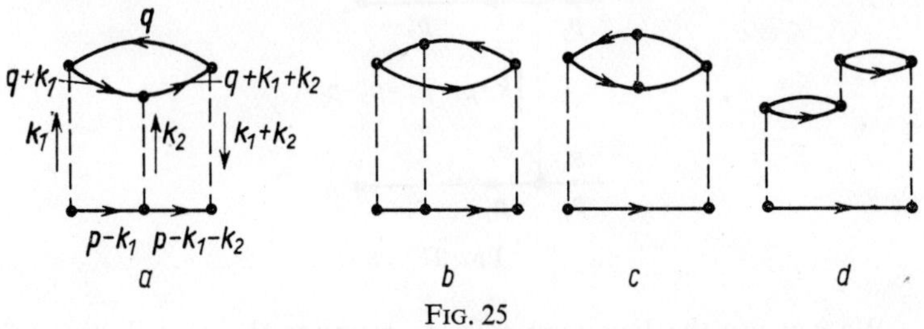

FIG. 25

16.5. We will now evaluate the higher-order diagrams. We will consider third-order diagrams of the self-energy (Fig. 25). In comparison with the second-order diagram for Σ, one interaction line and two particle lines (or a particle and a hole) have been added.

We will show that the diagram in Fig. 25a is of basic importance for dilute systems. This follows from the fact that this diagram has a minimum number of holes (equal to one). Since, so far as the momenta of the holes are concerned, the integration is over small values of p_0 (compared with k), it is evident that the remaining diagrams will give a small contribution (for small η).

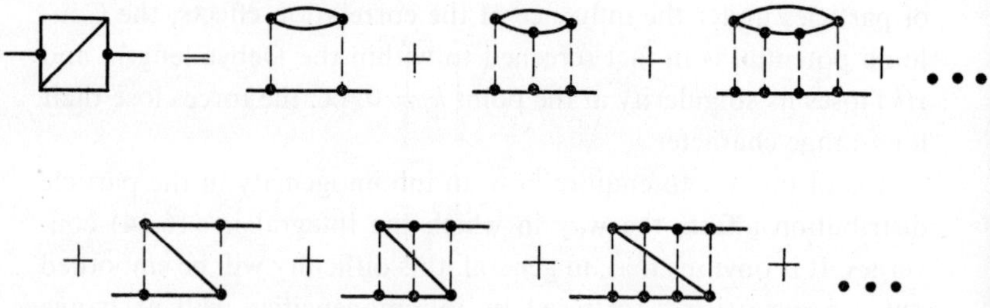

FIG. 26

If we consider higher-order diagrams, then, from a similar reasoning, diagrams with a minimum number of holes will play the most important part. Thus, in dilute systems, it is sufficient to include only the following infinite set of diagrams for the self-energy Σ (Fig. 26). The general property of these diagrams is that all the interaction lines connect the same pair of particle lines.

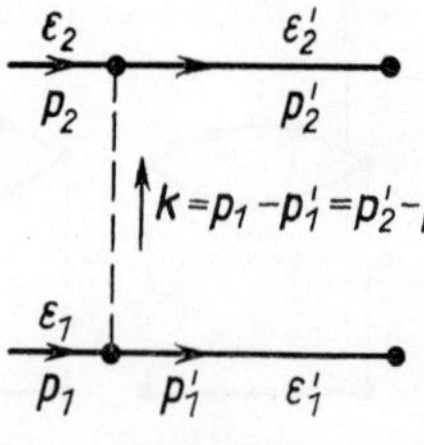

FIG. 27

We can use the Feynman rules to estimate the contribution of each of these diagrams. In the transition from the nth-order to the $(n+1)$th-order diagram, the additional components shown in

Fig. 27 are added. According to the Feynman rules, we must also introduce the following combination into the expression for Σ:

$$Q_1 \sim \int d^4p_1'\, d^4p_2'\, \delta^4(p_1 + p_2 - p_1' - p_2')\, v(\boldsymbol{p}_1 - \boldsymbol{p}_1') \times G_0(\boldsymbol{p}_1', \varepsilon_1')\, G_0(\boldsymbol{p}_2', \varepsilon_2').$$

Integrating in terms of ε_1' and ε_2', we find

$$Q_1 \sim M \int \frac{d^3\boldsymbol{p}_1'\, d^3\boldsymbol{p}_2'\, v(\boldsymbol{p}_1 - \boldsymbol{p}_1')}{p_1^2 + p_2^2 - p_1'^2 - p_2'^2} (1 - n_{p_1'})(1 - n_{p_2'}) \times \delta(\boldsymbol{p}_1 + \boldsymbol{p}_2 - \boldsymbol{p}_1' - \boldsymbol{p}_2'). \qquad (16.15)$$

Changing to the variable $\boldsymbol{k} = \boldsymbol{p}_1 - \boldsymbol{p}_1'$ and remembering that $k \gg p_0$, we find

$$Q_1 \sim M \int_{p_0}^{\infty} d^3\boldsymbol{k}\, \frac{v(k)}{k^2} \sim \alpha/\eta \sim \alpha',$$

where α' is the parameter in the Born expansion. Thus the contribution from the diagrams in which we are interested is of the form

$$\beta \sim \alpha \sum_{n=1}^{\infty} C_n(\alpha')^n, \qquad (16.16)$$

where the C_n are certain numerical coefficients. The contributions from the discarded diagrams are a factor η^{-1} smaller for each order of perturbation theory.

We must now go into the physical interpretation of the results we have obtained. It is evident from Fig. 26 that one needs to consider the correlations only for each of a pair of particles, since all the interaction lines connect two particle lines. In such a case, we speak of pair correlation of the particles. This has a perfectly clear physical interpretation. Owing to the smallness of the parameter η, the probability of three, or more particles coming within the sphere of action of the forces (and this is an essential requirement for triple, or more complicated correlations) is very small.

16.6. Diagrams of the type illustrated in Fig. 25d are the most important in dense systems. In a similar way, the diagrams in

Fig. 28 contribute the most amongst the higher-order diagrams. A general feature of these diagrams is the presence of a maximum number of closed loops, as a result of which all the momentum transfers are equal.

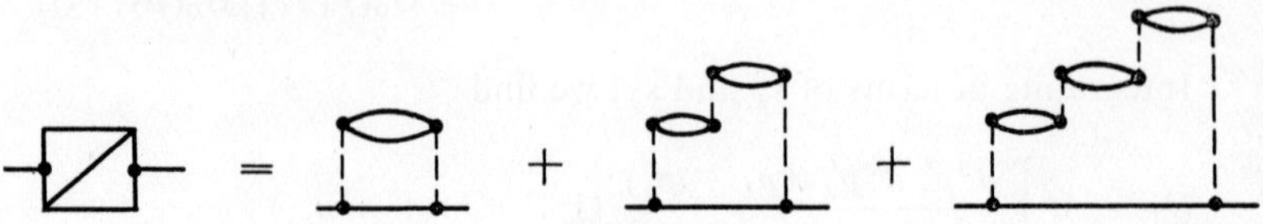

FIG. 28

The special nature of these diagrams is explained by the fact that the momentum transfer k is small compared with the Fermi momentum p_0 in dense systems. The Fourier transform $\nu(k)$ decreases as k increases: this decrease is of the form $\nu(k) \sim k^{-2}$ for $k \gg 1/R$ and, in particular, for $k \sim p_0$ (see § 16.4). Hence the largest contribution is given by diagrams with the smallest value for the momentum transfer. The diagrams in Fig. 28 have this property: all the momentum transfers are the same, and are equal to k. On the other hand, the diagram in Fig. 25b, for example, has one of the transmitted momenta of the order $p_0 \gg k$.

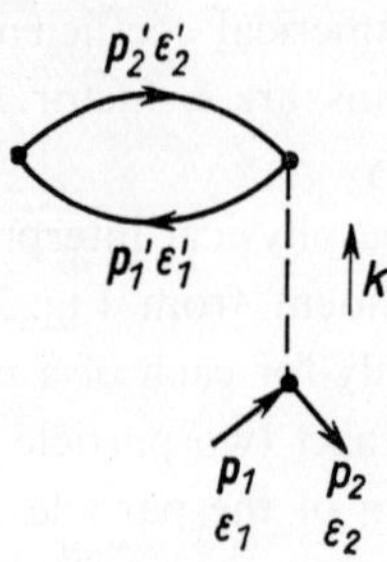

FIG. 29

We will make a quantitative comparison of the contributions from the diagrams in Fig. 28. In the transition from the nth to the $(n + 1)$th diagram, the component shown in Fig. 29 is added. The corresponding expression, after integration over the energy, has the form

$$Q_2 \sim M \int d^3p_1' \, d^3p_2' \frac{v(p_1' - p_2') \, n_{p_1'} (1 - n_{p_2'})}{p_1^2 + p_2'^2 - p_2^2 - p_1'^2} \times \delta(p_1 + p_2' - p_2 - p_1'). \tag{16.17}$$

Introducing the variable $\mathbf{k} = \mathbf{p}_1' - \mathbf{p}_2' = \mathbf{p}_1 - \mathbf{p}_2$, and proceeding as in § 16.4, with $k \ll p_0$, we obtain

$$Q_2 \sim Mp_0 v(k).$$

Diagrams of this type therefore give the following overall contribution

$$\beta \sim \frac{M^2}{p_0^2} \int_0^{p_0} dk \, k^3 |v(k)|^2 f[Mp_0 v(k)], \tag{16.18}$$

where $f(x) = \sum_{n=0}^{\infty} C_n x^n$ (the C_n are certain coefficients); an explicit expression for $f(x)$ will be presented in Chapter 4. Jumping ahead a little, we may note that when $x \to \infty$, $f(x) \to x^{-1}$, and when $x \to 0$, $f(x) \to$ const.

We will now introduce a quantity k_0 defined by the quality

$$M^2 p_0^2 |v(k_0)|^2 \sim 1. \tag{16.19}$$

We must distinguish here between certain cases. The function $|v(k)|^2$ generally decreases as k increases. Hence if $M^2 p_0^2 |v(1/R)|^2 \ll 1$, the equality (16.19) cannot in general hold, so this inequality will be true always. In this case the sum in the expression for $f(x)$ effectively reduces to its first term. Therefore, when

$$\alpha\eta^3 \ll 1,$$

we need only consider the first correlation correction, which leads us to the results in § 16.4.

The second highly important case occurs when k_0, defined by (16.19), lies between $1/R$ and p_0. Since $v^2(k) \sim V_0^2 R^2 k^4$, we have

$$k_0 \sim (MV_0 p_0 R)^{1/2} \sim (\alpha\eta)^{1/2} p_0. \tag{16.20}$$

The requirement $1/R \ll k_0 \ll p_0$ takes the form

$$\alpha\eta^3 \gg 1; \quad \alpha\eta \ll 1.$$

We will estimate the value of β in this case, remembering that in the region $0 < k < 1/R$, $v(k) \sim V_0 R^3$, $Mp_0 v \sim \alpha\eta^3 \gg 1$, $f \sim 1/(\alpha\eta^3)$; in the region $1/R < k < k_0$, $v(k) \sim V_0 R/k^2$, $Mp_0 v \sim k_0^2/k^2$, $f \sim k^2/k_0^2$; in the region $k_0 < k < p_0$, $Mp_0 v \sim k_0^2/k^2$, $f \sim 1$. The contribution of the first region to the integral (16.18) is $\sim\alpha/\eta$, the second is $\sim\alpha^2\eta^2$, and the third $\sim\alpha^2\eta^2 \ln(\alpha\eta)$. The most important is obviously the contribution by the region $k_0 < k < p_0$, which is

$$\beta \sim \alpha^2\eta^2 \ln(\alpha\eta). \tag{16.21}$$

This is small compared with unity.

In principle it is also possible to have

$$k_0 > p_0; \quad \alpha\eta > 1. \tag{16.22}$$

However, all our previous arguments then lose their force since they are based on k being small compared with p_0. In particular all perturbation theory diagrams must be taken into account. This signifies physically (see § 27) that the potential V, screened by the correlation interactions, ceases to be long range (as compared with the distance between the particles d). In other words the system with (16.22) holding does not belong to the class of dense systems.

16.7. We will next consider dense Coulomb systems. For them, $v(k) \sim e^2/k^2$ and $Mp_0 v \sim p_0/a_0 k^2$ for all values of k. Splitting the integral (16.18) into two regions, $k < k_0$ and $k > k_0$, where, from (16.19)

$$k_0 \sim (p_0/a_0)^{1/2}, \tag{16.23}$$

we have in the first region $Mp_0 v \sim k_0^2/k^2 \gg 1, f \sim k^2/k_0^2$ and

$$\beta_1 \sim (a_0 p_0)^{-2} \sim \alpha^2.$$

We see that the divergence for small k has indeed disappeared.

In the second region $Mp_0 v \ll 1$ and $f \sim 1$, whence

$$\beta_2 \sim \alpha^2 \ln(p_0/k_0).$$

Hence, the important contribution is given by the region $k_0 < k < p_0$ and

$$\beta \sim \alpha^2 \ln(1/\alpha). \tag{16.24}$$

It can be seen that the situation is very reminiscent of that we considered in the previous section. This is hardly surprising: a dense system with short-range forces has a range of these forces which is large compared with the distance between the particles and this makes it approximately like a Coulomb system.

The quantity k_0 is called the Debye momentum, the inverse

$$R_0 \sim (a_0/p_0)^{1/2} \tag{16.25}$$

is called the Debye screening length.

We will now return to (16.18), and expand the function f formally as a series in terms of its argument. The nth-order term of the expansion gives a contribution

$$\beta_n \sim \frac{p_0^{n-2}}{a_0^{n+2}} \int \frac{dk}{k^{2n+1}} \sim \frac{p_0^{n-2}}{a_0^{n+2} k_{\text{eff}}^{2n}}; \quad \beta_0 \sim (a_0 p_0)^{-2} \ln (p_0/k_{\text{eff}}).$$

Here k_{eff} is the effective lower limit in the integration over k. This evidently coincides with the Debye momentum k_0: $k_0 \sim (p_0/a_0)^{1/2}$. If this is substituted in the expression for β_n, it can be seen that all the terms in the expansion of β introduce an equal contribution. It is a necessary requirement that the divergences of the β_n compensate each other, that is, that an effective "cut-off" appears in the integral (16. 18).

Instances are possible when this integral actually "cuts-off" for some other reason. This is true, for example, of a spatially bounded system. The momentum transfers $k < 1/r$ (where r is the size of the system) are suppressed; k_{eff} is therefore equal to the larger of the quantities $1/r$ and k_0. If the dimensions of the system are smaller than the Debye screening length R, i.e. $1/r \gg k_0$, then $\beta_n \sim \alpha^2 (k_0 r)^{2n}$ and the term with $n = 0$ is of most importance. Hence, if the condition

$$\eta \sim (a_0/d)^{1/2} \gg N^{1/3} \tag{16.26}$$

is satisfied (where N is the number of particles in the system: $r \sim dN^{1/3}$), the lowest-order diagram is of basic importance, and it provides a contribution

$$\beta \sim \alpha^2 \ln N. \tag{16.27}$$

The contribution from the nth-order diagram is $\beta_n \sim \alpha^2 (N^{1/3}/\eta)^n$.

As distinct from dilute systems, for which pair correlation between particles is of fundamental importance, we now have multiparticle, collective correlations dominant. This is evident from the diagram in Fig. 28: in each order of perturbation theory, a maximum number of particles interact. This is explained by the large value for the range of the forces as compared with the distance between the particles.

16.8. We will now return to dilute systems. We have seen in § 16.5 which are the principal diagrams for such systems and what order of magnitude the correlation effects have.

In practice it is more convenient to take another approach: to replace the true interaction potential by the pseudopotential (introduced in § 1) right at the start†

$$\hat{V}_{\text{eff}}(\boldsymbol{r}) = \frac{4\pi l}{M}\delta(\boldsymbol{r})\left(1 + r\frac{\partial}{\partial r}\right) \tag{16.28}$$

and to select the principal diagrams corresponding to this pseudopotential. This approach, besides reducing considerably the amount of work, also makes it possible to treat cases for which the analysis of § 16.5 is incorrect.

The interaction parameter α, for the potential (16.28), is equal to

$$\alpha \sim l/d \sim lp_0 \tag{16.29}$$

from the results of § 1. The condensation parameter is in this case equal to zero, and the system is described by the single parameter α.

We will now consider the diagrams corresponding to the potential (16.28), using eqns. (16.9), (16.15) and (16.17). Particular care is required here, however.

We will rewrite eqn. (16.9) in the following form:

$$\beta \sim \frac{M^2}{p_0^5}\int d^3\boldsymbol{p}_1\, d^3\boldsymbol{p}_2 n_{p_1} n_{p_2} \int d^3\boldsymbol{r}_1\, d^3\boldsymbol{r}_2 V_{\text{eff}}(r_1)\, V_{\text{eff}}(r_2)\, I(\boldsymbol{r}_1 - \boldsymbol{r}_2),$$

† Here l is the scattering length, connected with the scattering amplitude by the relation

$$a = -l/(1 + ikl).$$

A summation over the diagrams singled out in § 16.5 leads, in fact, to an immediate replacement of the true potential by the pseudopotential (16.28). This will be demonstrated in § 26.

where

$$I(\boldsymbol{r}) = \int d^3\boldsymbol{k} \exp\left[-i(\boldsymbol{k}\cdot\boldsymbol{r})\right] \frac{(1 - n_{\boldsymbol{p}_1+\boldsymbol{k}})(1 - n_{\boldsymbol{p}_2-\boldsymbol{k}})}{(\boldsymbol{p}_1 + \boldsymbol{k})^2 + (\boldsymbol{p}_2 - \boldsymbol{k})^2 - p_1^2 - p_2^2}.$$

Here we have replaced the Fourier transform by the potential itself. We will consider the integral

$$I_0(\boldsymbol{r}) = \int \frac{d^3\boldsymbol{k} \exp\left[-i(\boldsymbol{k}\cdot\boldsymbol{r})\right]}{(\boldsymbol{p}_1 + \boldsymbol{k})^2 + (\boldsymbol{p}_2 - \boldsymbol{k})^2 - p_1^2 - p_2^2}$$

for small r. A straightforward analysis shows that I_0 consists of two terms which do not disappear as $r \to 0$; one has a singularity $1/r$, and the other is proportional to $(\boldsymbol{p}_1 - \boldsymbol{p}_2 \cdot \boldsymbol{r})/r$.

If we now substitute I_0 for I in the expression for β, we obtain a zero result. This is connected with the particular form of the potential $\hat{V}_{\text{eff}}$ which, as has already been remarked in § 1, eliminates terms of the order $1/r$. The second term of I_0 disappears after integrating over the angles. We may therefore simply replace I by $I - I_0$. But this difference is a regular function of $|\boldsymbol{r}_1 - \boldsymbol{r}_2|$, which permits us to omit the operator $1 + r\partial/\partial r$ in the expression for V_{eff}. We have finally

$$\beta \sim \frac{l^2}{p_0^5} \int d^3\boldsymbol{p}_1\, d^3\boldsymbol{p}_2\, d^3\boldsymbol{k}\, \frac{n_{\boldsymbol{p}_1} n_{\boldsymbol{p}_2}[(1 - n_{\boldsymbol{p}_1+\boldsymbol{k}})(1 - n_{\boldsymbol{p}_2-\boldsymbol{k}}) - 1]}{(\boldsymbol{p}_1 + \boldsymbol{k})^2 + (\boldsymbol{p}_2 - \boldsymbol{k})^2 - p_1^2 - p_2^2}.$$

Characteristic values of $p_{1,2}$ and k in this integral are of the order p_0, so we finally have for the lowest-order perturbation theory

$$\beta \sim l^2 p_0^2 \sim \alpha^2. \tag{16.30}$$

We will estimate the magnitude of the integral in (16.15): it can be put in the form

$$Q_1 \sim M \int_{p_0}^{\infty} \frac{d^3k}{k^2} v(k) \sim \int d^3\boldsymbol{r}\, l\delta(\boldsymbol{r}) \left(1 + r\frac{\partial}{\partial r}\right) \int_{p_0}^{\infty} \frac{d^3\boldsymbol{k}}{k^2} \left[\exp -i(\boldsymbol{k}\cdot\boldsymbol{r})\right].$$

The integral over k in this can be written as $1/r - (2/\pi)\, p_0 + O(r)$ for small r. From the properties of the operator $1 + r\partial/\partial r$, we find

$$Q_1 \sim p_0 l \sim \alpha. \tag{16.31}$$

We can estimate the integral in (16.17) in a similar way. In the region $k < p_0$, which is of chief importance,

$$Q_2 \sim Mp_0v \sim p_0l \sim \alpha \tag{16.32}$$

(when $k \gg p_0$, $Q_2 \sim \alpha(p_0^2/k^2)$).

We thus conclude that the pseudopotential does not select preferred diagrams. All the diagrams of a given order in perturbation theory introduce the same contribution. This result is hardly surprising, since there is only one dimensionless parameter: α.

We see, moreover, that for $\alpha \ll 1$ and the scattering length and amplitude small compared with the distance between the particles we can limit ourselves to the lowest-order diagram. This leads to the same results as were obtained in summing the diagrams in § 16.5.

The situation is extremely complicated when $\alpha \gtrsim 1$, with $d/R \gg 1$ and $l/d \gtrsim 1$. This corresponds to resonance, which was examined in § 1. Here all the perturbation theory diagrams are essential, and this makes the solution of the problem very difficult.†

The results obtained in this section (see Table 1) are important in the solution of specific problems in many-body theory and will be used in the following to deduce approximate quantitative relations. In those instances where it is not sufficiently obvious that the parameters characterizing the system are small, it is impossible to limit consideration only to the appropriate principal diagrams. However, the summation of these diagrams provides a convenient initial approximation on the basis of which better results may be obtained. In the next stage one can again limit oneself to a particular subgroup of diagrams. Thus, for dilute systems one can consider diagrams with one line representing a hole; for dense systems, there are the diagrams in which one of the interaction lines has a momentum transfer of the order p_0.

We must emphasize in conclusion that the above analysis is only immediately applicable to systems with repulsive forces which are

† The fact that the analysis of § 16.5 is inapplicable in this case is connected with the divergence (or poor convergence) of those series which represent the sum of the contributions from the various perturbation theory diagrams corresponding to the true interaction potential.

kept in equilibrium by external constraints. The presence of attractive forces necessitates a special investigation of the stability of the system against "bunching" of the constituent particles.

We should note, first of all, the possibility of an absolute type of instability that leads to the collapse of the system (see § 7.7). This effect is connected with the predominance of the effective energy of attraction over the kinetic energy, and appears already in the Hartree–Fock approximation. This type of instability disappears so long as there are additional, sufficiently strong, repulsive forces (the model considered in §§ 7 and 18), or if the phase volume of the attractive forces is small (the model for superconductivity in metals).

Even if these conditions are satisfied, there may be weaker instabilities which do not lead to so radical a change in the bulk properties of the system. Owing to the correlation part of the attractive interaction, bound complexes can be formed (the Cooper pairs in superconductivity), as a result of which the "normal" state of a system becomes unstable. The resultant rearrangement of this state, which has little effect on the overall energy of the system, is an essential element in explaining numerous phenomena—including superconductivity.

17. Application to the Theory of Two-Electron Atoms

17.1. We will look at the simplest many-body system—the two-electron atom or ion (He, Li^{+}, Be^{++}, B^{+++}, etc.). When the experimental data on the energy of the electron shell are compared with the results of the appropriate calculations in the Hartree–Fock approximation, a certain difference is observed which may be put down to correlation effects.

We will not consider the methods used in spectroscopy to calculate this difference (there are, in particular, methods whereby the configurations are coupled and the variables are not completely separated), but we will demonstrate how the perturbation theory developed in the previous sections can be used to this end [89].

We will consider the general member of the isoelectronic sequence in which we are interested (number of electrons $N = 2$, nuclear charge Z), and will estimate the value of the condensation and interaction parameters. The average distance between the particles is of the order a_0/Z and from (1.27) and (1.28)

$$\eta \sim Z^{1/2}; \quad \alpha \sim Z^{-1}. \tag{17.1}$$

Thus if Z is large, this is a dense system with weak interaction. A general property of such systems, as we have remarked in § 16, is that the correlation effects contribute little. This conclusion applies in fact to the case $Z = 2$, which is evidently explained by the favourable values of the corresponding numerical coefficients.

This system is one of a class of bounded systems which satisfy the condition (16.26): $d/a_0 \sim Z^{-1} \ll N^{-2/3}$, for large Z. We can therefore limit ourselves to the lowest-order diagram: the contribution from diagrams of the nth order is $\sim Z^{-n}$.

17.2. We will start from eqn. (15.16), and will limit ourselves to considering the atom in the ground state, for which $\chi_{\mu_1} = \delta_{\sigma,1/2} \times \chi_0(\boldsymbol{r})$; $\chi_{\mu_2} = \delta_{\sigma,-1/2}\chi_0(\boldsymbol{r})$. The exchange terms disappear identically, since the particles are in states with opposite spins. We then obtain

$$E - E_0 = -\sum_{\substack{n,\, m \\ n \neq 0,\, m \neq 0}} \frac{|\langle 0, 0|e^2/r|m, n\rangle|^2}{\varepsilon_m + \varepsilon_n - 2\varepsilon_0}, \tag{17.2}$$

where

$$\langle i, j|e^2/r|m, n\rangle = e^2 \int \frac{d^3x\, d^3x'}{|\boldsymbol{x} - \boldsymbol{x}'|} \chi_i^*(x)\, \chi_j^*(x')\, \chi_m(x)\, \chi_n(x').$$

The sum with both of the states $n = 0$ and $m = 0$ omitted can be written in the form

$$\sum_{\substack{n \neq 0 \\ m \neq 0}} = \sum_{m,n}{}' - \sum_{\substack{n \\ m = 0}} - \sum_{\substack{m \\ n = 0}},$$

where the dash denotes that only the states with $m = n = 0$ are omitted. Then

$$E - E_0 = E_{\text{Coul}} - 2E_{\text{Self}}, \tag{17.3}$$

where

$$E_{\text{Coul}} = -\sum_{n,m}{}' \frac{|\langle 0, 0|e^2/r|m, n\rangle|^2}{\varepsilon_m + \varepsilon_n - 2\varepsilon_0}$$

is the energy correction due purely to the Coulomb interaction according to the usual perturbation theory;

$$E_{\text{Self}} = -\sum_{n \neq 0} \frac{|\langle 0|B|n\rangle|^2}{\varepsilon_n - \varepsilon_0} \tag{17.4}$$

is the energy correction due to the self-consistent potential

$$B(r) = e^2 \int \frac{d^3\boldsymbol{x}'}{|\boldsymbol{x} - \boldsymbol{x}'|} |\chi_0(x')|^2 .$$

It follows from these relations that

$$E_{\text{Coul}} < 0, \quad E_{\text{Self}} < 0; \quad E_{\text{Coul}} < E - E_0 < 0, \tag{17.5}$$

which confirms the well-known result, obtained from the variational principle, that the correlation energy has a negative sign.

17.3. We may, with sufficient accuracy, replace the wave functions in the Hartree approximation by those in the hydrogen-like approximation, where a specially selected screening of the nucleus is substituted for the self-consistent interaction between the electrons. These wave functions are eigenfunctions for the single-electron problem in the field of a nucleus of charge $Z^* = Z - 5/16$ [38], with

$$\chi_0(r) = \frac{1}{\sqrt{\pi}} (Z^*/a_0)^{3/2} \exp(-Z^* r/a_0), \quad \varepsilon_0 = -\frac{(Z^*)^2}{2} e^2/a_0.$$

So far as the excited states are concerned

$$\chi_n(r) = (Z^*/a_0)^{3/2} f_n(Z^* r/a_0), \quad \varepsilon_n \sim (Z^*)^2 e^2/a_0,$$

where f_n is some known function. We will not require any more accurate expressions.

Making the substitution $r \to r/Z^*$, we can easily confirm that the matrix element $\langle i, j|e^2/r|m, n\rangle$ is proportional to Z^*, and the energy denominator to $(Z^*)^2$. Hence, all the quantities $E - E_0$, E_{Coul} and E_{Self} are independent of Z^* and, consequently, of Z. Thus the correlation correction to the energy is the same for all terms of the isoelectronic sequence. This has been established previously by a semi-empirical method [90].

Moreover, it follows from the fact that E_{Coul} is independent of Z^*, that this quantity is equal to the correction to the Coulomb interaction between two electrons which are situated only in the field of the nucleus, i.e. which have the zero-approximation wave function $(1/\pi)^{\frac{1}{2}}(Z/a_0)^{3/2}\exp(-Zr/a_0)$. This correction has been calculated by Hylleraas [91] and is

$$E_{\text{Coul}} = -0{\cdot}1574e^2/a_0. \tag{17.6}$$

17.4. Before calculating E_{Self}, we must introduce several important relations.

We will suppose that we are interested in the second-order correction to the energy, and the first-order correction to the wave function in perturbation theory, where the perturbing function $V(r)$ has radial symmetry:

$$[-\nabla^2/2M + V_0(r) + V(r) - E]\,\chi(r) = 0.$$

Introducing $\chi(r) = [1 + \varkappa(r)]\,\chi_0(r)$, where χ_0 is the solution of the unperturbed equation with energy $E_0(\varkappa \ll 1)$, we obtain

$$\varkappa'' + 2(\chi_0'/\chi_0 + 1/r)\varkappa' = 2M[V(r) - \langle V\rangle],$$

where we have introduced the notation $\langle\ldots\rangle \equiv \int d^3r(\ldots)\,\chi_0^2(r)$. In particular, from the normalization condition, $\langle\varkappa\rangle = 0$.

The above equation can be easily solved, and gives

$$\left.\begin{aligned}\varkappa(r) &= \int_0^r \Phi(\xi)\,D^{-1}(\xi)\,d\xi - \int_0^\infty dr D(r)\int_0^r \Phi(\xi)D^{-1}(\xi)\,d\xi,\\ E - E_0 &= -\frac{1}{2}\int_0^\infty dr\,\Phi^2(r)\,D^{-1}(r),\\ \text{where}\quad \Phi(r) &= 2M\int_0^r d\xi\,D(\xi)\,[V(\xi) - \langle V\rangle]\end{aligned}\right\} \tag{17.7}$$

and the radial density $D(r) = 4\pi r^2\chi_0(r)^2$.

Thus, for a radial perturbing function,† the wave function and energy can be determined by straightforward quadrature [89, 92].

† The results are, in fact, applicable to the ground state of the system, so long as we suppose that $\chi_0^* = \chi_0$ and that χ_0 is independent of angle.

We will apply these relations to the determination of E_{Self}. The perturbation will be provided by

$$V(r) = B(r) = \frac{Z^* e^2}{a_0 \zeta} [1 - \exp(-2\zeta)(1 + \zeta)],$$

where $\zeta \equiv Z^* r/a_0$. The function Φ is equal to

$$\frac{Z^* e^2}{4a_0} \exp(-2\zeta) [10\zeta^2 - 6\zeta - 3 + \exp(-2\zeta)(8\zeta^2 + 12\zeta + 3)]$$

and

$$E_{\text{Self}} = -\frac{26 + 243 \ln(4/3)}{1728} e^2/a_0 = -0{\cdot}0555 e^2/a_0. \quad (17.8)$$

17.5. From eqns. (17.3), (17.6) and (17.8), it is easy to obtain the following estimate for the correlation energy of a two-electron atom:

$$E - E_0 = -0{\cdot}046 e^2/a_0. \quad (17.9)$$

This value, which is independent of Z, is in satisfactory agreement with the results obtained from comparing the values computed in the Hartree approximation with the experimental data. The corresponding difference varies from $-0{\cdot}042$ to $-0{\cdot}054 e^2/a_0$ as Z goes from 2 to 6. The residual deviation can be explained partly by the inaccuracy of the hydrogen-like approximation, and partly by the influence of the higher-order diagrams.

The surprisingly small value for the correlation correction deserves special notice. The kinetic energy of the electrons, $\overline{T}$ (and also the energy in the field of the nucleus $\overline{U}$) in a two-electron atom is, on the average, of the order Z^2; the self-consistent interaction energy $\sim Z$; the correlation energy, as has been established above, is completely independent of Z. Hence, for large values of Z, the situation is quite comprehensible—it can be explained simply by the fact that the system is dense. It is very surprising that the correlation energy is small for helium ($Z = 2$). In this case $\overline{T} + \overline{U}$ is $-4e^2/a_0$, the self-consistent part of the energy is about $+e^2/a_0$ (the positive sign is connected with the electron repulsion), whilst the correlation energy is $-0{\cdot}046 e^2/a_0$. Hence, when there are no obvious small parameters, the correlation and the self-consistent parts of the energy differ by almost two orders of magnitude.

17.6. A peculiarity of this example is that, under the influence of the strong external nuclear field, the electron system is sufficiently condensed so that not only the distance between the particles, but even the overall dimensions of the system are smaller than the Debye screening length. It is this fact that made it possible to restrict consideration to the lowest-order diagram.

Another example of the same sort is a multi-electron atom ($N \gg 1$), condensed by external forces, or an ion with $N \ll Z$, condensed by the Coulomb field of the nucleus.

In this first case, when $N = Z \gg 1$, the requirement (16.26) imposes the following limitations on the radius of the atom $r \sim dZ^{1/3}$:

$$r \ll a_0/Z^{1/3} \tag{17.10}$$

and on the density

$$\varrho \gg Z^2/a_0^3 . \tag{17.11}$$

The equivalent pressures evidently correspond to regions II and III (see § 6). The correlation energy of such an atom is of the order

$$E - E_0 \sim (e^2/a_0)\, Z \ln Z \tag{17.12}$$

and is independent of the density of the system.

However, it is impossible to use the cell model here for calculating the correlation contribution to the pressure: since $R_0 \gg r$ (where r is the cell radius and R_0 the Debye length), the correlation binding of the electrons in a given cell with their neighbours is not included. Hence a compressed isolated atom in regions II and III by no means imitates a real compressed substance.

The condensed ion provides another example. We have from (16.26)

$$Z - N \gg N^{1/3} . \tag{17.13}$$

We have allowed here for the fact that the size of the ion r is of the order $a_0/(Z - N)$ (where $Z - N$ is the effective nuclear charge near the periphery of the ion). The correlation energy of the ion is of the order

$$E - E_0 \sim \frac{e^2}{a_0} N \ln N . \tag{17.14}$$

It is independent of the nuclear charge Z.

18. Application to the Theory of the Atomic Nucleus

18.1. It was noted in § 7, when we were considering nuclear matter in the Hartree–Fock approximation, that it is impossible to obtain a stable state of the nucleus for finite values of the density in this approximation. This is connected with the incomplete incorporation of the repulsive forces between the nuclei: in fact nucleons which interact like solid spheres can not approach nearer to each other than their diameter. It appears that a sufficiently complete account of this effect could be made within the framework of perturbation theory using only a small number of the corresponding terms in this expansion.

As has been pointed out in § 7, nuclear matter can, with sufficient accuracy, be described by the following model: nucleons, which have their dispersion law changed when long-range attractive forces, which are taken into account in the Hartree–Fock approximation, are considered, are bound only by repulsive forces of the hard-core type. The justification for this model from the point of view of the interaction potential (1.11) will be given in the next chapter.

The Hamiltonian for a system corresponding to this model can be written as

$$\hat{H} = \int dq\, \hat{\psi}^+(q)\,(\hat{T} + \hat{W})\,\hat{\psi}(q) + C + \tfrac{1}{2}\int dq\, dq'\, \hat{\psi}^+(q)\,\hat{\psi}^+(q') \times V_{(c)}\hat{\psi}(q')\,\hat{\psi}(q), \tag{18.1}$$

where $\hat{T} + \hat{W}$ is given by (7.18) with the momentum $\boldsymbol{p}$ replaced by the operator $\hat{\boldsymbol{p}}$, and the last term in (7.20) subtracted. We may write symbolically

$$\hat{T} + \hat{W} = \int_0^{\hat{p}} \frac{p\, dp}{M_0(p)} - \frac{\pi}{12M(a-c)^2}[y^3 + 9\mathrm{Si}y - 9\sin y], \tag{18.2}$$

where M_0 is given by (7.19); $y = ap_0$. The second term on the right-hand side of (18.2), together with C, acts as the potential of the external field; this is a general attractive potential well in which the nucleons are situated.

The potential $V_{(c)}$, can be replaced by the pseudopotential which, from eqns. (1.17) to (1.19) and (7.16) has the form of a sum of two components. The first, of order c, takes the form

$$\hat{V}_{\text{eff}} = \frac{4\pi c}{M_{\text{eff}}} \delta(\boldsymbol{r}) \left(1 + r\frac{\partial}{\partial r} + \beta r\right). \tag{18.3}$$

The second is made up of the corresponding term for the p-scattering

$$\delta \hat{V}_{\text{eff}(1)} = -\frac{4\pi c^3}{M}\left[\sum_{\alpha} \nabla_a \,\delta(\boldsymbol{r})\, \nabla_a + \delta(\boldsymbol{r})\, \nabla^2\right] \tag{18.4}$$

and a term or order c^3, which derives from the series expansion (7.16) in c:

$$\delta \hat{V}_{\text{eff}(0)} = -\frac{4\pi c^3}{3M} \delta(\boldsymbol{r})\, \nabla^2. \tag{18.5}$$

In the latter two expressions we have replaced M_{eff} by M since $M/M_{\text{eff}} = 1 + O(cp_0)$.

If we now go over to the Hartree–Fock approximation, we return to the results of § 7. In this case the pseudopotential (18.3), owing to its δ-character, does not alter the dispersion law for the nucleons: its Fourier transform is independent of $\boldsymbol{p}$. Hence, the quantity $M/M_0(p)$ does not change when we change to the Hartree–Fock approximation for the potential $V_{(c)}$; only the constant term in the operator (18.2) is changed.

18.2. We did not allow for the contribution of the pseudopotentials of (18.4) and (18.5) in § 7 (they only need to be considered in the Hartree–Fock approximation). They depend on the momentum operator: we must therefore use the general relations (4.30) and (4.31). We must consider in more detail how much the potential of (18.4) contributes to the quantity $\hat{W}$ and to the energy of the system.

If we substitute the following expressions into (4.30) and (4.31),

$$f(\boldsymbol{x}, \boldsymbol{p}' + \hat{\boldsymbol{p}}) = \theta[p_0^2(x) - (\boldsymbol{p}' + \hat{\boldsymbol{p}})^2],$$

$$V\left[\boldsymbol{x} - \boldsymbol{x}', \pm\frac{1}{2}(i\boldsymbol{p}' + \nabla_{x'})\right] = -\frac{4\pi c^3}{M}\left[\pm\left(\frac{i\boldsymbol{p}'}{2}\cdot\nabla\right)\delta(\boldsymbol{r}) - \frac{1}{4}\delta(\boldsymbol{r})p'^2\right]$$

(it is unnecessary to differentiate the function f in the quasi-classical approximation), we find, after a straightforward integration,

$$\hat{A} = -\frac{1}{4}\hat{B}; \quad \hat{W} = \frac{c^3}{2\pi M}\left(p_0^5 + \frac{5}{3}p_0^3\hat{p}^2\right). \tag{18.6}$$

In this example, $\hat{B}$, as well as $\hat{A}$, depends on the momentum operator.

The corresponding contribution to the energy can be found directly from the general equation (4.36)

$$\delta \mathscr{E} = 2 \int d^3\boldsymbol{p}\, \theta(p_0^2 - p^2)\, \hat{W}(p),$$

whence

$$\delta \mathscr{E} = \frac{c^3 p_0^8}{(3\pi^3 M)}.$$

Putting the volume of the system in the form $A/2\varrho = 3\pi^2 A/2p_0^3$, where A is the mass number, we have

$$\delta E = \frac{c^3 p_0^5 A}{2\pi M}. \tag{18.7}$$

The contribution of the potential (18.5) can be calculated in a similar way.

18.3. We will substitute the pseudopotential (18.3) in the interaction Hamiltonian (18.1). The corresponding value of the parameter $p_0 l$ (see § 16.8) is equal to $p_0 c < 1$. We may therefore limit ourselves to the lower-order diagrams in Fig. 26.

We will consider the second-order diagrams. The corresponding expression for the energy is given by (15.22). It is necessary, however, to be careful when substituting the Fourier transform of the potential, $v(k)$, in this equation.

We will show that we can put $v(k) = 4\pi c/M_{\text{eff}}$ in (15.22) (as if the operator $1 + r\partial/\partial r + \beta r$ was absent), and simultaneously replace the combination $(1 - n_{p_1+k})\,(1 - n_{p_2-k})$ by $(1 - n_{p_1+k}) \times (1 - n_{p_2-k}) - 1$. A similar change must be made in the higher orders of perturbation theory, too.

In actual fact this operator reduces to unity when acting on a function which is regular at the point $r = 0$ (in view of the subsequent multiplication by the δ-function), and goes to zero when acting on the expression $\mathrm{P} \int [d^3\boldsymbol{p} \sin pr/pr\Delta\varepsilon(\boldsymbol{p})]$. The integral (15.22) contains the product of two functions of v. When the pseudopotential (7.16) is substituted in one of them, the operator $1 + r\partial/\partial r + \beta r$ has to be included; for the other, the pseudopotential is already acting on a regular function—on the right-hand bracket of the matrix element. Hence we may immediately put in one of the factors $\nu = 4\pi c/M_{\text{eff}}$. This gives

$$E - E_0 = -\left(\frac{4\pi c}{M_{\text{eff}}}\right)^2 3(2\pi)^3 \int d^3\boldsymbol{p}_1\, d^3\boldsymbol{p}_2\, d^3\boldsymbol{k}$$

$$\times \frac{n_{p_1}n_{p_2}(1 - n_{p_1+k})(1 - n_{p_2-k})}{\varepsilon_{p_1+k} + \varepsilon_{p_2-k} - \varepsilon_{p_1} - \varepsilon_{p_2}}$$

$$\times \int d^3\boldsymbol{r}\, \delta(\boldsymbol{r}) \left(1 + r\frac{\partial}{\partial r} + \beta r\right) \exp[-i(\boldsymbol{k}\cdot\boldsymbol{r})].$$

Since the states $\boldsymbol{p}_{1,2}$ lie below, and $\boldsymbol{p}_1 + \boldsymbol{k}$ and $\boldsymbol{p}_2 - \boldsymbol{k}$ lie above, the Fermi surface, we must in fact understand the integral over $\boldsymbol{k}$ as implying its principal value. Thus we find for the integral in $E - E_0$:

$$\mathrm{P}\int d^3\boldsymbol{k} \exp i(\boldsymbol{k}\cdot\boldsymbol{r}) \frac{(1 - n_{p_1+k})(1 - n_{p_2-k})}{\Delta\varepsilon(\boldsymbol{k})},$$

which is acted on by the pseudopotential. If $(1 - n)(1 - n)$ is replaced by one, this integral gives a contribution of zero. Hence if we replace $(1 - n)(1 - n)$ by $(1 - n)(1 - n) - 1$, we do not alter the result. However, the integral over $\boldsymbol{k}$ is now a regular function of r at the point $r = 0$, since the integral converges for any r. We may then discard the operator $1 + \partial/\partial r + \beta r$ and arrive at our final expression

$$E - E_0 = -3(2\pi)^3 \left(\frac{4\pi c}{M_{\text{eff}}}\right)^2 \int d^3\boldsymbol{p}_1\, d^3\boldsymbol{p}_2\, d^3\boldsymbol{k}$$

$$\times \frac{n_{p_1}n_{p_2}[(1 - n_{p_1+k})(1 - n_{p_2-k}) - 1]}{\varepsilon_{p_1+k} + \varepsilon_{p_2-k} - \varepsilon_{p_1} - \varepsilon_{p_2}}. \tag{18.8}$$

This is not an easy integral to evaluate, and it must therefore be simplified. In fact the integration really covers a fairly narrow region near the Fermi surface. As a result, the value of the integral changes very little if we replace ε_p by $p^2/2M_0$, where M_0 refers to the Fermi surface. Using the results of § 7, we obtain ($y = ap_0$):

$$\frac{M}{M_0} = 1 + \frac{3}{2\pi}\frac{V_0 M a^2}{y}\left(1 + \frac{\sin 2y}{2y} - \frac{2\sin^2 y}{y^2}\right).$$

Substituting this expression in (18.8), we have†

$$E - E_0 = -\frac{3}{4\pi^4}\cdot\frac{c^2}{M}\left(\frac{M}{M_{\text{eff}}}\right)^2\frac{M_0}{M}I, \tag{18.9}$$

where

$$I = \int d^3p_1\, d^3p_2\, d^3k \frac{[(1 - n_{p_1+k})(1 - n_{p_2-k}) - 1]\, n_{p_1} n_{p_2}}{(p_1 + k)^2 + (p_2 - k)^2 - p_1^2 - p_2^2}.$$

The final expression for the lowest-order correlation correction has the form

$$E - E_0 = \frac{6(11 - 2\ln 2)\, c^2 p_0^4}{35\pi^2 M}\left(\frac{M}{M_{\text{eff}}}\right)^2\left(\frac{M_0}{M}\right)A. \tag{18.10}$$

18.4. We will not calculate the correlation correction to the third order in cp_0. There are data in the literature [39] on the total value of this correction and the contribution of the pseudopotential (18.5) for the case of a quadratic dispersion relation. The correction has the form

$$E - E_0 = \frac{0{\cdot}13 c^3 p_0^5}{M}A. \tag{18.11}$$

A departure from a quadratic dispersion relation might appear as an additional factor $(M/M_{\text{eff}})^3\,(M_0/M)^2$, depending on the presence of two energy denominators and three pseudopotentials in the expression for the third-order correction. This factor, however, is close to one, and we can retain (18.11).

Thus the total energy of nuclear matter, when the correlation due to the repulsive forces is included, can be written as

$$E = E_0 + \frac{6(11 - 2\ln 2)\, c^2 p_0^4}{35\pi^2 M}\left(\frac{M}{M_{\text{eff}}}\right)^2\frac{M_0}{M}A + \frac{0{\cdot}13 c^3 p_0^5}{M}A + \frac{c^3 p_0^5}{2\pi M}A, \tag{18.12}$$

where E_0 is given by (7.22).

† The calculation of I is described in detail in ref. [45].

E, as distinct from E_0, has a minimum for reasonable values of p_0. This is connected with the rapid growth of the correlation terms as $p_0 \to \infty$. We will give the final result for equilibrium values of p_0 and E [46].

$$\left.\begin{aligned} p_0 &\approx 1{\cdot}4 \text{ fm}^{-1}, \\ E/A &\approx -11 \text{ MeV}. \end{aligned}\right\} \tag{18.13}$$

The separate energy terms have the following values: the kinetic energy—25 MeV; the attraction energy—68 MeV; the repulsion energy—32 MeV. This last quantity is made up of four figures equal to 21, 6, 2·5 and 2·5 MeV, respectively. The first three correspond to the self-consistent part and the two first correlation corrections describing the s-scattering. It is obvious that the corresponding series converges with sufficient rapidity. The last term gives the p-scattering contribution.

A comparison of the data in (18.13) with the empirical data indicates satisfactory agreement. The magnitude of p_0 is right, and E/A is close to the coefficient C_1 in Weizsäcker's formula (see § 7) which has a value -15 MeV.

18.5. If we wish to obtain values for the remaining coefficients in Weizsäcker's formula, it is necessary to consider besides the expressions used in § 7, the correlation terms obtained on the

TABLE 2. *A Comparison of the Calculated and Empirical Characteristics of the Atomic Nucleus*

Characteristic of the atomic nucleus	Calculated magnitude	Empirical magnitude
Nuclear radius, $R/A^{1/3}$	1·1 fm	1·1 fm
Width of the surface layer, d	3 fm	2·4 fm
Volume energy, C_1	−11 MeV	−15 MeV
Coulomb energy, C_2	0·7 MeV	0·7 MeV
Surface energy, C_3	23 MeV	18 MeV
Energy of symmetry, C_4	33 MeV	24 MeV

condition that the numbers of neutrons and protons are not equal and that the nucleus is finite. The inequality in the number of neutrons and protons can be allowed for by inserting different limiting momenta in the corresponding equations: $p_{0(p)} \neq p_{0(n)}$. The inhomogeneity in the particle distribution can be allowed for by referring all quantities in the surface layer to a value of p_0 half the size of that at the centre of the nucleus. It seems from the resulting calculations [80] that the correlation effects have most influence on the surface part of the energy appreciably decreasing its magnitude. The remaining quantities, apart from the volume-proportional part of the energy, are little changed (Table 2).

Thus there seems to be a more or less good agreement with experiment.† This evidently indicates that three-body forces in the atomic nucleus are, in general, less important that might have been supposed from general considerations (see § 1).

† So far as the deviation in the value for C_4 is concerned, we must note the following. We have completely ignored in our calculations interference between the surface effects and the symmetry effects. A current variant [93] of Weizsäcker's empirical formula, which takes this interference into account, obtains a value of $C_4 \approx 30$ MeV, $C_3 \approx 21$ MeV, in closer agreement with those derived above. Hence, final conclusions can only be drawn when this interference has been included.

CHAPTER 4

The Method of Green Functions in Quantum Mechanics

19. The Single-Particle Green Function

19.1. We will consider in this chapter approaches to the many-body problem which are not based on a perturbation series expansion. The need to get away from the perturbation theory approach is dictated mainly by the wish to study systems with strong interactions between particles. Solving the corresponding "exact" field equations is of the same order of difficulty as solving the exact Schrödinger equation. The situation of greatest interest, however, is that which was considered in § 16, when a certain, distinct, infinite subgroup of perturbation terms is of fundamental importance. These terms can be summed in a very simple and compact way by field theory methods.

The most important task of many-body theory—determining the spectrum of the excited states of a system—is also impossible on perturbation theory. It cannot even be set up in a complete form within the framework of the theory. Some kinds of excitation, due to interactions between particles, are completely ignored.

The approach we will now consider has as its most important element the Green functions; the simplest of these represents a direct generalization of the operator contraction. Green functions contain considerable physical information. Once the one- and two-particle Green functions have been evaluated, we dispose of a practically complete description of the many-body system.

Those aspects of the description in § 15, which refer to the distribution of the single-particle properties of the system (the density matrix) and the energy of the system, can be obtained from the self-energy part of the S-matrix Σ. This quantity is directly connected with the single-particle Green function.

We can use the two-particle Green function to obtain a more detailed knowledge of the two-particle properties of a system: such as correlation functions, the kinetic coefficients describing the reaction of a system to external perturbation, etc.

Finally, the Green functions provide the possibility of determining the excitation spectrum of a system. The characteristics of the elementary excitations (the quasi-particles) can be obtained from Green functions by relatively simple mathematical manipulations.

19.2. We have used the operator contraction several times in the previous chapter.

$$\overbrace{\hat{\psi}(1)\,\hat{\psi}^{+}(2)} = \langle\Psi_0|T[\hat{\psi}(1)\,\hat{\psi}^{+}(2)]|\Psi_0\rangle = iG_0(1, 2).$$

It forms an important part of the diagram technique. The operator contraction describes the propagation of particles uncorrelated with other particles in the system.

If we are to go outside the framework of perturbation theory, we must generalize the concept of operator contraction and consider particle propagation with all possible correlation interactions with other particles in the system included. These correlations reduce to describing that the particle excites the system in every possible way, i.e. some of the particles undergo a transition through the Fermi surface, while later this excitation is removed.

In graphical language, the propagation function under consideration must combine all the diagrams which connect two given points. Representing this generalized function by a heavy line, we can write down the diagramatic equality (Fig. 30), which signifies that the propagation of a particle allowing for interaction is equivalent to the propagation of a free particle plus the propagation of a particle with all possible self-energy insertions. The generalized propagation function is made up of the diagrams shown in Fig. 31.

We have denoted the simple propagation function by $iG_0(1, 2)$ where G_0 is the free-particle Green function. Similarly, we will denote the generalized propagation function by $iG(1, 2)$, where G

is the exact single-particle Green function. Using the Feynman rules, and omitting the factor i, the diagramatic equality of Fig. 30 can be written in the form

$$G(1, 2) = G_0(1, 2) + \int d3\, d4\, G_0(1, 3)\, \Sigma(3, 4)\, G_0(4, 2). \qquad (19.1)$$

We will show that the Green function G can be written as an expectation value

$$G(1, 2) = -i \frac{\langle \Psi_0 | T[\hat{\psi}_{\mathrm{Int}}(1)\, \hat{\psi}^{+}_{\mathrm{Int}}(2)\, \hat{S}] | \Psi_0 \rangle}{\langle \Psi_0 | \hat{S} | \Psi_0 \rangle}, \qquad (19.2)$$

where the T-product of operators containing the S-matrix, should be understood as

$$\hat{S}(\infty, t_1)\, \hat{\psi}_{\mathrm{Int}}(1)\, \hat{S}(t_1, t_2)\, \hat{\psi}^{+}_{\mathrm{Int}}(2)\, \hat{S}(t_2, -\infty)$$

when $t_1 > t_2$; similarly for the converse case.

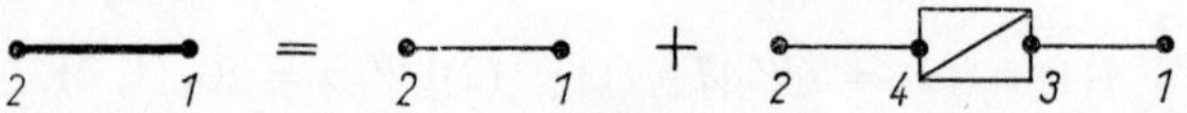

FIG. 30

This can be proved by substituting the S-matrix expansion (14.1) in the expression (19.2). Only the first two terms of this expansion provide a contribution which differs from zero. Proceeding in precisely the same way as in § 15.2, we obtain (19.1).

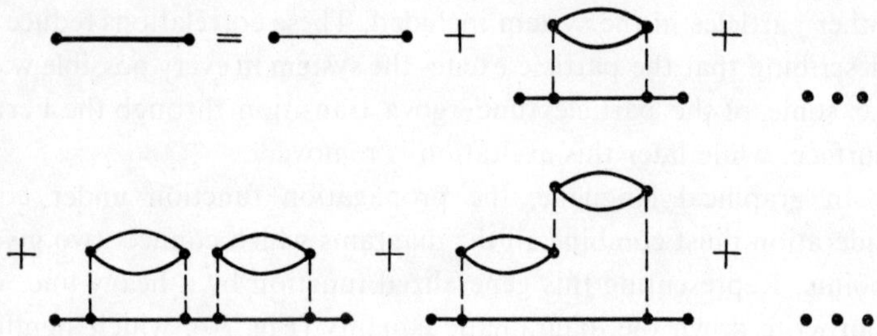

FIG. 31

The Green function can be expressed in terms of the operators in the Heisenberg representation by means of (15.5):

$$G(1, 2) = -i \langle \Psi | T[\hat{\psi}_H(1)\, \hat{\psi}^{+}_H(2)] | \Psi \rangle. \qquad (19.3)$$

This expression differs from (19.2) in its simpler structure, and is therefore suitable for investigations of a general nature. It contains, however, the Heisenberg operators and the wave functions including interaction effects which can only be found by solving the interaction problem.

19.3. Equation (19.1) relating the Green function to the self-energy, Σ, can be replaced by a more useful relationship called Dyson's equation.

We will first of all analyse the self-energy in terms of the diagrams which compose it. Figure 21 indicates that Σ contains both what are known as compact diagrams, i.e. they cannot be crossed by a line which intersects only one line of the diagram (see Figs. 21b, 21c and 21d) and non-compact diagrams, which do not have this property (see Fig. 21e). Σ can be reduced to a sum of the compact diagrams only: this is denoted by $M(1, 2)$ and is called the mass operator. It is represented graphically in Fig. 32.

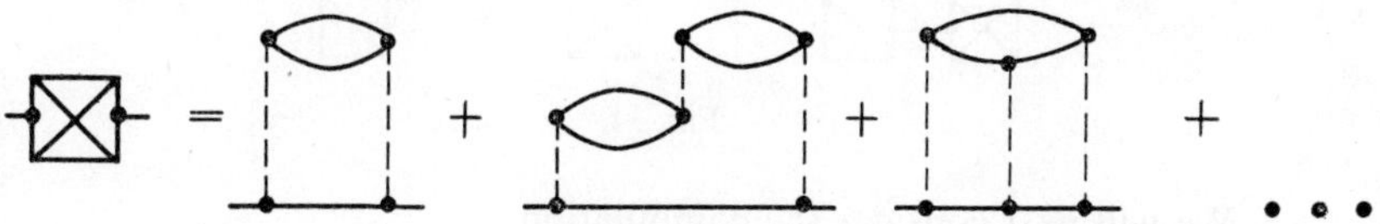

FIG. 32

We can establish a connection between Σ and M by the following straightforward argument. The sum of all the diagrams (compact and non-compact) which form Σ can be obtained if we first take the sum of the compact diagrams, then the sum of the non-compact which are formed from two compact, then the sum of the non-compact formed from three compact, and so on (Fig. 33).

FIG. 33

When we use the Feynman rules to obtain the corresponding analytical relation, we must remember that, unlike vacuum diagrams (§ 14), there are now no additional numerical factors. This is due to the presence of a preferred direction of propagation, and makes a further change of variables impossible, if we require that the form of a given element be retained. We therefore obtain the simple relationship:

$$\Sigma(1,2) = M(1,2) + \int d3\, d4\, M(1,3)\, G_0(3,4)\, M(4,2) + \int d3\, d4\, d5\, d6\, M(1,3)\, G_0(3,4)\, M(4,5)\, G_0(5,6)\, M(6,2) + \cdots. \tag{19.4}$$

This infinite series may be thought of as the solution of the following integral equation:

$$\Sigma(1,2) = M(1,2) + \int d3\, d4\, M(1,3)\, G_0(3,4)\, \Sigma(4,2). \tag{19.5}$$

Indeed if we iterate this equation successively, i.e. substitute M for Σ on the right-hand side, then obtain a corrected value for Σ and so on, we get back to eqn. (19.4). This integral equation is shown diagramatically in Fig. 34:

FIG. 34

We will next consider the combination

$$\int d3\, \Sigma(1,3)\, G_0(3,2) = \int d3\, M(1,3) \times [G_0(3,2) + \int d4\, d5\, G_0(3,4)\, \Sigma(4,5)\, G_0(5,3)]. \tag{19.6}$$

According to (19.1), we can introduce the Green function G on the right-hand side, giving (Fig. 35)

$$\int d3\, \Sigma(1,3)\, G_0(3,2) = \int d3\, M(1,3)\, G(3,2). \tag{19.6'}$$

FIG. 35

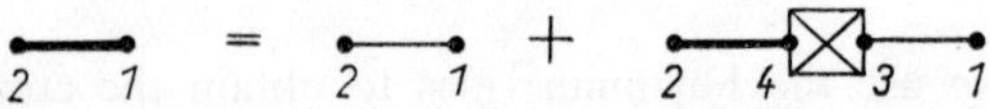

FIG. 36

We can now change from (19.1) to Dyson's integral equation (Fig. 36):

$$G(1,2) = G_0(1,2) + \int d3\, d4\, G_0(1,3)\, M(3,4)\, G(4,2). \tag{19.7}$$

If we operate on Dyson's equation from the left with $i\partial/\partial t_1 - \hat{T} - \hat{W}$, then, from (10.6), we can obtain the equation

$$\left(i\frac{\partial}{\partial t_1} - \hat{T} - \hat{W}\right) G(1, 2) - \int d3\, M(1, 3)\, G(3, 2) = \delta(1 - 2). \tag{19.8}$$

Thus the exact Green function also appears as the Green function of an equation which describes the propagation of a particle. In the Hartree–Fock approximation this equation simply coincided with the Schrödinger equation for an individual particle. In the case under consideration we must also include the mass operator which describes the way in which the particle motion changes due to the correlation interaction.

19.4. We will now turn to the energy representation, writing the Green function in the form

$$G(1, 2) = \sum_{\mu\nu} \int \frac{d\varepsilon}{2\pi} \exp\left[-i\varepsilon(t_1 - t_2)\right] \chi_\mu^*(q_2)\, \chi_\nu(q_1)\, G_{\mu\nu}(\varepsilon). \tag{19.9}$$

If we introduce a similar representation for the mass operator too, we can rewrite (19.8) as

$$(\varepsilon - \varepsilon_\nu)\, G_{\mu\nu}(\varepsilon) - \sum_\sigma M_{\mu\sigma}(\varepsilon)\, G_{\sigma\nu}(\varepsilon) = \delta_{\mu\nu}. \tag{19.10}$$

These relations look even simpler for a spatially uniform system. Putting

$$G(1, 2) = \int d^4p\, G(\boldsymbol{p}, \varepsilon) \exp\left[i(\boldsymbol{p}\cdot \boldsymbol{x}_1 - \boldsymbol{x}_2) - i\varepsilon(t_1 - t_2)\right] \tag{19.11}$$

and using a similar representation for M, we obtain

$$G(\boldsymbol{p}, \varepsilon) = G_0(\boldsymbol{p}, \varepsilon)\left[1 + M(\boldsymbol{p}, \varepsilon)\, G(\boldsymbol{p}, \varepsilon)\right]$$

instead of (19.7), or, applying (10.8) for $G_0(\boldsymbol{p}, \varepsilon)$

$$G(\boldsymbol{p}, \varepsilon) = \left[\varepsilon - \varepsilon_p - M(\boldsymbol{p}, \varepsilon)\right]^{-1}. \tag{19.12}$$

The rules for integrating around the corresponding singularities of $G(\boldsymbol{p}, \varepsilon)$ will be considered in § 23.

When the interaction between particles was included in the Hartree–Fock approximation, the dispersion law was altered (the quantity $p^2/2M$ was replaced by ε_p). When the correlation interaction is allowed for, further changes of this type take place, as is shown by (19.12). These latter changes are more far-reaching in nature.

19.5. We can use the Green function to express the physical quantities considered in § 15—the density matrix and the ground state energy of a system.

We will consider first the single-particle density matrix. It follows from (15.3) that this quantity is directly related the Green function:

$$R(q_1, q_2) = -i \lim_{t_2 - t_1 \to +0} G(1, 2). \tag{19.13}$$

If we remember that G depends only on the combination $t_1 - t_2$, and not on the times t_1 and t_2 individually, and expand G into a Fourier integral in terms of this difference

$$G(1, 2) = \int \frac{d\varepsilon}{2\pi} \exp\left[-i\varepsilon(t_1 - t_2)\right] G(q_1, q_2, \varepsilon),$$

we have, similarly to the results of § 15,

$$R(q_1, q_2) = -i \int_C \frac{d\varepsilon}{2\pi} G(q_1, q_2, \varepsilon) \tag{19.14}$$

(the contour C is shown in Fig. 24).

It is now easy to find an expression for the exact distribution function, $f(\boldsymbol{x}, \boldsymbol{p})$, expanding $G(q_1, q_2, \varepsilon)$ into a Fourier integral in the difference $\boldsymbol{x}_1 - \boldsymbol{x}_2$

$$G(q_1, q_2, \varepsilon) = \int d^3\boldsymbol{p}\, G(\boldsymbol{x}_1, \boldsymbol{p}, \varepsilon) \exp\left[i(\boldsymbol{p} \cdot \boldsymbol{x}_1 - \boldsymbol{x}_2)\right],$$

whence

$$f(\boldsymbol{x}, \boldsymbol{p}) = -i \int_C \frac{d\varepsilon}{2\pi} G(\boldsymbol{x}, \boldsymbol{p}, \varepsilon). \tag{19.15}$$

For a spatially uniform system, $G(\boldsymbol{x}, \boldsymbol{p}, \varepsilon)$ is independent of $\boldsymbol{x}$ and is simply the Fourier transform $G(\boldsymbol{x}_1 - \boldsymbol{x}_2, \varepsilon)$ in terms of $\boldsymbol{x}_1 - \boldsymbol{x}_2$. The quantity

$$\varrho(\boldsymbol{p}) = -i \int_C \frac{d\varepsilon}{2\pi} G(\boldsymbol{p}, \varepsilon) \tag{19.16}$$

now gives the momentum distribution of the particles.

We next consider the energy of the system. It follows from the results obtained in § 15.3 that

$$E = -\frac{i}{2}\int dq_1 \lim_{\substack{2\to 1 \\ t_2-t_1\to +0}} \left(i\frac{\partial}{\partial t_1} + \hat{T}\right) G(1, 2). \tag{19.17}$$

For a spatially uniform system

$$E = -\frac{i\Omega}{2}\,\mathrm{Tr}_{\sigma\tau}\int_C d^4p(\varepsilon + p^2/2M)\, G(\boldsymbol{p}, \varepsilon). \tag{19.18}$$

We can also use eqn. (19.18) to write

$$E = -i\int dq_1 \lim_{\substack{2\to 1 \\ t_2-t_1\to +0}} \Big\{(\hat{T} + \hat{W}/2)\, G(1, 2) + \frac{1}{2}\int d3\, M(1, 3)\, G(3, 2)\Big\}. \tag{19.19}$$

For a spatially uniform system we have

$$E = -\frac{i\Omega}{2}\,\mathrm{Tr}_{\sigma\tau}\int_C d^4p[\varepsilon_p + (p^2/2M) + M(\boldsymbol{p}, \varepsilon)]\, G(\boldsymbol{p}, \varepsilon). \tag{19.20}$$

This differs from the one introduced, for example, by Galitskii and Migdal [94] only in its form: the mass operator which figures here does not include the self-consistent part, equal to $\varepsilon_p - p^2/2M$.

The expression for the energy can be obtained in another form if the parameter λ is introduced (see § 15). We find for the correlation energy:

$$E - E_0 = \frac{i}{2}\int_0^1 \frac{d\lambda}{\lambda}\int dq_1 \times \lim_{\substack{2\to 1 \\ t_2-t_1\to +0}} \Big\{\int d3\, M(1, 3)\, G(3, 2) - \lambda \hat{W}_{q_1}[G(1, 2) - G_0(1, 2)]\Big\}. \tag{19.21}$$

For a spatially uniform system

$$E - E_0 = \frac{i\Omega}{2}\,\mathrm{Tr}_{\sigma\tau}\int_0^1 \frac{d\lambda}{\lambda}\int_C d^4p \times [1 - \lambda(\varepsilon_p - p^2/2M)\, G_0(\boldsymbol{p}, \varepsilon)]\, M(\boldsymbol{p}, \varepsilon)\, G(\boldsymbol{p}, \varepsilon). \tag{19.22}$$

20. The Two-Particle Green Function

20.1. Besides the single-particle Green function, which describes the propagation of one particle (allowing for its correlation interactions with the system), we can also introduce higher-order Green functions. The most important of these is the pair (or two-particle) Green function, defined by the relation

$$G(1, 2, 1', 2') = (-i)^2 \frac{\langle \Psi_0 | T[\hat{\psi}_{\mathrm{Int}}(1)\, \hat{\psi}_{\mathrm{Int}}(2)\, \hat{\psi}^+_{\mathrm{Int}}(2')\, \hat{\psi}^+_{\mathrm{Int}}(1')\, \hat{S}] | \Psi_0 \rangle}{\langle \Psi_0 | \hat{S} | \Psi_0 \rangle} \tag{20.1}$$

In terms of the Heisenberg representation, we can write

$$G(1, 2, 1', 2') = (-i)^2 \langle \Psi | T[\hat{\psi}_H(1)\, \hat{\psi}_H(2)\, \hat{\psi}^+_H(2')\, \hat{\psi}^+_H(1')] | \Psi \rangle \tag{20.2}$$

This new Green function describes the propagation of two particles (or two holes, or a particle and a hole) allowing for their correlation interaction with one another and with the remaining particles of the system. The Green function in the form (20.2) describes the propagation of two particles from the points 1′, 2′ to 1, 2. If we rewrite (20.2) as

$$G(1, 2, 1', 2') = \langle \Psi | T[\hat{\psi}_H(1)\, \hat{\psi}^+_H(2')\, \hat{\psi}_H(2)\, \hat{\psi}^+_H(1')] | \Psi \rangle, \tag{20.2'}$$

it is evident that it simultaneously describes the propagation of a particle and a hole from 1′ to 2 and from 1 to 2′. In the same way it may be confirmed that the two-particle Green function also includes the propagation of two holes (from 1, 2 to 1′, 2′). A graphical representation of the two-particle Green function appears in Fig. 37a.

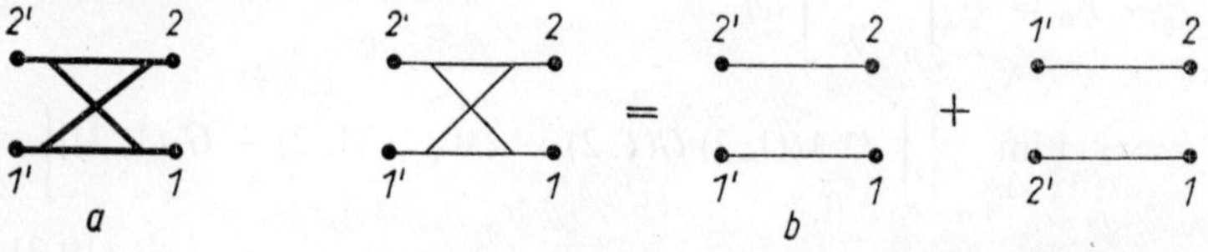

FIG. 37

The two-particle Green function has the following symmetry properties:

$$G(1, 2, 1', 2') = -G(2, 1, 1', 2') = -G(1, 2, 2', 1').$$

In the Hartree–Fock approximation ($\hat{S} = 1$), the two-particle Green function can be decomposed into an anti-symmetrical product of single-particle functions. Applying Wick's theorem to (20.1), we find

$$G_0(1, 2, 1', 2') = G_0(1, 1')\, G_0(2, 2') - G_0(1, 2')\, G_0(2, 1'). \quad (20.3)$$

We are dealing, in this case, with the uncorrelated propagation of two particles, including only the exchange effects (see Fig. 37b). The two lighter lines with the cross indicate the function $G_0(1, 2, 1', 2')$.

20.2. We will substitute the S-matrix expansion (14.1) in (20.1). The first term gives simply the two-particle function in the Hartree–Fock approximation $G_0(1, 2, 1', 2')$. The second term of the expansion takes the form

$$\int d3\, d4\, \Sigma(3, 4)\, \{G_0(1, 3)\, G_0(2, 4, 2', 1') + G_0(2, 3)\, G_0(1, 4, 1', 2')\} \quad (20.4)$$

after Wick's theorem has been applied. Finally, we find from the third term, which contains the vertex part,

$$\int d3\, d4\, d5\, d6\, G_0(1, 2, 3, 4)\, \Gamma(3, 4, 5, 6)\, G_0(5, 6, 1', 2'). \quad (20.5)$$

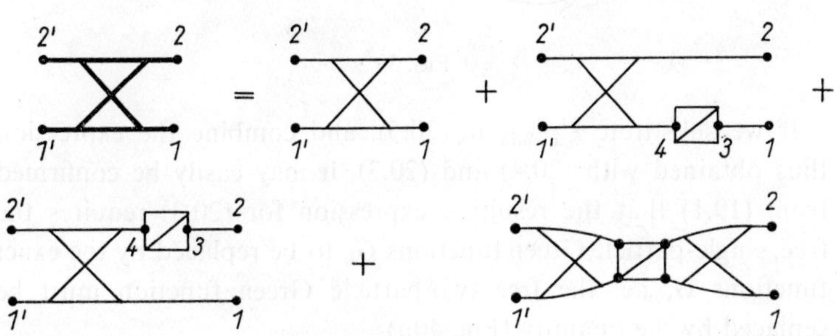

FIG. 38

The remaining terms of (14.1) do not contribute to the expression under consideration. The equality obtained is given graphically in Fig. 38. The first term describes the propagation of the uncorrelated particles; the second and third, the correlation of the particles with the system (but not with each other), and the last the correlation of the particles with each other (and also, as we will see, with the system).

A further consideration must be given to the appearance of certain peculiarities in the structure of the vertex part which represents the sum of all possible diagrams containing four external lines. These necessarily include unconnected diagrams (which can be crossed by a line which does not intersect any of the lines in the diagram—see Fig. 39a). The sum of the unconnected diagrams can be expressed in terms of the self-energy Σ (see Fig. 39b):†

$$\Gamma_{\text{unconn}}(1, 2, 3, 4) = \tfrac{1}{2}\Sigma(1, 3)\,\Sigma(2, 4). \qquad (20.6)$$

The coefficient $\frac{1}{2}$ comes from the possibility of changing the variables in both unconnected parts.

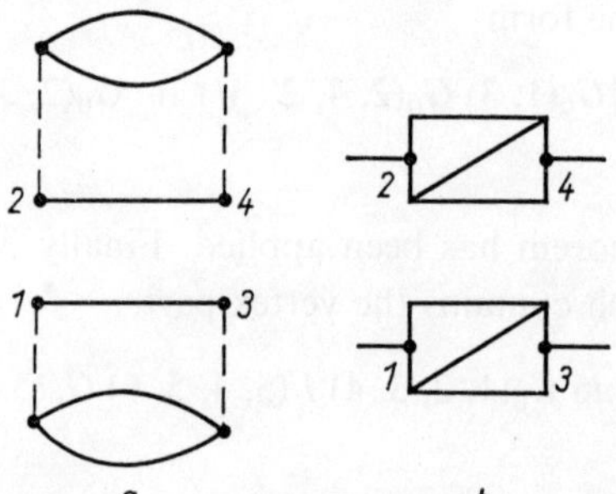

a b

FIG. 39

If we substitute Γ_{unconn} in (20.5), and combine the expression thus obtained with (20.4) and (20.3), it may easily be confirmed from (19.1) that the resulting expression for (20.3) requires the free, single-particle Green functions G_0 to be replaced by the exact functions G, i.e. the free two-particle Green function must be replaced by the quantity (Fig. 40a):

$$\tilde{G}(1, 2, 1', 2') = G(1, 1')\,G(2, 2') - G(1, 2')\,G(2, 1'). \quad (20.7)$$

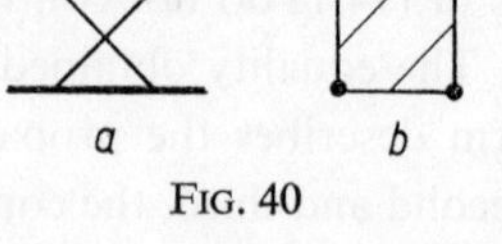

a b

FIG. 40

† The expression for Γ_{unconn} (1, 2, 3, 4) is in unsymmetrized form.

Hence, the equation we have obtained for the two-particle Green function can be written in the form

$$G(1, 2, 1', 2') = \tilde{G}(1, 2, 1', 2') + \int d3\, d4\, d5\, d6\, G_0(1, 2, 3, 4) \\ \times \Gamma_{\text{conn}}(3, 4, 5, 6)\, G_0(5, 6, 1', 2'). \qquad (20.8)$$

The first terms here describe the motion of a pair of particles independent of one another (but correlated with the system); the last term takes into account also their mutual correlation effects (Γ_{conn} is the connected vertex part, see Fig. 40b). Figure 41 represents eqn. (20.8) diagrammatically.

FIG. 41

20.3. Equation (20.8) is similar to eqn. (19.1) for the single-particle Green function. Both equations relate the accurate and the free-particle Green function. The free-particle function in (20.8) may be replaced by the function G if the vertex part is further transformed.

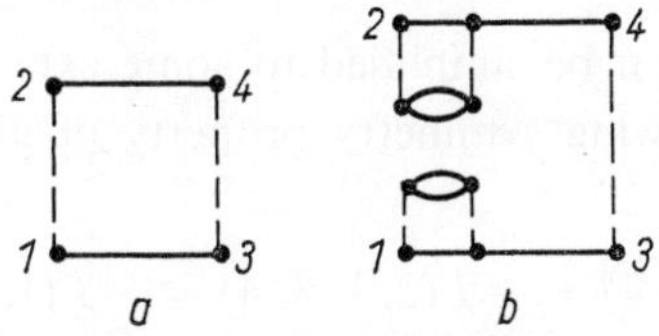

FIG. 42

To this end, we note that the connected vertex part $\Gamma_{\text{conn}}(1, 2, 3, 4)$, may be composed of two different types of diagram. The first type consists of diagrams in which each of the points 1, 2, 3, 4 is connected with the others (Fig. 42a). The second type refers to diagrams in which at least one of these points is not so related (see Fig. 42b). If we denote the sum of all diagrams of the first type by Δ (Fig. 43a), we can write down a diagramatic equality relating Γ_{conn} to Δ (see Fig. 43b). The points here correspond to the analogous diagrams containing all possible permutations of vertices. The corresponding analytical equation, which can easily be derived from the Feynman rules, will not be introduced here owing to its complexity.

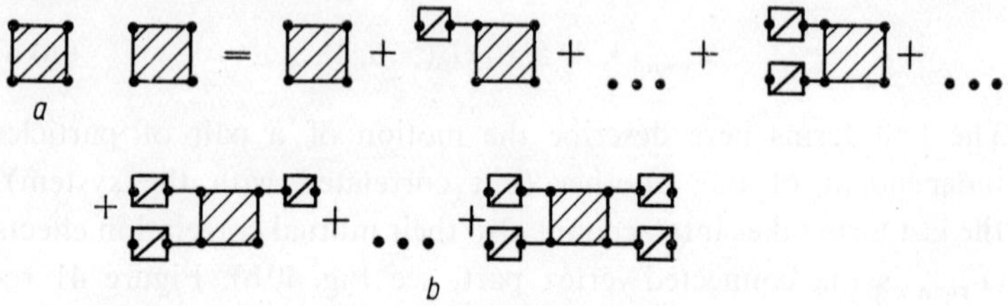

FIG. 43

It can be shown without difficulty that the combination of the free, two-particle Green functions which enter into eqn. (20.8) with the self-energies (see Fig. 43b), leads to functions of the type $G_0(1, 2, 3, 4)$ in this equation being replaced by others of the type $\tilde{G}(1, 2, 3, 4)$. To do this, it is necessary to substitute (19.1) into this latter expression and then to construct the corresponding diagrams. We may thus write (Fig. 44a):

$$G(1, 2, 1', 2') = \tilde{G}(1, 2, 1', 2') + \int d3\, d4\, d5\, d6\, \tilde{G}(1, 2, 3, 4) \times \Delta(3, 4, 5, 6)\, \tilde{G}(5, 6, 1', 2'). \tag{20.9}$$

This expression can be simplified to some extent, if we take into account the following symmetry property of all the vertex parts (see § 14.5):

$$\Gamma(1, 2, 3, 4) = -\Gamma(2, 1, 3, 4) = -\Gamma(1, 2, 4, 3). \tag{20.10}$$

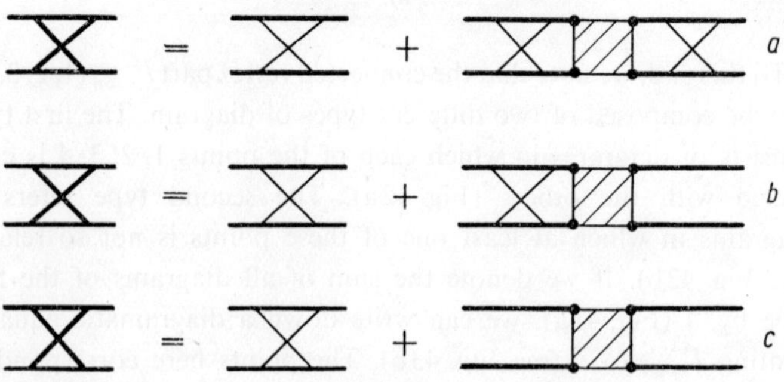

FIG. 44

Therefore (see Fig. 44b)

$$G(1, 2, 1', 2') = \tilde{G}(1, 2, 1', 2') + 2 \int d3\, d4\, d5\, d6\, G(1, 3)\, G(2, 4)$$
$$\times \Delta(3, 4, 5, 6)\, \tilde{G}(5, 6, 1', 2'). \qquad (20.11)$$

A similar procedure (see Fig. 44c) gives

$$G(1, 2, 1', 2') = G'(1, 2, 1', 2') + 4 \int d3\, d4\, d5\, d6\, G(1, 3)\, G(2, 4)$$
$$\times \Delta(3, 4, 5, 6)\, G(5, 1')\, G(6, 2'). \qquad (20.12)$$

The diagrams contributing to Δ include the non-compact diagrams, which can be crossed by a line intersecting only two continuous lines of the diagram. It might be possible in principle to replace the kernel in the preceding equation by a compact kernel and so produce an integral equation, analogous to Dyson's equation for the single-particle Green function

$$G(1, 2, 1', 2') = G'(1, 2, 1', 2') + \int d3\, d4\, d5\, d6$$
$$\times K(1, 2, 1', 2'; 3, 4, 5, 6)\, G(3, 4, 5, 6), \qquad (20.13)$$

where the kernel K can be expressed in terms of the compact part of Δ, and G' in terms of the Green function $G(1, 2)$, etc. It is by no means easy, however, to establish this relationship.

Equation (20.12) connecting the two-particle Green function with the vertex part Δ plays an important part in Landau's semi-empirical theory of the Fermi fluid [25, 95]. The quantity Δ is closely related to the process of the scattering of quasi-particles by one another and under certain conditions coincides with the corresponding scattering amplitude.

20.4. There exist certain relationships which connect the single-particle and the two-particle Green function. We will now derive one of these.

We will operate with $i\partial/\partial t_1 - \hat{T}_{q_1}$ on the expression for the Green function (19.3). Using the equation of motion for a field operator in the Heisenberg representation (3.28), the relation (9.12′) and the commutation rules for operators at coincident times (3.27), we obtain

$$\left(i\frac{\partial}{\partial t_1} + \hat{T}\right) G(1, 2) = \delta(1 - 2)$$

$$- i \int d3 \, \langle \Psi | T [\hat{\psi}_H^+(3) \, \hat{V}(1, 3) \, \hat{\psi}_H(3) \, \hat{\psi}_H(1) \, \hat{\psi}_H^+(2)] | \Psi \rangle . \qquad (20.14)$$

We will reduce the mean value of the operators to the form of a two-particle Green function. Since the times t_1 and t_3 coincide, the corresponding operators under the T-product sign cannot be permuted. The possibility of such an interchange was justified in § 9 by the fact that ordering in terms of time is the same as arranging the operators in the required order. Obviously, if the operators refer to the same time, this argument falls down.

When reducing a four-particle expectation value to a two-particle Green function we should temporarily introduce unequal times into the first three operators of this average value. It is then possible to write

$$\langle \Psi | T [\hat{\psi}_H^+(3) \, \hat{\psi}_H(3) \, \hat{\psi}_H(1) \, \hat{\psi}_H^+(2)] | \Psi \rangle = \lim_{\substack{4 \to 3 \\ t_4 > t_3 > t_1, \; t_4 \to t_1}} G(3, 1, 4, 2) . \qquad (20.15)$$

The operators here do indeed acquire the correct order. Hence the desired equation may be written in the form

$$\left(i\frac{\partial}{\partial t_1} - \hat{T}\right) G(1, 2) + i \int d3 \, V(1, 3) \lim_{\substack{4 \to 3 \\ t_4 > t_3 > t_1, \; t_4 \to t_1}} G(3, 1, 4, 2)$$
$$= \delta(1 - 2). \qquad (20.16)$$

If the left-hand side of this equation contained the operator $i\partial/\partial t - \hat{T} - \hat{W}$ instead of $i\partial/\partial t - \hat{T}$, then we could divide this into both sides of (20.16), and so express $G(1, 2)$ in terms of G_0 and the two-particle Green function. Equation (20.16) can be reduced to this form, if we use the relationship

$$\hat{W}_{q_1} G(1,2) = \int d3 \, V(1,3) \, [R(q_3, q_3) \, G(1,2) - R(q_3, q_1) \, G(q_3, t_1, 2)],$$

which follows from the definition of the operator $\hat{W}$ (see § 4). If we introduce the Green function G_0 instead of the density matrix and subtract the resulting expression from (20.16), we find

$$\left(i\frac{\partial}{\partial t_1} - \hat{T} - \hat{W}\right) G(1, 2) + i \int d3\, V(1, 3) \lim_{\substack{4\to 3\\ t_4>t_3>t_1,\ t_4\to t_1}} \{G(3, 1, 4, 2)$$

$$- G_0(3, 4)\, G(1, 2) + G_0(1, 4)\, G(3, 2)\} = \delta(1 - 2).$$

This can be written more compactly in the following way:

$$G(1, 2) - G_0(1, 2) = -i \int d3\, d4\, G_0(1, 3)\, V(3, 4)$$

$$\times \lim_{\substack{5\to 4\\ t_5>t_4>t_3,\ t_5\to t_3}} \{G(4, 3, 5, 2) - G_0(4, 5)\, G(3, 2) + G_0(3, 5)\, G(4, 2)\} \tag{20.17}$$

This relationship may be verified most easily by operating on both sides with $i\partial/\partial t_1 - \hat{T} - \hat{W}$ and taking into account eqn. (10.6); (20.17) then reduces to the previous equation.

Relationships of another type, which are expressible in terms of Ward's theorem will be considered in § 20.8.

In a similar way, if we operate with $i\partial/\partial t - \hat{T}$ on the two-particle Green function, it would be possible to find a connection between the two- and three-particle Green functions, etc. An infinite number of equations of this sort, relating two Green functions of adjacent order, would in principle be sufficient to determine them. It is obvious, however, that the solution of this system of coupled equations requires that it should be simplified. We will not be using this system in the following.

20.5. It is possible from a comparison of (20.17) and (19.7) to express the mass operator M in terms of the two-particle Green function and then in terms of the vertex part. This expression will be used to obtain the single-particle Green function. Subtracting (20.17) from (19.7), and operating on the result from the left with $i\partial/\partial t - \hat{T} - \hat{W}$, we obtain

$$\int d4\, M(1, 4)\, G(4, 2) = -i \int d4\, V(1, 4)$$

$$\times \lim_{\substack{5\to 4\\ t_5>t_4>t_1,\ t_5\to t_1}} \{G(4, 1, 5, 2) - G_0(4, 5)\, G(1, 2) + G_0(1, 5)\, G(4, 2)\}.$$

Substituting (20.12) in the right-hand side of this equality, we can reduce it to the form

$$-i \int d4\, V(1,4) \{[G(4,5) - G_0(4,5)]\, G(1,2) - [G(1,5) - G_0(1,5)] \times G(4,2) + 4 \int d6\, d7\, d8\, d9\, G(4,6)\, G(1,7)\, \Delta(6,7,8,9) \times G(8,5)\, G(9,2)\}.$$

Now altering the notation appropriately the expression will contain the function $G(4, 2)$ to the right. We can isolate the expression for the mass operator, itself:

$$M(1,2) = M_1(1,2) + M_2(1,2) - 4i \int d3\, d4\, d5\, d6 \times \lim_{\substack{7\to 3 \\ t_7 > t_3 > t_1}} V(1,3)\, G(3,4)\, G(1,5)\, \Delta(4,5,6,2)\, G(6,7), \tag{20.18}$$

where

$$\left.\begin{aligned} M_1(1,2) &= -i\delta(1-2) \int d3\, V(1,3) \lim_{\substack{4\to 3 \\ t_4 > t_3 > t_1}} [G(3,4) - G_0(3,4)], \\ M_2(1,2) &= i \int d3\, V(1,3) \lim_{\substack{4\to 3 \\ t_4 > t_3 > t_1}} [G(1,4) - G_0(1,4)]. \end{aligned}\right\} \tag{20.19}$$

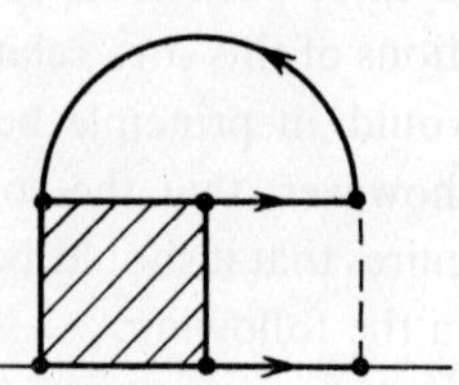

FIG. 45

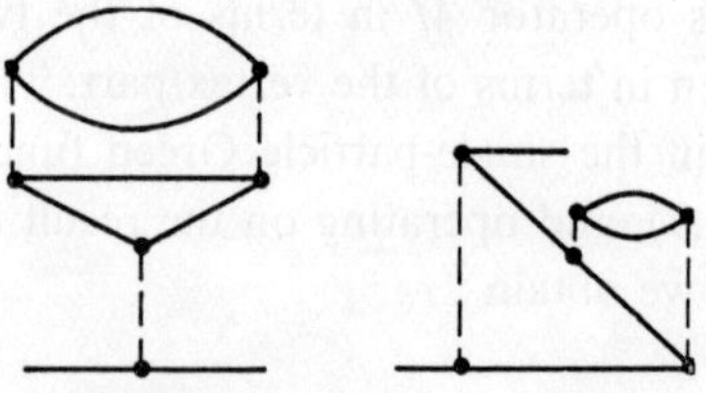

FIG. 46

Figure 45 represents a diagram corresponding to the third term of (20.18). Typical diagrams for M_1 and M_2 are given in Figs. 46a and 46b, respectively. In the examples, which we will be considering later, the contribution of M_1 and M_2 to the mass operator is negligibly small.

20.6. Besides the relations so far mentioned, the two-particle Green function also contains information about the system, which cannot be obtained from the single-particle function.

This is true, in particular, of the correlations in the system which indicate how two particles in the system influence each other. A natural measure of the correlation effect for a pair of particles is provided by the difference

$$K(1, 2, 3, 4) = G(1, 2, 3, 4) - \tilde{G}(1, 2, 3, 4). \qquad (20.20)$$

This quantity describes the dynamical correlation only, and in the Hartree–Fock approximation goes to zero. All the possible two-body correlations of a system can be obtained from (20.20). We will consider in detail the coordinate correlation.

We introduce the operator for the number density of the particles $\hat{n}(q) = \hat{\psi}^+(q)\,\hat{\psi}(q)$ (see § 3). The average value of the operator product $\langle\Psi|\hat{n}(q_1)\,\hat{n}(q_2)|\Psi\rangle$ can be put in the form

$$\lim_{\substack{3\to 1,\ 4\to 2\\ t_3>t_1>t_4>t_2,\ t_3\to t_2}} G(1, 2, 3, 4).$$

It is to be distinguished from the product of the average values of the operators $\hat{n}(q)$, each of which can be written in the form

$$\langle\Psi|\hat{n}(q_1)|\Psi\rangle = i \lim_{\substack{2\to 1\\ t_2>t_1}} G(1, 2) = \lim_{q_2\to q_1} R(q_1, q_2).$$

The corresponding difference can also serve as a measure of the correlation of the particle coordinates; it can be written as

$$\langle\Psi|\hat{n}(q_1)\,\hat{n}(q_2)|\Psi\rangle - \langle\Psi|\hat{n}(q_1)|\Psi\rangle\,\langle\Psi|\hat{n}(q_2)|\Psi\rangle$$
$$\equiv \langle\Psi|\hat{n}(q_1)|\Psi\rangle\,\langle\Psi|\hat{n}(q_2)|\Psi\rangle f(q_1 - q_2), \qquad (20.21)$$

where the function f represents the probability of finding a particle in the coordinate state q_1 (i.e. at the point $\boldsymbol{x}_1$ with known values of the discrete coordinates), given another particle at the point q_2. The average here is over the states of the remaining particles.

The left-hand side of (20.21) may be expressed without difficulty in terms of the two-particle Green function. Using (20.20), we find

$$\lim_{\substack{3\to 1,\, 4\to 2 \\ t_3>t_1>t_4>t_2,\, t_3\to t_2}} K(1, 2, 3, 4) - R(q_1, q_2)\, R(q_2, q_1). \qquad (20.22)$$

The first term of this expression describes the dynamical part of the correlation, the second the exchange part.

In the Hartree–Fock approximation only the second part is left. We will calculate the corresponding expression for the function f, confining ourselves to the case of a homogeneous system:

$$R_0(q_1, q_2) = \delta_{\varrho_1 \varrho_2} \int d^3\boldsymbol{p}\, \theta(p_0^2 - p^2) \exp[i(\boldsymbol{p} \cdot \boldsymbol{x}_1 - \boldsymbol{x}_2)]$$

$$= -\frac{p_0^3}{2\pi^2} \frac{1}{\xi} \frac{\partial}{\partial \xi} \left(\frac{\sin \xi}{\xi} \right) \delta_{\varrho_1 \varrho_2},$$

$$\langle \Psi | \hat{n}(q) | \Psi \rangle = p_0^3 / 6\pi^2,$$

where $\xi = p_0 |\boldsymbol{x}_1 - \boldsymbol{x}_2|$. We hence obtain for the function f

$$f(\xi) = -\frac{9}{\xi^6} (\sin \xi - \xi \cos \xi)^2\, \delta_{\varrho_1 \varrho_2}. \qquad (20.23)$$

This falls off rapidly as ξ increases, corresponding to a weakening of the statistical influence of the particles as their distance apart increases. On the other hand, for $\xi < 1$, i.e. for distances smaller than the average particle wave length, the value of f is close to -1, tending toward this value as ξ decreases, while the average value $\langle \Psi | \hat{n}(q_1)\, \hat{n}(q_2) | \Psi \rangle$ tends to zero. This corresponds to the impossibility of finding two Fermi particles at the same point. This only applies to particles which have the same values for the discrete coordinates and obviously agrees with Pauli's principle.

20.7. There is one more important piece of information which may be obtained from the two-particle Green function. This is the way in which a system changes when it is acted on by some weak, external agent (in particular, external fields of force).

Suppose the Hamiltonian for the interaction of the system with this external agent has the form

$$\delta\hat{H} = \int dq\, \hat{\psi}^+(q)\, \delta\hat{U}\, \hat{\psi}(q), \tag{20.24}$$

i.e. has the nature of a single-particle operator. The operator $\delta\hat{U}$ represents the energy density of the corresponding interaction. For example, if we are considering an external magnetic field $\boldsymbol{H}$, then $\delta\hat{U} = -(\boldsymbol{M}\cdot\boldsymbol{H})$, where $\boldsymbol{M}$ is the magnetization of the system.

We will try to find the change in the single-particle Green function due to $\delta\hat{H}$, supposing it to be small. We will start from the general expression for the Green function in the interaction representation:†

$$G(1, 2) = -i\,\frac{\langle \Psi_0 | T[\hat{\psi}_{\text{Int}}(1)\, \hat{\psi}^+_{\text{Int}}(2)\, \hat{S}] | \Psi_0 \rangle}{\langle \Psi_0 | \hat{S} | \Psi_0 \rangle}. \tag{20.25}$$

Here the S-matrix is defined by the equation

$$i\,\frac{\partial \hat{S}}{\partial t} = (\hat{H}'_{\text{Int}} + \delta\hat{H}_{\text{Int}})\, \hat{S} \tag{20.26}$$

and describes the evolution of a system under the action of both the Hamiltonian for the interaction between particles $\hat{H}'$ and of the Hamiltonian $\delta\hat{H}$.

If we denote the solution of (20.26), for $\delta\hat{H} = 0$, by $\hat{S}_0$, then, putting $\hat{S} = \hat{S}_0 + \delta\hat{S}$, we obtain the equation

$$i\,\frac{\partial\, \delta\hat{S}}{\partial t} = \hat{H}'_{\text{Int}}\, \delta\hat{S} + \delta\hat{H}_{\text{Int}}\, \hat{S}_0.$$

Its solution, as may be easily verified by substitution, has the form

$$\begin{aligned} \delta\hat{S}(t, -\infty) &= -i\int_{-\infty}^{t} d\tau\, \hat{S}_0(t, \tau)\, \delta\hat{H}_{\text{Int}}(\tau)\, \hat{S}_0(\tau, -\infty) \\ &= -i\int_{-\infty}^{t} d\tau\, T[\delta\hat{H}_{\text{Int}}(\tau)\, \hat{S}_0(t, -\infty)]. \end{aligned} \tag{20.27}$$

† Strictly speaking, this expression, which is based on stability condition for Ψ_0, is true only for a time-independent $\delta\hat{U}$.

If we write the correction to $G(1, 2)$ due to external influence as $\delta G(1, 2)$, we have

$$\delta G(1, 2) = \int_{-\infty}^{\infty} d\tau \, \langle \Psi_0 | \hat{S}_0 | \Psi_0 \rangle^{-1}$$
$$\times \{ i \langle \Psi_0 | T[\delta \hat{H}_{\text{Int}}(\tau) \, \hat{S}_0] | \Psi_0 \rangle \, G(1, 2)$$
$$- \langle \Psi_0 | T[\hat{\psi}_{\text{Int}}(1) \, \hat{\psi}^{+}_{\text{Int}}(2) \, \delta \hat{H}_{\text{Int}}(\tau) \, \hat{S}_0] | \Psi_0 \rangle \} .$$

We may then write, from the definition of the Green function,

$$\delta G(1, 2) = \int d3 \, \delta \hat{U}(3) \lim_{4 \to 3} \{ G(1, 2) \, G(3, 4) - G(1, 3, 2, 4) \} . \tag{20.28}$$

We can also, from (20.12), express $\delta G(1, 2)$ in terms of the vertex part Δ:

$$\delta G(1, 2) = \int d3 [G(1, 3) \, \delta \hat{U}(3) \, G(3, 2) + 4 \int d4 \, d5 \, d6 \, d7 \, G(1, 4)$$
$$\times \, \delta \hat{U}(3) \, G(3, 5) \Delta(4, 5, 6, 7) \, G(6, 2) \, G(7, 3)] . \tag{20.29}$$

We will now introduce the inverse Green function $G^{-1}(1, 2)$, which is defined by

$$\int d2 \, G^{-1}(1, 2) \, G(2, 3) = \int d2 \, G(1, 2) \, G^{-1}(2, 3) = \delta(1 - 3).$$

It is obvious that, in the momentum representation,

$$G^{-1}(p, \varepsilon) = [G(p, \varepsilon)]^{-1} .$$

The inverse Green function is closely related to the mass operator:

$$G^{-1}(1, 2) = G_0^{-1}(1, 2) + M(1, 2).$$

This relationship may in fact be obtained by multiplying Dyson's equation from the left by $G_0^{-1}(3, 1)$ and from the right by $G^{-1}(2, 4)$, and integrating over 2 and 3. The definition of the inverse Green function gives the relationship

$$\delta G^{-1}(1, 2) = - \int d3 \, d4 \, G^{-1}(1, 3) \, \delta G(3, 4) \, G^{-1}(4, 2).$$

If we now multiply (20.29) from the left by $G^{-1}(\ldots 1)$ and from the right by $G^{-1}(2, \ldots)$, and integrate over 1 and 2, then

$$\delta G^{-1}(1, 2) = \delta G_0^{-1}(1, 2) + \delta M(1, 2) = - \delta \hat{U}(1) \, \delta(1 - 2)$$
$$- 4 \int d3 \, d4 \, d5 \, \delta \hat{U}(3) \, G(3, 4) \, \Delta(1, 4, 2, 5) \, G(5, 3) . \tag{20.30}$$

This actually contains two independent relationships, since the first term on the right-hand side of (20.30) is equal to the change in G_0^{-1}. We may therefore simply put

$$\delta M(1, 2) = -4 \int d3\, d4\, d5\, \delta\hat{U}(3)\, G(3, 4)\, \Delta(1, 4, 2, 5)\, G(5, 3).$$

20.8. Besides the importance for kinetic effects of the change in the Green function due to a given perturbation, these equations can also be applied to find certain general relationships. To this end, we will choose an operator $\delta\hat{U}$, for which the change in the Green function is known beforehand.

We will select specifically $\delta\hat{U} = \alpha = \text{const.}$; the operator $\delta\hat{H}$ is then equal to $\alpha\hat{N}$, where $\hat{N}$ (the operator representing the total number of particles in the system) commutes with the total Hamiltonian of the system. For this reason, the exact wave function of the system Ψ does not change at all when $\delta\hat{H}$ is added (except for an unimportant phase factor), but the operator $\hat{\Psi}_H(1)$ acquires a factor $\exp(-i\alpha t_1)$. This may be confirmed if we remember that the equation for the perturbation operator $\hat{\psi}_H$ has the form

$$\left(i\frac{\partial}{\partial t_1} - \hat{T} - \alpha\right)\hat{\psi}_H(1) = \int d2\, \hat{\psi}_H^+(2)\, \hat{V}(1, 2)\, \hat{\psi}_H(2)\, \hat{\psi}_H(1).$$

Hence the overall change in the Green function (19.3) is equal to

$$\delta G(1, 2) = -i\alpha(t_1 - t_2)\, G(1, 2).$$

Similarly,

$$\delta G^{-1}(1, 2) = -i\alpha(t_1 - t_2)\, G^{-1}(1, 2).$$

The substitution of this expression in (20.30) gives

$$(t_1 - t_2)\, G^{-1}(1, 2) = -i\delta(1 - 2) - 4i \int d3\, d4\, d5\, G(3, 4) \times \Delta(1, 4, 2, 5)\, G(5, 3). \tag{20.31}$$

In the momentum representation† this relation, which reflects yet another connection between the single-particle and the two-

† When changing to the momentum representation it is essential to remember that the expression under the integral sign is not unique [94, 96]. The left-hand side of (20.31) can be written (in the momentum representation) as

$$-i\frac{\partial}{\partial \varepsilon} G^{-1}(p, \varepsilon)$$

particle Green functions, gives one of the Ward identities. Its difference from (20.17) is that it contains only the effective interaction Δ, whilst (20.17) also contains the true interaction potential V.

An analogous Ward equation can also be written for the quantity $(\boldsymbol{x}_1 - \boldsymbol{x}_2)\, G^{-1}$.

20.9. Another expression for $G(1, 2)$ also exists in the literature. It may be derived in a way which avoids the assumption that the operator $\delta\hat{U}$ is independent of time.

Suppose, therefore, that the Hamiltonian representing the external perturbation has the form of (20.24) with $\delta\hat{U}$ time dependent. We must then alter (20.25) to give the more general expression

$$G(1, 2) = -i\,\langle\Psi_0|\hat{S}^+T[\hat{\psi}_{\mathrm{Int}}(1)\,\hat{\psi}^+_{\mathrm{Int}}(2)\,\hat{S}]|\Psi_0\rangle.$$

The correction to the S-matrix may be calculated, as before, from (20.27). The following expression is obtained for the correction to the Green function:

$$\begin{aligned}\delta G(1, 2) = \int_{-\infty}^{\infty} d\tau\, \{&\langle\Psi_0|\hat{S}_0^+(\tau, -\infty)\,\delta\hat{H}(\tau)\,\hat{S}_0^+(\infty, \tau)\\ &\times T[\hat{\psi}_{\mathrm{Int}}(1)\,\hat{\psi}^+_{\mathrm{Int}}(2)\,\hat{S}_0]|\Psi_0\rangle\\ &- \langle\Psi_0|\hat{S}^+T[\hat{\psi}_{\mathrm{Int}}(1)\,\hat{\psi}^+_{\mathrm{Int}}(2)\,\delta\hat{H}(\tau)\,\hat{S}_0]|\Psi_0\rangle\}.\end{aligned}$$

We can change to the Heisenberg representation with the aid of the general relationships given in Chapter 3. This gives

$$\begin{aligned}\delta G(1,2) = \int_{-\infty}^{\infty} d\tau\, \{&\langle\Psi|\delta\hat{H}_H(\tau)\,T[\hat{\psi}_H(1)\,\hat{\psi}_H^+(2)]|\Psi\rangle\\ &- \langle\Psi|T[\hat{\psi}_H(1)\,\hat{\psi}_H^+(2)\,\delta\hat{H}_H(\tau)]|\Psi\rangle\}. \qquad (20.32)\end{aligned}$$

It is evident from this expression that the region of integration over τ from t_1 to ∞ (for $t_1 > t_2$), or from t_2 to ∞ (for $t_1 < t_2$), automatically disappears.

We will consider the change in the density matrix due to the action of an external perturbing force. If we put $t_2 - t_1 \to +0$ in (20.32), we have

$$\delta R(q_1, q_2, t_1) = i\int_{-\infty}^{t_1} d\tau\,\langle\Psi|[\delta\hat{H}_H(\tau), \hat{\psi}_H^+(q_2, t_1)\,\hat{\psi}_H(q_1, t_1)]_-|\Psi\rangle.$$

This commutator cannot, of course, be calculated explicitly, since the operator refers to different moments of time.

We will reduce this expression to a two-particle Green function by first of all writing the average value as

$$-A(q_1, q_2) + A^*(q_2, q_1),$$

where

$$A(q_1, q_2) \equiv \langle \Psi | \hat{\psi}_H^+(q_2, t_1)\, \hat{\psi}_H(q_1, t_1)\, \delta H_H(\tau) | \Psi \rangle .$$

This latter quality may easily be rewritten in the form

$$\int dq_3\, \delta \hat{U}_{q_3}(t_3) \lim_{\substack{4 \to 3 \\ t_2 > t_1 > t_4 > t_3 \\ t_2 \to t_1}} G(1, 3, 2, 4).$$

Hence we have finally

$$\delta R(q_1, q_2, t_1) = \int_{-\infty}^{t_1} d3\, \delta \hat{U}(3) \lim_{\substack{4 \to 3 \\ t_2 > t_1 > t_4 > t_3,\ t_2 \to t_1}} [G(1,3,4,2) - G^*(1,3,4,2)]. \tag{20.33}$$

Knowing the two-particle Green function for the unperturbed system, it becomes possible to calculate the single-particle characteristics of a weakly perturbed system. The above equation determines many of the transport properties of the system: in particular the conductivity.

21. The Excited States of a System (Hartree–Fock Approximation)

21.1. The previous sections have considered almost exclusively the ground state of a system only. Problems concerning excited states of systems, where the energy differs from the minimum, are also of great importance in many-body theory.

Besides systems with excited states, which have the same number of particles (N) as the ground state, the alternative case, where this number differs from N, is of considerable importance.† In particular

† From now on we will indicate the number of particles in the argument of the wave function. Excited states of systems containing N particles will be called pair states; if they contain $N \pm 1$ particles, they will be called single-particle (or single-hole) states.

there are systems where the excited states have one extra particle or hole, i.e. which contain $N \pm 1$ particles.

An important class of problems in many-body theory concerns the way in which external particles will interact with a system of particles of the same sort. On entering such a system, a particle can give up an appreciable amount of energy, thus exciting the system, and remain in it for a greater or lesser length of time. The resultant state of the system obviously belongs to the class of problems we are considering.

21.2. It is expedient to begin our consideration of excited states, as we did for the ground state, with the Hartree–Fock approximation. If we denote the corresponding wave function and energy of the excited state by Ψ_{0n} and E_{0n}, we can write

$$H_0\Psi_{0n} = E_{0n}\Psi_{0n}, \tag{21.1}$$

where it is assumed that the zero-approximation Hamiltonian $\hat{H}_0$ coincides with (4.29). With this choice of $\hat{H}_0$, the self-consistency is limited to the configuration corresponding to the ground state of the system. Hence only this state is described in the best way (see § 4). It would be possible, in principle, to construct a zero-approximation Hamiltonian for each state, with the requirement that the quantity $\langle\Psi_{0n}|(\hat{H} - \hat{H}_0)^2|\Psi_{0n}\rangle$ be a minimum. But such a way of approaching the problem could lead to a quite unjustified complication of the work. We will therefore always select a single Hamiltonian in the Hartree–Fock approximation for all configurations: one which describes the ground state best of all. In the following chapter, which is devoted to quantum statistics, the Hamiltonian $\hat{H}_0$ is determined by the condition $\langle(\hat{H} - \hat{H}_0)^2\rangle_0 = \min$, where the symbol $\langle\ldots\rangle_0$ denotes a statistical averaging over an ensemble. This gives the best average description of a system.

We may now retain all our previous methods when we describe the excited states of a system. In particular the wave functions of the basis χ_ν, the expression for the field operators, etc., retain their form. There are, however, two factors which are somewhat different for the description of excited states. The first is that, when an N-product of the field operators is averaged in the state Ψ_{0n}, the result differs from zero. The second concerns the degeneracy of excited energy levels.

21.3. We next consider some properties of the solutions of (21.1) for states with fixed occupation numbers. A particular solution of (21.1) for a particle–hole excited state can always be chosen in the form

$$\Psi_{0n}(N) = \prod_{i,j=1}^{k} \hat{a}_{\nu_i}^{+}\hat{b}_{\mu_j}^{+}\Psi_0. \tag{21.2}$$

It corresponds to the presence of k pairs, with particles in the states ν_i and holes in the states μ_j, where $\varepsilon_{\nu_i} > \varepsilon_F$, $\varepsilon_{\mu_j} < \varepsilon_F$. Similarly, for a single-particle excited state of the system, we can write

$$\Psi_{0n}(N+1) = \hat{a}_{\nu}^{+} \prod_{i,j=1}^{k} \hat{a}_{\nu_i}^{+}\hat{b}_{\mu_j}^{+}\Psi_0. \tag{21.3}$$

The number k is not fixed in either case but must be less than the total number of particles N.

As a confirmation that (21.2) and (21.3) do indeed satisfy (21.1), and as a way of calculating the corresponding energy E_{0n}, we will consider the commutators:

$$[\hat{H}_0, \hat{a}_{\nu}^{+}]_- = \varepsilon_\nu \hat{a}_{\nu}^{+}; \quad [\hat{H}_0, \hat{b}_{\mu}^{+}]_- = -\varepsilon_\mu \hat{b}_{\mu}^{+}.$$

We have used here (4.3) and the equality $\hat{A}_{\nu}^{+}\hat{A}_{\nu} = \hat{b}_{\nu}\hat{b}_{\nu}^{+} (\varepsilon_\nu < \varepsilon_F)$ (see § 8). Hence, transposing the operators $\hat{H}_0$ and $\hat{a}^{+}$ and $\hat{b}^{+}$ consecutively in the expression for Ψ_{0n}, we can confirm that these latter are eigenfunctions of $\hat{H}_0$. A similar procedure gives their eigenvalues:

$$E_{0n} = E_0 + \sum_{\nu} \varepsilon_\nu - \sum_{\mu} \varepsilon_\mu,$$

where the summation is over all occupied states of particles and holes; the quantity E_0 is the energy of the ground state (see § 4). We can, in other words, write down the following expression for the energy of the excited state of a system:

$$E_{0n} - E_0 = \sum_{\nu} \varepsilon_\nu (n_\nu^{(p)} - n_\nu^{(h)}), \tag{21.4}$$

where $n_\nu^{(p)}$ is the occupation number for the particles and $n_\nu^{(h)}$ is the occupation number for the holes in the given state.

The difference $E_{0n} - E_0$ is given a special name—the excitation energy—and is denoted by ΔE_{0n}. This quantity can be written in a somewhat different way for a single-particle excited state so long as N is sufficiently large. We will write down ΔE_{0n} for this case, indicating explicitly the number of particles:

$$\Delta E_{0n} = E_{0n}(N+1) - E_0(N).$$

Adding and subtracting the ground state energy for a system of $N+1$ particles, $E_0(N+1)$, we have

$$\Delta E_{0n} = \Delta' E_{0n} + \mu, \tag{21.5}$$

where $\Delta' E_{0n} = E_{0n}(N+1) - E_0(N+1)$ represents the excitation energy of a system of $N+1$ particles (or a system of N particles since N is large); $\mu = \partial E_0(N)/\partial N \approx E_0(N+1) - E_0(N)$ is the chemical potential of the system. For a single-hole state, the sign of μ in this expression has to be changed.

21.4. The energy levels corresponding to (21.1) are, as a rule, degenerate.† The given value of the excitation energy ΔE_{0n} corresponds to a whole series of different excited states of the system, which can be differentiated by the number of pairs and by the particle and hole states that go to make up these pairs.

We will begin by considering these particle-hole excited states. The simplest state of this sort corresponds to the presence of a single pair; the corresponding excitation energy can be written as $\varepsilon_\nu - \varepsilon_\mu$, where ν is the particle state and μ the hole state. It is evident that, if the relationship

$$\Delta E_{0n} = \varepsilon_\nu - \varepsilon_\mu$$

(ΔE_{0n} is the given excitation energy) can be satisfied for different states μ and ν, then the degeneracy of this particular level appears immediately. The spectrum of the single-particle states of a system is most frequently continuous (or almost continuous) in character. In this case, the equation has a set of solutions. The wave function for a state with a single pair, of excitation energy ΔE_{0n}, is represented

† We have avoided considering degeneracy of the ground state by limiting ourselves to systems with filled shells.

by a superposition:

$$\Psi_{0n}(N) = \sum_{\mu,\nu} C_{\mu\nu}\delta(\varepsilon_\mu - \varepsilon_\nu - \Delta E_{0n})\, \hat{a}_\nu^+ \hat{b}_\mu^+ \Psi_0, \qquad (21.6)$$

where $C_{\mu\nu}$ are the appropriate numerical coefficients, and the δ-function ensures the constancy of the excitation energy.

The simplest wave function for a single-particle excited state is

$$\Psi_{0n}(N+1) = \sum_{\nu} C_\nu \delta(\varepsilon_\nu - \Delta E_{0n})\, \hat{a}_\nu^+ \Psi_0. \qquad (21.7)$$

Pairs are absent here and degeneracy arises from the possibility of changing the state of a single particle without altering the value of the energy; in particular, by changing the direction of the spin, momentum, etc.

The other reason for degeneracy arises from the possibility of selecting a different number of pairs k. The general expression for the wave function of a state with excitation energy ΔE_{0n} consequently takes the form

$$\Psi_{0n}(N) = \sum_{\nu,\mu,k} C_{\nu_i,\mu_j,k}\delta(\Sigma\varepsilon_{\nu_i} - \Sigma\varepsilon_{\mu_j} - \Delta E_{0n}) \prod \hat{a}_{\nu_i}^+ \hat{b}_{\mu_j}^+ \Psi_0, \qquad (21.8)$$

and similarly for $\Psi_{0n}(N+1)$.

21.5. The above expansion for the wave function of an excited state can be put in a more convenient form:

$$\begin{aligned}\Psi_{0n}(N) = \{&\textstyle\int dq_1\, dq_2\, \Phi_{0n}(q_1, q_2)\, N[\hat{\psi}^+(q_1)\, \hat{\psi}(q_2)] \\ &+ \textstyle\int dq_1\, dq_2\, dq_3\, dq_4\, \Phi_{0n}(q_1, q_2, q_3, q_4) \\ &\times N[\hat{\psi}^+(q_1)\, \hat{\psi}^+(q_3)\, \hat{\psi}(q_2)\, \hat{\psi}(q_4)] + \cdots\}\, \Psi_0, \qquad (21.9)\end{aligned}$$

$$\begin{aligned}\Psi_{0n}(N+1) = \{&\textstyle\int dq_1\, \Phi_{0n}(q_1)\, \hat{\psi}^+(q_1) + \int dq_1\, dq_2\, dq_3 \\ &\times \Phi_{0n}(q_1, q_2, q_3)\, N[\hat{\psi}^+(q_1)\, \hat{\psi}^+(q_3)\, \hat{\psi}(q_2)] + \cdots\}\, \Psi_0. \qquad (21.10)\end{aligned}$$

We have introduced here the normal products of the operators, so that only the creation part of the operators should remain. The appearance of even a single annihilation operator for a particle or hole would produce a zero result.

The functions Φ_{0n}, which determine the weight with which a given configuration of particles and holes contributes to Ψ_{0n}, need to satisfy several conditions. We have previously denoted by 1, 3, ... those arguments of Φ_{0n} which correspond to particle-creation operators, and by 2, 4, ... hole-creation operators. If we expand Φ_{0n} in terms of the set of functions, χ_ν, i.e. change to the energy representation, it is easy enough to see that the corresponding coefficients of the expansion can only be non-zero for the region above the Fermi surface for the coordinates 1, 3, ..., and below the Fermi surface for the coordinates 2, 4, ... Other regions of the spectrum automatically disappear from (21.9) and (21.10).

Apart from this, the functions Φ_{0n} must be such that only states corresponding to the same excitation energies E_{0n} enter into the required superposition. For this to be true, it is essential that the coefficients of the expansion of Φ_{0n}, in terms of the set χ_ν, contain the function $\delta(\varepsilon_{\nu_1} + \varepsilon_{\nu_3} + \cdots - \varepsilon_{\nu_2} - \varepsilon_{\nu_4} - \cdots - \Delta E_{0n})$, where the + sign represents the energies corresponding to the coordinates of the particles $q_1, q_3, \ldots$, and the − sign to the coordinates of the holes $q_2, q_4, \ldots$ In particular,

$$\Phi_{0n}(q_1) = \sum_\nu \alpha_\nu (1 - n_\nu)\,\delta(\varepsilon_\nu - \Delta E_{0n})\,\chi_\nu(q_1),$$

$$\Phi_{0n}(q_1, q_2) = \sum_{\mu\nu} \alpha_{\mu\nu} n_\mu (1 - n_\nu)\,\delta(\varepsilon_\nu - \varepsilon_\mu - \Delta E_{0n})\,\chi_\mu^*(q_2)\,\chi_\nu(q_1),$$

where the α are certain numerical coefficients.

These pecularities of the functions Φ_{0n} lead to some important relationships. From the expansion of the density matrix

$$R_0(q_1, q_2) = \sum_\nu n_\nu \chi_\nu^*(q_2)\,\chi_\nu(q_1),$$

we find

$$\int dq_2\, R_0(q_1, q_2)\,\Phi_{0n}(q_2) = 0. \tag{21.11}$$

Convenient relationships can also be derived for the functions $\Phi_{0n}(q_1, q_2)$:

$$\left.\begin{aligned} &\int dq_2\, R_0(q_1, q_2)\,\Phi_{0n}(q_2, q_1) = 0, \\ &\int dq_1\, R_0(q_1, q_2)\,\Phi_{0n}(q_2, q_1) = \Phi_{0n}(q_2, q_2). \end{aligned}\right\} \tag{21.12}$$

From these relations, and the expansions (21.9) and (21.10) it is an easy task to show that the weight functions Φ_{0n} can be represented by matrix elements in the following way:

$$\left.\begin{aligned} \Phi_{0n}(q_1) &= \langle \Psi_0 | \hat{\psi}(q_1) | \Psi_{0n}(N+1) \rangle, \\ \Phi_{0n}(q_1, q_2) &= \langle \Psi_0 | \hat{\psi}^+(q_2)\, \hat{\psi}(q_1) | \Psi_{0n}(N) \rangle. \end{aligned}\right\} \tag{21.13}$$

21.6. It is useful to introduce time-dependent weight functions

$$\left.\begin{aligned} \Phi_{0n}(1) &= \langle \Psi_0 | \hat{\psi}_{\mathrm{Int}}(1) | \Psi_{0n}(N+1) \rangle, \\ \Phi_{0n}(1, 2) &= \langle \Psi_0 | T[\hat{\psi}^+_{\mathrm{Int}}(2)\, \hat{\psi}_{\mathrm{Int}}(1)] | \Psi_{0n}(N) \rangle, \end{aligned}\right\} \tag{21.14}$$

etc. These transform to the functions in (21.13) when $t_1 \to 0$ and $t_1, t_2 \to 0$ respectively. In point of fact the T-product symbol in the latter equality will only be written for convenience in subsequent calculations. The difference between $T(\hat{\psi}^+\hat{\psi})$ and $\hat{\psi}^+\hat{\psi}$ reduces to a c-number and, allowing for the orthogonality of the states Ψ_0 and Ψ_{0n} vanishes.

From the equation of motion for an operator in the interaction representation, the following equations are easily obtained:

$$\left.\begin{aligned} \left(i\frac{\partial}{\partial t_1} - \hat{T} - \hat{W}\right)\Phi_{0n}(1) &= 0, \\ \left(i\frac{\partial}{\partial t_1} - \hat{T} - \hat{W}\right)\Phi_{0n}(1, 2) &= 0, \end{aligned}\right\} \tag{21.15}$$

and an associated equation for the derivative with respect to t_2.

It is a straightforward task to find the explicit dependence of these quantities on time. From (2.8) we have

$$\Phi_{0n}(1) = \Phi_{0n}(q_1) \exp(-i\Delta E_{0n} t_1). \tag{21.16}$$

In a similar way we can show that

$$\Phi_{0n}(1, 2) = \Phi_{0n}(q_1, 0; q_2, \tau) \exp(-i\Delta E_{0n} t_1). \tag{21.16'}$$

Here $\tau = t_2 - t_1$. Thus the matrix elements we are considering, unlike the average values, are not only dependent on the difference in the times, but also on their absolute values.

It is possible to write a series of relationships, similar to (21.11) and (21.12), for the time-dependent functions Φ_{0n}. We note the expansion of the free-particle Green function (10.5)

$$G_0(1, 2) = - i \sum_{\nu} \chi_{\nu}^*(q_2)\, \chi_{\nu}(q_1) \exp\,[-i\varepsilon_{\nu}(t_1 - t_2)] \begin{cases} 1 - n_{\nu} & t_1 < t_2 \\ - n_{\nu} & t_1 < t_2 \end{cases}$$

and the time dependence of (21.16) and (21.16′). Then, from the expansions of the functions $\Phi_{0n}(q_1)$, etc., we have

$$\int dq_2\, G_0(1, 2)\, \Phi_{0n}(2) = -i\,\theta(t_1 - t_2)\, \Phi_{0n}(1). \quad (21.17)$$

Similarly, we can show that

$$\left.\begin{aligned} \int dq_2\, G_0(1, 2)\, \Phi_{0n}(2, 3) &= 0 \quad t_1 < t_2, \\ \int dq_2\, \Phi_{0n}(3, 2)\, G_0(2, 1) &= 0 \quad t_1 < t_2, \\ \int dq_1\, dq_2\, G_0(1, 2)\, \Phi_{0n}(2, 1) &= 0. \end{aligned}\right| \quad (21.18)$$

We can next express the wave functions of an excited state of a system in terms of the time-dependent functions $\Phi_{0n}(q_1)$ etc.

$$\Psi_{0n}(N) = \int dq_1\, dq_2\, \Phi_{0n}(1, 2)\, N[\hat{\psi}^+_{\mathrm{Int}}(1)\, \hat{\psi}_{\mathrm{Int}}(2)]\, \Psi_0 + \cdots \quad (21.19)$$

$$\Psi_{0n}(N + 1) = \int dq_1\, \Phi_{0n}(1)\, \hat{\psi}^+_{\mathrm{Int}}(1)\, \Psi_0 + \cdots \quad (21.20)$$

It comes down, in fact, to replacing $\Phi_{0n}(q_1)$ by $\Phi_{0n}(1)$, etc., and replacing the operators in the Schrödinger representation by the operators in the interaction representation. To demonstrate the truth of this assertion, we can substitute (21.14) in (21.20) and expand the field operators. This gives

$$\Psi_{0n}(N + 1) = \int dq_1 \sum_{\mu, \nu} \langle \Psi_0 | \hat{a}_{\nu} | \Psi_{0n} \rangle\, \hat{a}^+_{\mu}\, \chi_{\nu}(q_1)\, \chi^*_{\mu}(q_1) \times \exp\,[i(\varepsilon_{\mu} - \varepsilon_{\nu})\, t_1]\, \Psi_0.$$

We can use the orthogonality of the functions χ_{ν} to show that the time t_1 disappears from this expression, i.e. that we return to (21.10). In the same way we can demonstrate that (21.19) is correct.

21.7. We will now examine the physical meaning of the quantities Φ_{0n}. We have already noted above that they determine the amount which a given configuration of particles and holes contributes to the wave function of an excited state. They also have a more immediate physical interpretation. As is evident from the expansion in terms of the set χ_ν, the function $\Phi_{0n}(q_1)$ simply represents the wave function of a "surplus" particle which is responsible for the difference between the state $\Psi_{0n}(N + 1)$ of the system and the ground state.

In a similar way the function $\Phi_{0n}(q_1, q_2)$ can be treated as the wave function of a pair which is excited from the "background" of the ground state. $\chi_\nu(q_1)$ here represents the wave function of a particle, and $\chi_\nu^*(q_2)$ represents the wave function of a hole. It should be remarked in this connection that a hole must be described by a conjugate wave function. This follows from a comparison of the parts of the field operators corresponding to the creation (or annihilation) of particles and of holes [see eqn. (8.4)].

More complicated Φ_{0n} functions can be interpreted in a similar sort of way. One can say, in general, that Φ_{0n} is the wave function of the particle–hole complex, responsible for the difference between Ψ_{0n} and Ψ_0.

We must now make a few comments on the normalization of the wave functions. If we remember that the terms in the expansions (21.9) and (21.10) are mutually orthogonal (this follows from the fact that the corresponding scalar product contains the average value of an unequal number of creation and annihilation operators, and therefore reduces to zero), it is obvious that the condition $\langle\Psi_{0n}\Psi_{0n}\rangle = 1$ gives

$$\left.\begin{aligned} \int dq_1 |\Phi_{0n}(q_1)|^2 + \cdots &= 1, \\ \int dq_1\, dq_2 |\Phi_{0n}(q_1, q_2)|^2 + \cdots &= 1. \end{aligned}\right\} \tag{21.21}$$

We have written out here only the first terms of the various expansions. The extension to higher terms of (21.9) and (21.10) is obvious.

We will next consider certain of the physical characteristics of an excited state: first of all its density matrix which is determined by

$$R_{0n}(q_1, q_2) = \langle\Psi_{0n}|\hat{\psi}^+(q_2)\,\hat{\psi}(q_1)|\Psi_{0n}\rangle. \tag{21.22}$$

Substituting the expansions (21.9), (21.10) in this, we find for the single-particle excited state of a system

$$R_{0n}(q_1, q_2) = \int dq_3\, dq_4\, \Phi_{0n}(q_3)\, \Phi^*_{0n}(q_4)$$
$$\times \langle \Psi_0 | \hat{\psi}(q_4)\, \hat{\psi}^+(q_2)\, \hat{\psi}(q_1)\, \hat{\psi}^+(q_3) | \Psi_0 \rangle .$$

Averaging this by the methods described in Appendix A, and using (21.11) and (21.17), we obtain

$$R_{0n}(q_1, q_2) = R_0(q_1, q_2) + \Phi^*_{0n}(q_2)\, \Phi_{0n}(q_1). \qquad (21.23)$$

This gives the density matrix of the excited state as the sum of the density matrix of the ground state and of a complex whose wave function is Φ_{0n}.

Tedious, but straightforward, calculations for the particle–hole excited state of a system give

$$R_{0n}(q_1, q_2) = R_0(q_1, q_2) + \int dq_3\, [\Phi^*_{0n}(q_2, q_3)\, \Phi_{0n}(q_1, q_3)$$
$$- \Phi^*_{0n}(q_3, q_1)\, \Phi_{0n}(q_3, q_2)]. \qquad (21.24)$$

The first term inside the square brackets is the density matrix for a particle, the second is the density matrix for a hole; the minus sign indicates the absence of a particle in that state.

The following matrix element

$$M_{0n} = \langle \Psi_0 | \hat{A} | \Psi_{0n} \rangle$$

also represents an important characteristic of an excited state of a system of N particles. $\hat{A}$ is some operator describing an external force capable of exciting the system. The excitation probability can be estimated from the quantity M_{0n}. The operator $\hat{A}$ is most frequently single-particle in character. In this case

$$M_{0n} = \int dq_1 \langle \Psi_0 | \hat{\psi}^+(q_1)\, \hat{A}_{q_1}\, \hat{\psi}(q_1) | \Psi_{0n} \rangle$$
$$= \int dq_1 \lim_{q_2 \to q_1} \hat{A}_{q_1}\, \Phi_{0n}(q_1, q_2). \qquad (21.25)$$

It is therefore directly related to the weight function Φ_{0n}.

22. The Excited States of a System (Allowing for the Correlation Interaction)

22.1. We have so far been considering the excited states of a system in terms of the Hartree–Fock approximation. We will now take into account the correlation interaction between particles and relate the precise wave function of an excited state Ψ_n to the quantity Ψ_{0n} introduced above. We will write the latter in the condensed form:

$$\Psi_{0n} = \sum_{\alpha} C_{\alpha} \Psi_{0n}^{(\alpha)},$$

where $\Psi_{0n}^{(\alpha)}$ are the wave functions of excited states with the same excitation energy and with fixed occupation numbers.

We now repeat the discussion of § 9 concerning the connection between Ψ and Ψ_0, with allowance for the degeneracy of the considered state of the system. This makes (9.18) inapplicable, and it must now be replaced by the general relationship†

$$\hat{S}\Psi_{0n} = \sum_{\alpha,\beta} C_{\alpha} \langle \alpha | \hat{S} | \beta \rangle \Psi_{0n}^{(\beta)}. \tag{22.1}$$

The summation here, as for the previous equation, is taken over states with the same energy.

Equation (9.23) which is satisfied by the wave function in the interaction representation $\Psi_{\text{Int}\,n}(-\infty)$, still holds good, since the absence of degeneracy was not postulated in deriving it. It remains to discover under what conditions the function Ψ_{0n} satisfies the same equation.

We can use (2.1) and the equation $\hat{H}_0 \Psi_{0n}^{(\alpha)} = E_{0n} \Psi_{0n}^{(\alpha)}$ to show that

$$\hat{S}^{-1} \hat{H}^0 \hat{S} \Psi_{0n} = E_{0n} \Psi_{0n}.$$

Moreover,

$$\frac{i\delta}{2} \hat{S}^{-1} \frac{\partial \hat{S}}{\partial \lambda} \Psi_{0n} = \sum_{\alpha\,\beta} C_{\alpha} \sigma_{\alpha\beta} \Psi_{0n}^{(\beta)},$$

† We have introduced the notation

$$\langle \alpha | \hat{S} | \beta \rangle = \langle \Psi_{0n}^{(\alpha)} | \hat{S} | \Psi_{0n}^{(\beta)} \rangle.$$

We note that

$$\hat{S}^{-1} \Psi_{0n}^{(\alpha)} = \sum_{\beta} \langle \alpha | \hat{S}^{-1} | \beta \rangle \Psi_{0n}^{(\beta)}.$$

where

$$\sigma_{\alpha\beta} = \frac{i\delta}{2}\left\langle \alpha \middle| \hat{S}^{-1}\frac{\partial \hat{S}}{\partial \lambda} \middle| \beta \right\rangle.$$

Hence, substitution of Ψ_{0n} in eqn. (9.23) gives the set of equations

$$\sum_{\beta} C_{\beta}[(E_{0n} - E_n)\,\delta_{\alpha\beta} + \sigma_{\alpha\beta}] = 0. \qquad (22.2)$$

A solution of this set determines, first of all, the correlation contribution to the excitation energy, $\Delta E_n = E_n - E$, and secondly, the values of the coefficients C_α, for which the function $\Psi_{0n} = \Sigma\, C_\alpha \Psi_{0n}^{(\alpha)}$ satisfies the same equation as does $\Psi_{\mathrm{Int}\,n}(-\infty)$. Using the same reasoning as in § 9.7, we may conclude that the states $\Psi_{\mathrm{Int}\,n}(-\infty)$ and Ψ_{0n} have the same energy. However, owing to the degeneracy present here, we cannot conclude that the quantities $\Psi_{\mathrm{Int}\,n}(-\infty)$ and Ψ_{0n} coincide. Our further discussion therefore differs from that in § 9.7.

There is no unique solution to eqns. (22.2): there are r the number of substates) sets of numbers $C_\alpha^{(i)} (i = 1 \ldots r)$ for which the function

$$\Psi_{0n}^{(i)} = \sum_{\alpha} C_\alpha^{(i)} \Psi_{0n}^{(\alpha)}$$

satisfies the same equation as $\Psi_{\mathrm{Int}\,n}(-\infty)$. We can introduce r functions $\Psi_{\mathrm{Int}\,n}^{(i)}(-\infty)$, each of which is equal to $\Psi_{0n}^{(i)}$; this assertion does not contradict the degeneracy of the considered level: it comes down to choosing a specific basis, in terms of which the arbitrary state $\Psi_{\mathrm{Int}\,n}(-\infty)$ can be expanded. We can now write

$$\Psi_n^{(i)} = \hat{S}(0, -\infty)\, \Psi_{0n}^{(i)}. \qquad (22.3)$$

Thus a level, which is degenerate in the absence of correlation interaction, leads to a set of r accurate wave functions, when the interaction is included. There may be, amongst these, functions corresponding to the same value of the energy. This indicates that the correlation interaction does not completely remove the degeneracy.

22.2. These results indicate that each of the accurate wave functions for the excited state of a system is connected by the

S-matrix with a completely determinate and by no means arbitrary superposition of the functions $\Psi_{0n}^{(\alpha)}$. In other words if interaction is excluded, the functions $\Psi_n^{(i)}$ transform to the zero-approximation wave functions with completely determinate values of the expansion coefficients C_α, which can be found by solving eqns. (22.2).

A situation like this is typical of problems with degenerate energy levels. As is shown in courses on quantum mechanics, there are, among all the possible superpositions of states with equal energy, certain "stable" superpositions which are distinguished by the characteristic that, when a small perturbing potential is applied, they too, change only slightly. When interaction is excluded, the wave function for a stationary state of the system transforms to a stable zero-approximation wave function.

We will consider eqns. (22.2), which determine the coefficients C_β, on the supposition that the Hamiltonian $\hat{H}'$ is small, and with a limitation to first-order terms. If we use the expansion for the S-matrix

$$\hat{S} = 1 - i\lambda \int_{-\infty}^{\infty} dt \exp(-\delta|t|)\, \hat{H}'_{\text{Int}}(t) + \cdots$$

we have

$$\sigma_{\alpha\beta} = \frac{\delta}{2} \int_{-\infty}^{\infty} dt \exp(-\delta|t|)\, \langle\alpha|\hat{H}'_{\text{Int}}(t)|\beta\rangle.$$

But

$$\langle\alpha|\hat{H}'_{\text{Int}}|\beta\rangle = \langle\alpha|\exp(i\hat{H}_0 t)\, \hat{H}' \exp(-i\hat{H}_0 t)|\beta\rangle = \langle\alpha|\hat{H}'|\beta\rangle$$

owing to the identical energy of states α and β. Working out the elementary integral in t, we find

$$\sigma_{\alpha\beta} = \langle\alpha|\hat{H}'|\beta\rangle$$

and eqns. (22.2) take the form

$$\sum_\beta C_\beta[(E_{0n} - E_n)\, \delta_{\alpha\beta} + \langle\alpha|\hat{H}'|\beta\rangle] = 0. \tag{22.4}$$

These equations are precisely the same as the analogous equations for determining the coefficients of a stable state in the lowest order of degenerate perturbation theory [31].

The determinant of the set of eqns. (22.2), equated to zero, can have solutions for $E_n - E_{0n}$ which are not equal to each other.

This indicates that the initially degenerate energy level is split up under the influence of the correlation interaction. It follows from this that an arbitrarily chosen superposition of wave functions in the Hartree–Fock approximation transforms to a non-stationary state of the system when the correlation interaction is included.

In point of fact this superposition can always be decomposed into stable wave functions, each with the time evolution shown in (22.3), and each of which leads to a stationary state of the system with its own energy value. Hence the evolution of an arbitrarily chosen state of a system results in a superposition of stationary states with different values for the energy. Such a state is not, of course, stationary, and its energy is an indeterminate quantity. The corresponding energy dispersion is determined by the energies which go to make up the superposition.

When we speak of the stability of an excited state of a system, whose wave functions have expansion coefficients satisfying (22.2), we are using this term in the same way, essentially, as it was applied in § 9 to the ground state of a system. In other words if Ψ_{0n} is some kind of superposition, then

$$\left.\begin{aligned} \hat{S}\Psi_{0n} &= \langle\Psi_{0n}|\hat{S}|\Psi_{0n}\rangle\,\Psi_{0n}, \\ \Psi^*_{0n}\hat{S}^+ &= \langle\Psi_{0n}|\hat{S}|\Psi_{0n}\rangle^{-1}\,\Psi^*_{0n}. \end{aligned}\right\} \qquad (22.5)$$

In fact the solution of the set (22.2) implies simultaneously the reduction of the matrix $\sigma_{\alpha\beta}$ to the diagonal form. Hence

$$\hat{S}^{-1}\frac{\partial\hat{S}}{\partial\lambda}\Psi_{0n} = \left\langle\Psi_{0n}\left|\hat{S}^{-1}\frac{\partial\hat{S}}{\partial\lambda}\right|\Psi_{0n}\right\rangle\Psi_{0n},$$

which is equivalent to eqns. (22.5).

We may say in conclusion that an examination of the excited states of a system leads, due to their degeneracy, to the necessity for an additional, dynamical calculation, which reduces to finding the stable, zero-approximation wave functions. The direct solution of this problem is difficult. We will need to consider more round-about methods.

22.3. When the correlation interaction between particles is included, the expressions for the matrix elements Φ_{0n} take the form†

$$\left.\begin{aligned}\Phi_n(q_1) &= \langle\Psi|\hat{\psi}(q_1)|\,\Psi_n(N+1)\rangle,\\ \Phi_n(q_1, q_2) &= \langle\Psi|\hat{\psi}^+(q_2)\,\hat{\psi}(q_1)|\Psi_n(N)\rangle,\end{aligned}\right\}\tag{22.6}$$

etc. It is convenient here, as before, to introduce time coordinates and consider expressions of a more general type:

$$\left.\begin{aligned}\Phi_n(1) &= \langle\Psi|\hat{\psi}_H(1)|\Psi_n\rangle,\\ \Phi_n(1, 2) &= \langle\Psi|T[\hat{\psi}_H^+(2)\,\hat{\psi}_H(1)]|\Psi_n\rangle.\end{aligned}\right\}\tag{22.7}$$

The reverse transformation to the functions (22.6) can be made in the following way:

$$\Phi_n(q_1) = \lim_{t_1\to 0}\Phi_n(1),$$

$$\Phi_n(q_1, q_2) = \lim_{t_1,\, t_2\to 0}\Phi_n(1, 2).$$

It is easy to discover the time dependence of the matrix elements (22.7) in the general form. From the general relation (2.5), we have

$$\Phi_n(1) = \Phi_n(q_1)\exp(-i\Delta E_n t),\tag{22.8}$$

where ΔE_n is the exact excitation energy. For the functions $\Phi_n(1, 2)$, we obtain, analogously to (2.16′)

$$\Phi_n(1, 2) = \Phi_n(q_1, 0;\, q_2, \tau)\exp(-i\Delta E_n t),\tag{22.9}$$

where $\tau = t_1 - t_2$.

We can therefore write

$$\lim_{\tau\to 0}\Phi_n(1, 2) = \Phi_n(q_1, q_2)\exp(-i\Delta E_n t).\tag{22.10}$$

The expressions in (22.7) can be written in an interaction representation, similar to the results of § 15:

$$\left.\begin{aligned}\Phi_n(1) &= \frac{\langle\Psi_0|T[\hat{\psi}_{\text{Int}}(1)\,\hat{S}]|\Psi_{0n}(N+1)\rangle}{\langle\Psi_0|\hat{S}|\Psi_0\rangle},\\ \Phi_n(1, 2) &= \frac{\langle\Psi_0|T[\hat{\psi}_{\text{Int}}^+(2)\,\hat{\psi}_{\text{Int}}(1)\,\hat{S}]|\,\Psi_{0n}(N)\rangle}{\langle\Psi_0|\hat{S}|\Psi_0\rangle}.\end{aligned}\right\}\tag{22.11}$$

† These relations, like subsequent ones, can be written down for each of the functions $\Psi_n^{(i)}$, The index i will be omitted for simplicity.

We have used here (22.3) together with the stability condition for the ground state.

In the same sort of way, we can also introduce matrix elements of the type

$$\tilde{\Phi}_n(1) = \langle \Psi | \hat{\psi}_H^+(1) | \Psi_n(N-1) \rangle, \tag{22.12}$$

which correspond to states with one particle missing or, which is the same thing, with one excess hole.

22.4. When correlation interaction in the excited states of a system is taken into account, it becomes necessary to have some sort of information on the stability of the zero-approximation wave function Ψ_{0n}. The direct solution of eqns. (22.2) is by no means easy, although it is, in principle, possible if the parameters are small.

There is, however, a simple way of avoiding the direct solution of the problem, which is in many cases successful. One makes some very simple suppositions about the form which the stability of the function Ψ_{0n} takes; a subsequent check shows how far these suppositions are true. If the check is negative, a more complicated expression for Ψ_{0n} is chosen, and so on.

The simplest assumption is that the expansion of the function Ψ_{0n} consists of the first term

$$\left.\begin{aligned} \Psi_{0n}(N+1) &= \int dq_1\, \Phi_{0n}(1)\, \hat{\psi}_{\mathrm{Int}}^+(1)\, \Psi_0, \\ \Psi_{0n}(N) &= \int dq_1\, dq_2\, \Phi_{0n}(1,2)\, N[\hat{\psi}_{\mathrm{Int}}^+(2)\, \hat{\psi}_{\mathrm{Int}}(1)]\, \Psi_0. \end{aligned}\right\} \tag{22.13}$$

When the correlation interaction is included, we do not arrive at a stationary state of the system. It may turn out that the state is only slightly non-stationary (the measure of this will be accurately defined below). Hence we come to an approximate description of an excited state of a system; the accuracy being greater as the measure of its "non-stationariness" decreases.

We will now consider the single-particle excited state of a system all over again from this viewpoint. Substituting (22.13) in (22.11), we find

$$\Phi_n(1) = \int dq_2\, \Phi_{0n}(2) \frac{\langle \Psi_0 | T[\hat{\psi}_{\mathrm{Int}}(1)\, \hat{S}]\, \hat{\psi}_{\mathrm{Int}}^+(2) | \Psi_0 \rangle}{\langle \Psi_0 | \hat{S} | \Psi_0 \rangle}. \tag{22.14}$$

This expression can be reduced to the exact Green function by means of the following convenient approach, which we will also use later. We remember that, as has already been pointed out in § 21, the wave functions (22.13) do not actually depend on the time arguments. If, therefore, we substitute $t_2 = -\infty$ in the expression we have obtained for $\Phi_n(1)$, we arrive at an expression for the operator which is being averaged

$$T[\hat{\psi}_{\text{Int}}(1)\,\hat{S}]\,\hat{\psi}^+_{\text{Int}}(2) = T[\hat{\psi}_{\text{Int}}(1)\,\hat{\psi}^+_{\text{Int}}(2)\,\hat{S}].$$

Hence we have

$$\Phi_n(1) = i \int dq_2 \lim_{t_2 \to -\infty} G(1, 2)\,\Phi_{0n}(2). \tag{22.15}$$

This only holds for the approximate expression (22.13); the subsequent terms in the expansion (22.10) could lead to the appearance of higher-order Green functions in (22.15).

If we take into account the general equation for the Green function (19.8), act on (22.14) from the left with the operator $i\,\partial/\partial t_1 - \hat{T} - \hat{W}$, and remember that, for all finite t_1,

$$\int dq_2 \lim_{t_2 \to -\infty} \delta(1 - 2)\,\Phi_{0n}(2) = 0,$$

we find

$$\left(i\frac{\partial}{\partial t_1} - \hat{T} - \hat{W}\right)\Phi_n(1) - \int d2\,M(1, 2)\,\Phi_n(2) = 0. \tag{22.16}$$

The function $G(1, 2)$ is thus a Green function of the equation for the quantity $\langle\Psi|\hat{\psi}_H(1)|\Psi_n\rangle$.

If the function $\Phi_n(1)$ is indeed the exact expression for the given matrix element, then its time dependence should be given by (22.8). Expanding the mass operator M in a Fourier integral in terms of the time

$$M(1, 2) = \int \frac{d\varepsilon}{2\pi} \exp\left[-i\varepsilon(t_1 - t_2)\right] M(q_1, q_2, \varepsilon),$$

we may write (22.16) in the form

$$(\Delta E_n - \hat{T} - \hat{W})\,\Phi_n(q_1) = \int dq_2\, M(q_1, q_2, \Delta E_n)\,\Phi_n(q_2). \tag{22.17}$$

The solution of this equation with the corresponding boundary conditions gives the eigenvalues for the excitation energy ΔE_n. This is particularly obvious in the case of a spatially uniform system, for which the equation takes the form

$$\Delta E_n - \varepsilon_p - M(\boldsymbol{p}, \Delta E_n) = 0. \tag{22.18}$$

All permissible values of ΔE_n, whether referring to a discrete or a continuous spectrum, must satisfy this equation.

We should note that at the allowed values of ΔE_n the Fourier transform of the Green function $G(\boldsymbol{p}, \varepsilon)$, goes to infinity

$$G^{-1}(\boldsymbol{p}, \Delta E_n) = 0. \tag{22.19}$$

This fact will be especially emphasized in §§ 23 and 24.

This is specifically the situation when there is no correlation interaction; the function (22.13) is the exact wave function for a stationary state of the system. Putting $M = 0$ in eqn. (22.18), we obtain

$$\Delta E_{0n} = \varepsilon_p,$$

which evidently agrees entirely with the results of § 21.2. The quantity ΔE_{0n} (like ΔE_n) depends on the momentum of the "excess" particle $\boldsymbol{p}$, i.e. we will not find a single level for an excited state, but a whole series. The way in which ΔE_n depends on $\boldsymbol{p}$ (and on other characteristics of the state) is referred to as the excitation spectrum—in a restricted sense of the word.

The simple situation we have described only applies in exceptional cases: when eqns. (22.17) or (22.18) have a purely real solution. As a rule, the mass operator has an imaginary part which differs from zero, and ΔE_n is complex. This is important in examining the limits within which our assumed structure for the function Ψ_{0n} applies. The fact that the excitation energy has an imaginary part, differing from zero, reflects the non-stationary character of the state due to correlation (a stationary state corresponds to a purely oscillatory behaviour of a wave function). The relative magnitude of the imaginary part of ΔE_n, or, to be accurate, the ratio Im (ΔE_n)/Re (ΔE_n), also measures the extent to which a state is non-stationary.

We first justify the assertion that a low value for this ratio guarantees the correctness of our ΔE_n and $\Phi_n(q_1)$ as follows. The solution of (22.17) gives essentially not only the values of ΔE_n

and Φ_n, but also the structure of the stable zero-approximation wave function. If we substitute an arbitrary function $\Phi_{0n}(q_1)$ in eqn. (22.15), the function $\Phi_n(1)$, which it determines, will have an arbitrary time-dependence. In particular this function will not coincide with the solutions of eqn. (22.17), which describes a comparatively restricted class of "quasi-stationary" states of a system, whose wave functions depend exponentially on the time. To each of these states corresponds a completely determined function $\Phi_{0n}(q_1)$, which can be obtained from the solution of (22.17) by excluding the interaction. The function $\Phi_{0n}(q_1)$, thus obtained, after substitution in (22.13), gives the "most stable" wave function for that class of functions which contain one excess particle above the ground state.

We turn now to the assertion formulated above. If the imaginary part of ΔE_n is accurately equal to zero, then a superposition of the stable wave functions determined by $\Phi_{0n}(q_1)$ may consist either of a single state, or of states that remain degenerate even when the correlation interaction is included. In this latter case eqns. (22.2) have one multiple root, and the matrix $\sigma_{\alpha\beta}$ is proportional to the unit matrix. In either case the chosen function Ψ_{0n} is essentially stable. Hence if Im ΔE_n is small, this is a guarantee that the chosen function is close to the stable zero-approximation function.

22.5. We can consider, in a similar way, the quantities $\Phi_n(1, 2)$ which were obtained on the assumption that the zero-approximation wave function is given by (22.13). Substituting (22.13) in (22.11), with $t_1, t_2 \to -\infty$, we obtain

$$\Phi_n(1, 2) = \int dq_3\, dq_4 \lim_{t_3, t_4 \to -\infty} \Phi_{0n}(3, 4) \times \frac{\langle \Psi_0 | T[\hat{\psi}^+_{\text{Int}}(2)\, \hat{\psi}_{\text{Int}}(1)\, \hat{S}]\, N[\hat{\psi}^+_{\text{Int}}(3)\, \hat{\psi}_{\text{Int}}(4)] | \Psi_0 \rangle}{\langle \Psi_0 | \hat{S} | \Psi_0 \rangle}.$$

From the definition of a contraction (§ 9), we have

$$N[\hat{\psi}^+_{\text{Int}}(3)\, \hat{\psi}_{\text{Int}}(4)] = T[\hat{\psi}^+_{\text{Int}}(3)\, \hat{\psi}_{\text{Int}}(4)] + iG_0(4, 3).$$

Whence

$$\Phi_n(1, 2) = \int dq_3\, dq_4 \lim_{t_3, t_4 \to -\infty} \Phi_{0n}(3, 4) \times [G_0(4, 3)\, G(1, 2) - G(1, 4, 2, 3)].$$

Using (21.18), we have finally

$$\Phi_n(1, 2) = -\int dq_3\, dq_4 \lim_{t_3, t_4 \to -\infty} G(1, 4, 2, 3)\, \Phi_{0n}(3, 4). \tag{22.20}$$

The equations (22.15) and (22.20) are easily comprehensible; they determine the evolution of the quantities $\Phi_n(1)$, $\Phi_n(1, 2)$ from an infinite time in the past, when there was no correlation between particles, to the moment in which we are interested. The functions G are here demonstrating, the propagation properties inherent in Green functions.

Certain integral equations can be established for the functions $\Phi_n(1, 2)$, as for $\Phi_n(1)$. Substituting (20.12) in (22.20), we find

$$\begin{aligned}\Phi_n(1, 2) = -\int dq_3\, dq_4 \lim_{t_3, t_4 \to -\infty} [\tilde{G}(1, 4, 2, 3) \\ + 4 \int d5\, d6\, d7\, d8\, G(1, 5)\, G(4, 6)\, \Delta(5, 6, 7, 8) \\ \times G(7, 2)\, G(8, 3)]\, \Phi_{0n}(3, 4).\end{aligned} \tag{22.21}$$

In order to simplify this equation we note that the combination $\int dq_3, dq_4\, G(4, 3)\, \Phi_{0n}(3, 4)$ can be put equal to zero in the limit $t_3, t_4 \to -\infty$. In this case $G(4, 3) \to G_0(4, 3)$, and (22.18) gives a zero result.

We will also introduce the function

$$\overline{\Phi}_n(1, 2) = \int dq_3\, dq_4 \lim_{t_3, t_4 \to -\infty} G(1, 3)\, G(4, 2)\, \Phi_{0n}(3, 4), \tag{22.22}$$

which describes the correlation of particles and holes with the system, but not with each other. Then eqn. (22.21) takes the form

$$\begin{aligned}\Phi_n(1, 2) = \overline{\Phi}_n(1, 2) + 4 \int d3\, d4\, d5\, d6\, G(1, 3)\, G(5, 2) \\ \times \Delta(3, 4, 5, 6)\, \overline{\Phi}_n(6, 4).\end{aligned} \tag{22.21'}$$

This equation can be reduced to an integral equation for the function Φ_n, which can serve to give the spectrum. To this end, one should separate the compact vertex part as was done for the application to the two-particle Green function in § 20. We will examine these questions in more concrete terms in § 25; we note here that the corresponding equation for the determination of ΔE_n also has complex roots in the general case.

22.6. In the two preceding sections we have discussed the way to determine the spectrum of excited states of a system on the assumption that the stable zero-approximation wave functions can be described by the first terms of the expansions (21.9) and (21.10). It was explained that the magnitude of the excitation energy ΔE_n and the matrix element Φ_n are determined by equations which contain the mass operator and the vertex part, respectively. The imaginary part which appears in the solutions of these equations—the damping term—has, in this formulation of the problem, a purely formal connotation, and acts as a measure of the accuracy which the assumed form of the stable wave function possesses. If the magnitude of the damping is comparable with ΔE_n, this is simply a reflection that this assumption is very crude. If we then select, let us say, the first two terms of the expansions (21.9) and (21.10) as the stable function Ψ_{0n}, we obtain more complicated equations for the functions Φ_n, which contain in addition a vertex part (for single-particle excitation) and a six-pole term (for particle-hole excitation). The solution of these equations leads to diminished damping. If necessary, this process can be extended.

This kind of approach to the problem, with the aim of finding the stationary excited states of the system, is by no means the only one possible. An important class of problems in many-body theory examines a transition of a system to an excited state due to a definite excitation mechanism. In other words we must find the result when a given excited state of a system evolves in time in a prescribed way. There is, connected with this, the problem of describing the transition of the system back to its ground state. In particular we must consider the decay of a single-particle excited state with the emission of a particle. As will be explained below, the equations we have previously derived can be used here, but they now have a completely different significance. It is particularly important that the imaginary part of the root of these equations has a direct physical significance: it represents the rate at which the initial state disappears, or, more accurately, the rate at which the probability of observing this state decreases.

A characteristic peculiarity in this way of posing the problem is that the form of the wave function $\Psi_{\mathrm{Int}\,n}(-\infty)$ depends on the physical conditions imposed. The quantity $\Psi_{\text{Int}\,n}(-\infty)$ does not coincide with the stable zero-approximation wave function. The exact wave function for the excited state of a system is therefore a superposition of wave functions with different (though close) energy values. The system is now in a quasi-stationary state and the corresponding energy level has a finite width.

22.7. We will take, as an example, the excitation of a system (particularly the atomic nucleus) by an external particle of the same sort as the particles in the system. We suppose that the system is initially in the ground state. For a given case, we know the wave function of the system as $t \to -\infty$: it coincides in structure with the first term of the expansion (21.10),

$$\Psi_{\text{Int}\,n}(-\infty) = \int dq_2 \lim_{t_2 \to -\infty} \Phi_{0n}(2)\, \hat{\psi}^{+}_{\text{Int}}(2)\, \Psi_0, \qquad (22.23)$$

where $\Phi_{0n}(2)$ is some function. Additional pairs are necessarily absent.

We will substitute this expression for $\Psi_{\text{Int}\,n}(-\infty)$ in (22.11) which we will write in the form

$$\Phi_n(1) = \frac{\langle \Psi_0 | T[\hat{\psi}_{\text{Int}}(1)\, \hat{S}]|\, \Psi_{\text{Int}\,n}(-\infty)\rangle}{\langle \Psi_0 | \hat{S} | \Psi_0 \rangle}.$$

Repeating the discussion that led us to (22.15), we can confirm that it is now exactly correct. Equation (22.16) is now also exact:

$$\left(i\frac{\partial}{\partial t_1} - \hat{T} - \hat{W}\right)\Phi_n(1) - \int d2 M(1,2)\,\Phi_n(2) = 0; \qquad (22.24)$$

this can be treated as an effective Schrödinger equation. It describes the state of an external particle entering the system. Making the substitution $\Phi_n(1) = \Phi_n(q_1)\exp(-i\Delta E_n t_1)$, we arrive at an equation similar to (22.17):

$$(\Delta E_n - \hat{T} - \hat{W})\,\Phi_n(q_1) - \int dq_2\, M(q_1, q_2, \Delta E_n)\,\Phi_n(q_2) = 0 \qquad (22.25)$$

corresponding to the concept of the optical potential which is widely applied in the theory of nuclear reactions. The optical potential $\hat{U}$, equal to

$$\hat{U}\Phi_n(q_1) = \hat{W}\Phi_n(q_1) + \int dq_2\, M(q_1, q_2, \Delta E)\,\Phi_n(q_2), \qquad (22.26)$$

consists of two parts: the self-consistent potential $\hat{W}$, which depends on the momentum and determines a general potential well in which the particle moves, and the mass operator, which has a non-local structure, depends on the excitation energy, and contains an imaginary part corresponding to particle absorption.

We will find an expression for the imaginary (non-Hermitian) part of the optical potential in terms of the lowest-order perturbation theory, using an expression for the self-energy part Σ, which coincides in this approximation with M (see § 14). The relationship (14.9) gives:

$$\operatorname{Im} \Sigma_{\mu\nu}(\varepsilon) = -\pi \sum_{\mu_1\mu_2\mu_3} \langle \mu\mu_3|\hat{V}|\mu_1\mu_2\rangle \langle \mu_1\mu_2|\hat{V}(1-\hat{\mathscr{P}})|\nu\mu_3\rangle \times$$

$$[(1-n_{\mu_1})(1-n_{\mu_2})\, n_{\mu_3} - n_{\mu_1}n_{\mu_2}(1-n_{\mu_3})]\, \delta(\varepsilon + \varepsilon_{\mu_3} - \varepsilon_{\mu_1} - \varepsilon_{\mu_2}),$$

whence

$$\operatorname{Im} M(q_1, q_2, \Delta E) = \sum_{\mu\nu} \chi_\mu^*(q_2)\, \chi_\nu(q_1) \operatorname{Im} \Sigma_{\mu\nu}(\Delta E). \quad (22.27)$$

The optical potential in heavy nuclei has been considered for a model corresponding to the interaction potential (1.11) [97]. Excellent agreement with experiment was obtained in the low-energy region. The concept of an optical potential, and the field equations we have given above for determining it can also be usefully applied to the problem of electron scattering by an atom, particularly where it can be used to describe the polarization of the inner shells by an external electron.

The relationship (22.15) makes it possible to express the particle scattering cross-section of a system directly in terms of the single-particle Green function for the ground state of the system. For this purpose, $\Phi_{0n}(2)$ must be treated as the state of the incoming particle, and $\Phi_n(1)$, for $t_1 \to +\infty$, as the state of the scattered particle. We deal with them as is usual in scattering problems [98, 99].

22.8. The imaginary part of the excitation energy has a particularly obvious significance for the decay of an excited state of a system containing $N+1$ particles, when one particle is emitted and the remaining system returns to the ground state. In this case we know the wave function of the system as $t \to +\infty$; it coincides in structure with the first term of the expansion (21.10):

$$\Psi_{\mathrm{Int}\,n}(+\infty) = \int dq_2 \lim_{t_2\to\infty} \Phi(2)\, \hat{\psi}^+_{\mathrm{Int}}(2)\, \Psi_0 .$$

It is more convenient to consider here not the function $\Phi_n(1)$, but its conjugate

$$\Phi^*_n(1) = \langle \Psi_n | \hat{\psi}^+_H(1) | \Psi \rangle .$$

Substituting the relation

$$\Psi^*_n = \Psi^*_{\mathrm{Int}\,n}(\infty)\, \hat{S}(\infty, 0),$$

we find†

$$\Phi^*_n(1) = \int dq_2 \lim_{t_2\to+\infty} \langle \Psi_0 | T[\hat{\psi}_{\mathrm{Int}}(2)\, \hat{\psi}^+_{\mathrm{Int}}(1)\, \hat{S}] | \Psi_0 \rangle\, \Phi^*(2)$$

$$= i \int dq_2 \lim_{t_2\to+\infty} \Phi^*(2)\, G(2, 1).$$

If we wish to discuss the question further, we must first carry out a few preliminary calculations. We will show, first of all, that the following equation holds:

$$\left(-i\frac{\partial}{\partial t_2} - \hat{T}_{q_2} - \hat{W}_{q_2}\right) G_0(1, 2) = \delta(1 - 2).$$

To do this, we must start with the equation

$$\left(-i\frac{\partial}{\partial t_2} - \hat{T}_{q_2} - \hat{W}_{q_2}\right) \hat{\psi}^+_{\mathrm{Int}}(2) = 0$$

and argue as in § 10, allowing for the equality $\hat{W}^+ = \hat{W}$.

We want to obtain an analogous equation for the exact Green function. Let us suppose it has the form

$$\left(-i\frac{\partial}{\partial t_2} - \hat{X}\right) G(1, 2) = \delta(1 - 2),$$

where $\hat{X}$ is an unknown operator. From Dyson's equation, we have

$$\left(-i\frac{\partial}{\partial t_2} - \hat{X}\right) G_0(1, 2) + \int d3\, G_0(1, 3)\, M(3, 2) = \delta(1 - 2).$$

† We have omitted here the inessential phase factor $\langle \Psi_0 | \hat{S} | \Psi_0 \rangle$.

Comparing this equation for G_0 with the one given above, we find

$$\hat{X}f(2) = (\hat{T}_{q_2} + \hat{W}_{q_2})f(2) + \int d3\, f(3)\, M(3, 2).$$

Thus finally

$$\left(-i\frac{\partial}{\partial t_2} - \hat{T} - \hat{W}\right) G(1, 2) - \int d3\, G(1, 3)\, M(3, 2) = \delta(1 - 2).$$

Returning to the equation derived above for Φ_n^*, we find

$$\left(-i\frac{\partial}{\partial t_1} - \hat{T} - \hat{W}\right) \Phi_n^*(1) - \int d2\, \Phi_n^*(2)\, M(2, 1) = 0,$$

or the conjugate

$$\left(i\frac{\partial}{\partial t_1} - \hat{T} - \hat{W}\right) \Phi_n(1) - \int d2\, M^+(1, 2)\, \Phi_n(2) = 0.$$

Here $M^+(1, 2)$ denotes $M^*(2, 1)$. Looking for a solution of this equation in the form $\Phi_n(1) = \Phi_n(q_1) \exp(-i\Delta E_n t_1)$, we obtain

$$(\Delta E_n - \hat{T} - \hat{W})\, \Phi_n(q_1) - \int dq_2\, M^+(q_1, q_2, \Delta E_n)\, \Phi_n(q_2) = 0. \tag{22.28}$$

This equation, which has complex solutions, gives the spectrum of the emitting system, and the imaginary part gives the width of the levels. The quantity $\Phi_n(q_1) = \langle \Psi | \hat{\psi}(q_1) | \Psi_n \rangle$ represents the probability of observing a particle at the point q_1, in the excited state Ψ_n, which leaves the system in the ground state. In other words $\Phi_n(q_1)$ is the transition amplitude from the state Ψ_n to the state $\hat{\psi}^+(q_1)\, \Psi$.

We have so far been considering examples which refer to a single-particle excited state of a system. Pair excitation is similar. The imaginary part has a direct physical significance in this case, too: it describes the decay of an excited state of the system. We will leave a full consideration of these questions until § 28, simply noting the following. To determine whether we are dealing with a quasi-stationary, or a stationary, state of the system, we must exclude the correlation interaction between particles. If the state resulting from this is different from the stable zero-approximation state, then we are necessarily dealing with a quasi-stationary state. This is what happens, for example, for electromagnetic excitation of a system with Coulomb interactions. When correlation interaction is excluded, we obtain a state which is a superposition of states with one, two, etc., pairs. The contribution of states with n pairs is proportional to $(e^2/\hbar c)^n$. Since $e^2/\hbar c$ is small, we may suppose, in practice, that there is only the state with one pair, which does not coincide with the stable state. The contribution of states with n pairs to the latter is determined not by $(e^2/\hbar c)^n$, but by the appropriate power of the effective interaction constant between the particles of the system.

22.9. We will turn now to calculating the wave function of the excited state of a system Ψ_n. This quantity may, in principle, be related to the zero-approximation wave function by means of the operator $\hat{S}(0, -\infty)$. The practical execution of the corresponding calculations is, however, extremely difficult. Moreover, we do not normally need all the information which the function Ψ_n includes. We are most interested in the density matrix of the excited state $R_n(q_1, q_2)$, which gives the distribution of single-particle characteristics in this state, and which is connected in the usual way $[R_n(q_1, q_2) = -i \lim_{t_2 - t_1 \to +0} G_n(1, 2)]$ with the single-particle Green function of the excited state

$$G_n(1, 2) = -i\langle\Psi_n|T[\hat{\psi}_H(1)\,\hat{\psi}_H^+(2)]|\Psi_n\rangle. \tag{22.29}$$

There arises, in this connection, the following question: can the quantity $G_n(1, 2)$ be expressed in terms of the matrix elements Φ_n discussed above, or can it only be determined by solving some new equations? The function Φ_n, which is the matrix element connecting Ψ_n and Ψ, contains a certain amount of information about Ψ_n. What we are asking is whether this information is sufficient to describe the single-particle characteristics of the state Ψ_n.

This question can be transferred to another plane. We have seen in § 21 that the quantities $\Phi_{0n}(1)$, $\Phi_{0n}(1, 2)$ can be treated as the wave functions of a particle–hole complex, the appearance of which leads to excitation of the system. When the correlation interaction is included, $\Phi_{0n}(1)$ and $\Phi_{0n}(1, 2)$ transform to $\Phi_n(1)$ and

$\Phi_n(1, 2)$, respectively. If it appears (allowing for the assumptions made above about the stability of the function Ψ_{0n}) that $\Phi_n(1)$ and $\Phi_n(1, 2)$ can be treated, as before, as the wave functions of some complex, then the question can be solved without difficulty. Equations (21.23) and (21.24) hold in this case, as well as (21.21), in which the functions Φ_{0n} should be replaced by Φ_n:

$$R_n(q_1, q_2) - R(q_1, q_2) = \Phi_n^*(q_2)\,\Phi_n(q_1), \tag{22.30}$$

$$\begin{aligned} R_n(q_1, q_2) - R(q_1, q_2) = \int dq_3\,[\Phi_n^*(q_2, q_3)\,\Phi_n(q_1, q_3) \\ - \Phi_n^*(q_3, q_1)\,\Phi_n(q_3, q_2), \end{aligned} \tag{22.31}$$

$$\int dq_1\,|\Phi_n(q_1)|^2 = 1; \quad \int dq_1\,dq_2\,|\Phi_n(q_1, q_2)|^2 = 1. \tag{22.32}$$

We will now consider the problem we have set. We use the relations

$$\Psi_n = \hat{S}(0, -\infty)\,\Psi_{0n}, \quad \hat{S}\Psi_{0n} = \langle\Psi_{0n}|\hat{S}|\Psi_{0n}\rangle\,\Psi_{0n}$$

to rewrite (22.29) in the form

$$G_n(1, 2) = -i\,\frac{\langle\Psi_{0n}|T[\hat{\psi}_{\text{Int}}(1)\,\hat{\psi}^+_{\text{Int}}(2)\,\hat{S}]|\Psi_{0n}\rangle}{\langle\Psi_{0n}|\hat{S}|\Psi_{0n}\rangle}. \tag{22.33}$$

Substituting the expressions for Ψ_{0n} with $t_1 \to -\infty$, and for Ψ^*_{0n} with $t_1 \to +\infty$, we can reduce this to the Green functions of the ground state of the system:

$$G_n(1, 2) = \frac{i \int dq_3\,dq_4\,\Phi^*_{0n}(4)\,G(4, 1, 3, 2)\,\Phi_{0n}(3)}{i \int dq_3\,dq_4\,\Phi^*_{0n}(4)\,G(4, 3)\,\Phi_{0n}(3)},$$

where $t_3 \to -\infty$, $t_4 \to +\infty$. We note immediately that the denominator is actually equal to unity. In fact if we use (22.15), we can rewrite it in the form

$$\int dq_4 \lim_{t_4 \to \infty} \Phi^*_{0n}(4)\,\Phi_n(4).$$

But when $t_4 \to \infty$, the function Φ_n transforms to Φ_{0n} (as a result of excluding the interaction). Use of the normalization condition (21.21) then brings us to the desired result.

We will replace the Green function in the numerator of the expression for $G_n(1, 2)$ by its representation in terms of the vertex part (20.12). The functions Φ_{0n} may here be replaced by the functions Φ_n. This requires the use of (22.15), and also of the relation

$$\Phi_n^*(1) = -i \int dq_2 \lim_{t_2 \to \infty} G(2, 1)\, \Phi_{0n}^*(2),$$

which can be derived in the same way as eqn. (22.15). We then arrive at the following basic equation:

$$G_n(1, 2) - G(1, 2) = -i\Phi_n^*(2)\, \Phi_n(1)$$

$$- 4i \int d3\, d4\, d5\, d6\, \Phi_n^*(3)\, G(1, 4)\, G(6, 2)\, \Delta(3, 4, 5, 6)\, \Phi_n(5). \quad (22.34)$$

Changing to the density matrix, we obtain ($t_2 - t_1 \to +0$)

$$R_n(q_1, q_2) - R(q_1, q_2) = \Phi_n^*(q_2)\, \Phi_n(q_1)$$

$$+ 4 \int d3\, d4\, d5\, d6\, \Phi_n^*(3)\, G(1, 4)\, G(6, 2)\, \Delta(3, 4, 5, 6)\, \Phi_n(5). \quad (22.35)$$

The second term on the right-hand side of (22.35) violates (22.30). This has an obvious significance, since, as was pointed out in § 20, the term describes the scattering of one complex by another, which interferes with their independent existence. In the general case such scattering is of basic importance; dilute (non-resonance) and condensed systems, however, only have a comparatively small contribution from this source. Hence the concept of a wave function for single-particle excitation can be justified for these systems (to the required accuracy).

22.10. The question of pair excitation is more complicated. If we substitute the expressions for Ψ_{0n} with $t_1, t_2 \to -\infty$, and for Ψ_{0n}^* with $t_1, t_2 \to +\infty$, into the general equation (22.33), we find

$$G_n(1,2) = \frac{\begin{aligned}\int dq_3\, dq_4\, dq_5\, dq_6\, \Phi_{0n}^*(5, 6)\, \langle \Psi_0 | T[\hat{\psi}_{\mathrm{Int}}(5)\, \hat{\psi}_{\mathrm{Int}}(1)\, \hat{\psi}_{\mathrm{Int}}(4) \\ \times\, \hat{\psi}_{\mathrm{Int}}^+(6)\, \hat{\psi}_{\mathrm{Int}}^+(2)\, \hat{\psi}_{\mathrm{Int}}^+(3)\, \hat{S}] | \Psi_0 \rangle\, \Phi_{0n}(3, 4)\end{aligned}}{\begin{aligned}\int dq_3\, dq_4\, dq_5\, dq_6\, \Phi_{0n}^*(5, 6)\, \langle \Psi_0 | T[\hat{\psi}_{\mathrm{Int}}(5)\, \hat{\psi}_{\mathrm{Int}}(4)\, \hat{\psi}_{\mathrm{Int}}^+(3) \\ \times\, \hat{\psi}_{\mathrm{Int}}^+(6)\, \hat{S}] | \Psi_0 \rangle\, \Phi_{0n}(3, 4)\end{aligned}}, \quad (22.36)$$

where $t_5, t_6 \to \infty$; $t_3, t_4 \to -\infty$. Introducing the two-particle Green function in the denominator, and allowing for (22.20) and the normalization condition (21.21), it is easy to show that the denominator is equal to one.

If we introduce the three-particle Green function

$$G(1, 2, 3, 4, 5, 6) = (-i)^3 \frac{\langle \Psi_0 | T[\hat{\psi}_{\text{Int}}(1)\, \hat{\psi}_{\text{Int}}(2)\, \hat{\psi}_{\text{Int}}(3)\, \hat{\psi}^+_{\text{Int}}(6)\, \hat{\psi}^+_{\text{Int}}(5)\, \hat{\psi}^+_{\text{Int}}(4)\, \hat{S}] | \Psi_0 \rangle}{\langle \Psi_0 | \hat{S} | \Psi_0 \rangle}, \tag{22.37}$$

we can write

$$G_n(1, 2) = -i \int dq_3\, dq_4\, dq_5\, dq_6\, \Phi^*_{0n}(5, 6)\, G(5, 1, 4, 3, 2, 6)\, \Phi_{0n}(3, 4). \tag{22.38}$$

If we substitute the expansion of the S-matrix in terms of normal products into (22.37), then an additional term is required which contains the N-product of six operators. The corresponding function—the six-pole term referred to above—contains an unconnected part, and also a connected part which describes the mutual correlation of the three particles. The connected part of this term must be small if a two-particle wave function is to exist. This condition is satisfied under all circumstances in a dilute system, where only two-particle correlations need be considered. In dense systems triple correlations are as important as pair correlations, and they can only be neglected when they are small.

Even if triple correlations are ignored (22.38) does not reduce to (22.31). This could be achieved by the requirement that a three-particle Green function becomes a sum of integrals of the type

$$\int d7\, G(5, 1, 6, 7)\, G(7, 3, 2, 4),$$

where the connected six-pole term is ignored. In this case, (22.20) (and the analogous relation for Φ^*_n) would allow Φ_{0n} in (22.36) to be replaced by Φ_n, thus giving (22.31). This would be a very complicated business, however.†

† For single-particle excitation, the number of arguments of the Green function in the numerator of $G_n(1,2)$ is precisely twice as many as the number in (22.15). $G_n(1,2)$ has therefore been expressed as a product of two functions Φ_n. This purely arithmetical circumstance also distinguishes the single-particle from the pair excitation.

We now consider the results for a very dense electron gas [100]. In this case, the precise wave function of an excited state of the system can be put in the form [101].

$$\Psi_n(N) = \int dq_1\, dq_2\, \Phi_n'(q_1, q_2)\, N[\hat{\psi}^+(q_1)\, \hat{\psi}(q_2)]\, \Psi,$$

where Φ_n' is some function related to Φ_n by the equation†

$$\Phi_n(q_1, q_2) = \int dq_3\, [\Phi_n'(q_1, q_3)\, R(q_3, q_2) - R(q_1, q_3)\, \Phi_n'(q_3, q_2)].$$

The density matrix and the normalization conditions are expressed in terms of both of the functions Φ_n and Φ_n':

$$R_n(q_1, q_2) - R(q_1, q_2)$$
$$= \int dq_3\, [\Phi_n^*(q_2, q_3)\, \Phi_n'(q_1, q_3) - \Phi_n^*(q_3, q_1)\, \Phi_n'(q_3, q_2)],$$
$$\int dq_1\, dq_2\, \Phi_n'(q_1, q_2)\, \Phi_n^*(q_1, q_2) = 1.$$

Hence, in this case, there is no concept of a wave function for the complex. It may be shown that the application of (22.31) to the description of a uniform electron gas leads to a gross violation of more general theorems such as the virial theorem. The absence of any two-particle wave function can be explained in terms of the strong polarization of the medium, which makes it impossible to describe the excited state of a system using a single function of two coordinates. If vacuum polarization was important in quantum electrodynamics, the concept of a positronium wave function would lose its meaning in a similar way.

We may say in conclusion that the examination of the internal structure of this complex—the distribution of the coordinates, the momenta, etc., inside it—is a more complicated task than finding the excitation spectrum. The first problem requires the examination of higher-order Green functions than the second.

The question of introducing a wave function for this complex is even more involved. The assertion, found in the literature, that the wave function is Φ_n‡ can hardly be correct in all cases [86].

† If the correlation interaction is excluded, the function Φ_n' coincides with Φ_n.

‡ The function Φ_n, as before, determines the matrix element for the transition of a system to an excited state when correlation interaction is present.

23. Spectral Representations of Green Functions

23.1. This section will deal with a most important part of the Green functions method—their spectral (or parametric) representation [94, 102].†

This representation plays a double role in the theory. On the one hand, it helps us to solve finally the problem of finding the Green functions themselves. Even if the problem of finding the mass operator M is solved, the Green function is still not uniquely determined (see § 19). We require additional information on the interpretation of the function G at points where its denominator tends to zero. The spectral representation gives this information, or, in effect, defines the rules which we can use to integrate around the poles of the Green function.

In addition, the spectral representation makes it possible to introduce and interpret in a natural way the most important concept of contemporary many-body theory—the quasi-particle—and also to discover its basic characteristics: the dispersion law and the damping.

No assumptions about the nature of the system, or of the law of interaction between its particles, need be made in deducing the spectral representation. Although the approach is thus very general, the dependence of the Green function on the energy is specified by the spectral representation used.

23.2. We will derive the spectral representation of the single-particle Green function G. We will write it in the form

$$G(1, 2) = -i \begin{cases} \langle \Psi | \hat{\psi}_H(1)\, \hat{\psi}_H^+(2) | \Psi \rangle & t_1 > t_2 \\ -\langle \Psi | \hat{\psi}_H^+(2)\, \hat{\psi}_H(1) | \Psi \rangle & t_1 < t_2 \end{cases} \tag{23.1}$$

We may use the rule for multiplying matrices

$$\langle \Psi_i | \hat{a}\hat{b} | \Psi_k \rangle = \sum_l \langle \Psi_i | \hat{a} | \Psi_l \rangle \langle \Psi_l | \hat{b} | \Psi_k \rangle$$

which is true for any complete set of functions Ψ_l. The completeness condition may be written in the form

$$\sum_l \Psi_l^*(\alpha)\, \Psi_l(\beta) = \delta_{\alpha\beta},$$

† We will base our future work mainly on the paper by Galitskii and Migdal [94].

where α and β are the values of the variables on which Ψ_l depends. An approach using an intermediate set of functions was introduced into field theory by Kallén and Lehmann [103].

In our case it is convenient to use the exact eigenfunctions of the total Hamiltonian as the intermediate set

$$(\hat{H} - E_n)\,\Psi_n = 0.$$

We are speaking here of the eigenfunctions for all states (ground and excited). We ought to consider the eigenfunctions corresponding to any number of particles in the system (and not necessarily N), so long as the functions are eigenfunctions of the operators $\hat{H}$. Hence we should use all possible eigenfunctions of $\hat{H}$. The completeness of this set is one of the basic assumptions of any quantum-mechanical theory.

We will consider first the upper line of (23.1). It can be written in the form

$$-i \sum_n \langle \Psi | \hat{\psi}_H(1) | \Psi_n \rangle \langle \Psi_n | \hat{\psi}_H^+(2) | \Psi \rangle.$$

Only the single-particle intermediate state Ψ_n contributes to this expansion. In fact if we take into account the relation $\hat{\psi}_H(1) = \hat{S}(t_1, 0)\,\hat{\psi}(q_1)\,\hat{S}(0, t_1)$, we see that the operator $\hat{\psi}_H(1)$ diminishes the number of particles by one, whilst the S-matrix contains an equal number of creation and annihilation operators, and does not change the number of particles.

The Green function can now be expressed without difficulty in terms of the functions $\Phi_n(1)$ (22.7.). Specifically, when $t_1 > t_2$,

$$G(1, 2) = -i \sum_n \Phi_n^*(2)\,\Phi_n(1).$$

If we allow for the time dependence of the quantities in this equation, we have

$$G(1, 2) = -i \sum_n \Phi_n^*(q_2)\,\Phi_n(q_1) \exp\left[-i\Delta E_n(t_1 - t_2)\right].$$

The situation is similar for $t_1 < t_2$. In this case, however, the intermediate states have one fewer particle (or an excess hole). Introducing the matrix element $\tilde{\Phi}_n(1) = \langle \Psi | \hat{\psi}_H^+(1) | \Psi_n(N-1) \rangle$, the time dependence of which is determined by

$$\tilde{\Phi}_n(1) = \tilde{\Phi}_n(q_1) \exp\left(-i\Delta \tilde{E}_n t_1\right),$$

where

$$\Delta \tilde{E}_n = E_n(N-1) - E(N),$$

we can write finally

$$G(1, 2) = -i \sum_n \begin{cases} \Phi_n^*(q_2)\, \Phi_n(q_1) \exp\,[-i\Delta E_n(t_1 - t_2)] & t_1 > t_2 \\ -\tilde{\Phi}_n^*(q_1)\, \tilde{\Phi}_n(q_2) \exp\,[i\Delta \tilde{E}_n(t_1 - t_2)] & t_1 < t_2 \end{cases} \tag{23.2}$$

As in § 21, the quantities ΔE_n and $\Delta \tilde{E}_n$ can be expressed in terms of the excitation energy of a system containing the same number of particles (N), as the ground state. If we denote by μ the exact chemical potential of the system, then

$$\left.\begin{aligned} \Delta E_n &= \Delta E_n' + \mu, \\ \Delta \tilde{E}_n &= \Delta E_n' - \mu. \end{aligned}\right\} \tag{23.3}$$

The number of particles N is here supposed to be large.

If we now go over to the Fourier transform in terms of $t_1 - t_2$, and use the relationship given in Appendix C, we find that

$$G(q_1, q_2, \varepsilon) = \sum_n \left\{ \frac{\Phi_n^*(q_2)\, \Phi_n(q_1)}{\varepsilon - \Delta E_n' - \mu + i\delta} + \frac{\tilde{\Phi}_n^*(q_1)\, \tilde{\Phi}_n(q_2)}{\varepsilon + \Delta E_n' - \mu - i\delta} \right\}. \tag{23.4}$$

23.3. We now analyse this relationship. We will suppose that the summation over n proceeds in two stages: the first over all values of the excitation energy $\Delta E_n'$; the second over all substates corresponding to a given value of $\Delta E_n'$. It is convenient, in practice, to proceed as if the energy spectrum were continuous. If the system does, in fact, contain discrete levels, this is automatically reflected by the appearance of δ-functions of the energy in the expression under the integral sign.

If we consider, therefore, an infinitesimal interval of the excitation energy

$$E < \Delta E_n' < E + dE, \tag{23.5}$$

we can introduce new functions $A(q_1, q_2, E)$ and $B(q_1, q_2, E)$, which are defined by:

$$\left.\begin{aligned} A(q_1, q_2, E)\, dE &= \sum_n \Phi_n^*(q_2)\, \Phi_n(q_1), \\ B(q_1, q_2, E)\, dE &= \sum_n \tilde{\Phi}_n^*(q_1)\, \tilde{\Phi}_n(q_2), \end{aligned}\right\} \tag{23.6}$$

where the summation on the right-hand sides is taken over states which satisfy (23.5). The functions A and B thus determine the density of levels in the system. They can also be written in the form

$$A(q_1, q_2, E) = \sum_n \delta(E - \Delta E_n + \mu)\, \Phi_n^*(q_2)\, \Phi_n(q_1),$$
$$B(q_1, q_2, E) = \sum_n \delta(E - \Delta \tilde{E}_n - \mu)\, \tilde{\Phi}_n^*(q_1)\, \tilde{\Phi}_n(q_2). \quad (23.6')$$

We can now go from a summation over n, to an integration over E, using the general relation (which derives from (23.5) and (23.6)):

$$\sum_n \Phi_n^*(q_2)\, \Phi_n(q_1)\, X(\Delta E_n') \equiv \int dE\, A(q_1, q_2, E)\, X(E).$$

Here X is an arbitrary function. An analogous equation can also be written for B. Thus (23.4) can be rewritten as†

$$G(q_1, q_2, \varepsilon) = \int_0^\infty dE \left[\frac{A(q_1, q_2, E)}{\varepsilon - E - \mu + i\delta} + \frac{B(q_1, q_2, E)}{\varepsilon + E - \mu - i\delta}\right]. \quad (23.7)$$

This relation is also called the spectral representation of the Green function.

Note the useful relationship:

$$\int_0^\infty dE\, [A(q_1, q_2, E) + B(q_1, q_2, E)] = \delta(q_1 - q_2). \quad (23.8)$$

Its left-hand side can be written as

$$\sum_n [\Phi_n^*(q_2)\, \Phi_n(q_1) + \tilde{\Phi}_n^*(q_1)\, \tilde{\Phi}_n(q_2)]$$
$$= \sum_n [\langle \Psi | \hat{\psi}(q_1) | \Psi_n \rangle \langle \Psi_n | \hat{\psi}^+(q_2) | \Psi \rangle$$
$$+ \langle \Psi | \hat{\psi}^+(q_2) | \Psi_n \rangle \langle \Psi_n | \hat{\psi}(q_1) | \Psi_n \rangle].$$

The summation that comes into this can be calculated:

$$\langle \Psi | [\hat{\psi}(q_1), \hat{\psi}^+(q_2)]_+ | \Psi \rangle = \delta(q_1 - q_2).$$

Equation (23.8) is the neatest way of expressing the requirement that the set Ψ_n be complete.

† The quantity $E = \Delta E_n'$ is always positive, since the energy of the ground state corresponds to the lowest level of the spectrum.

We must now consider the asymptotic values of the Green function at large ε. Multiplying eqn. (23.7) by ε, and supposing $\varepsilon \to \infty$, we find

$$\varepsilon G(q_1, q_2, \varepsilon) \to \delta(q_1 - q_2). \tag{23.9}$$

Here we have made use of (23.8). The free-particle Green function, $G_0(q_1, q_2, \varepsilon)$, tends to the same limit, as is evident from the expressions for it (see § 10). This has an obvious physical interpretation: high-energy particles are only slightly affected by interaction with other particles.

The weight functions A and B may be related directly to the density matrix of the ground state of the system. Using the same arguments that led us to (23.8) we find that

$$R(q_1, q_2) = \int_0^\infty B(q_1, q_2, E)\, dE. \tag{23.10}$$

The integral $\int_0^\infty dE\, A(q_1, q_2, E)$ gives the corresponding density matrix describing the distribution of the holes.

23.4. In a spatially uniform system these relationships are somewhat simplified. In this case the momentum $\boldsymbol{p}$ of the excited state of the system is included amongst the state indices n in (23.6) (the momentum is considered to be zero in the ground state); also $\Phi_n(q_1) = a_{n'p} \exp i(\boldsymbol{p} \cdot \boldsymbol{x})$, where $a_{n'p}$ is some coefficient dependent on $\boldsymbol{p}$. Similarly, $\tilde{\Phi}_n(q_1) = \tilde{a}_{n'p} \exp[-i(\boldsymbol{p} \cdot \boldsymbol{x}_1)]$. Substituting these expressions into (23.6) and (23.7), and calculating the Fourier transform with respect to the difference $\boldsymbol{x}_1 - \boldsymbol{x}_2$, we find

$$G(\boldsymbol{p}, \varepsilon) = \int_0^\infty dE \left[\frac{A(\boldsymbol{p}, E)}{\varepsilon - E - \mu + i\delta} + \frac{B(\boldsymbol{p}, E)}{\varepsilon + E - \mu - i\delta}\right], \tag{23.11}$$

where the coefficient $A(\boldsymbol{p}, E)$ is proportional to $\sum_{n'} |a_{n'p}|^2$, $B(\boldsymbol{p}, E)$ is proportional to $\sum_{n'} |\tilde{a}_{n'p}|^2$, and the summation is taken over the interval (23.5) with fixed $\boldsymbol{p}$. It follows from this that A and B are positive quantities:

$$A \geqq 0; \quad B \geqq 0.$$

The expression for the free-particle Green function (see § 10)

$$G_0(\boldsymbol{p}, \varepsilon) = [\varepsilon - \varepsilon_p + i\delta \operatorname{sign}(\varepsilon_p - \varepsilon_F)]^{-1} \tag{23.12}$$

is obtained from the general spectral equation (23.11) with

$$\left.\begin{aligned} A(\mathbf{p}, E) &= \delta(E + \mu - \varepsilon_p), \\ B(\mathbf{p}, E) &= \delta(E - \mu + \varepsilon_p), \end{aligned}\right\} \tag{23.13}$$

where we have put $\mu = \varepsilon_F$.†

From a comparison of (23.11) and (23.12), we can interpret the spectral representation of the exact Green function as depending on the simultaneous propagation of free particles (or holes), which have an energy $E + \mu$ (or $-E + \mu$) and a weight A (or B). An object whose propagation is described by the function G is therefore a complicated superposition of particles and holes. If there is no interaction (or, more accurately, correlation) this superposition reduces to one term in accordance with (23.13).

23.5. We have remarked above that the spectral equation makes it possible to set up rules for integrating around the singularities of the Green function and for establishing its uniqueness. This is done by working out the connection between the real and imaginary parts of the Green function. With this in view, we will consider the symmetric and anti-symmetric parts of G, A and B (relative to transpositions of q_1 and q_2), denoting these parts by the suffixes, s and a, respectively. We note immediately that A_s and B_s are real, whilst A_a and B_a are purely imaginary. Indeed we have from (23.6):

$$\text{Im}\,[\Phi_n^*(q_2)\,\Phi_n(q_1) + \Phi_n^*(q_1)\,\Phi_n(q_2)] = 0,$$

$$\text{Re}\,[\Phi_n^*(q_2)\,\Phi_n(q_1) - \Phi_n^*(q_1)\,\Phi_n(q_2)] = 0,$$

† This equality is not limited to the case of a perfect gas (see § 4). Suppose we return to an inhomogeneous system, and remember that in the Hartree–Fock approximation $\Phi_n(q) = \chi_\nu(q)$, $\Delta E_n = \varepsilon_\nu$, $\tilde{\Phi}_n(q) = \chi_\nu^*(q)$, $\Delta \tilde{E}_n = -\varepsilon_\nu$. Then, according to (23.6′):

$$A(q_1, q_2, E) = \sum_\nu \delta(E - \varepsilon_\nu + \mu)\,\chi_\nu^*(q_2)\,\chi_\nu(q_1),$$

$$B(q_1, q_2, E) = \sum_\nu \delta(E + \varepsilon_\nu - \mu)\,\chi_\nu^*(q_2)\,\chi_\nu(q_1)$$

Substituting these expressions into (23.7), we find

$$G_0(q_1, q_2, \varepsilon) = \sum_\nu \chi_\nu^*(q_2)\,\chi_\nu(q_1)\,[\varepsilon - \varepsilon_\nu + i\delta\,\text{sign}\,(\varepsilon_\nu - \mu)]^{-1}.$$

Comparing this with (10.7), we can confirm that the chemical potential and the Fermi energy are equal in the Hartree–Fock approximation.

and similarly for the function B. Hence the imaginary part of the symmetrized Green function $G_s(q_1, q_2, \varepsilon)$, owes its appearance to the factor $\pm i\delta$ in the denominator. A similar assertion can be made about the real part $G_a(q_1, q_2, \varepsilon)$. If we take into account the relation

$$\operatorname{Im}(a \pm i\delta)^{-1} = \mp\pi\delta(a);$$

we find

$$\operatorname{Im} G_s = \pi \int_0^\infty dE\,\{-A_s(E)\,\delta(E - \varepsilon + \mu) + B_s(E)\,\delta(E + \varepsilon - \mu)\},$$

$$\operatorname{Re} G_a = i\pi \int_0^\infty dE\,\{-A_a(E)\,\delta(E - \varepsilon + \mu) + B_a(E)\,\delta(E + \varepsilon - \mu)\}.$$

When $\varepsilon > \mu$, the argument of the first term in the curly brackets goes to zero over the region of integration; when $\varepsilon < \mu$, the same happens effectively for the second term.

We find as a result

$$\left.\left.\begin{aligned} A_s(\varepsilon - \mu) &= -\frac{1}{\pi}\operatorname{Im} G_s(\varepsilon) \\ A_a(\varepsilon - \mu) &= \frac{i}{\pi}\operatorname{Re} G_a(\varepsilon) \end{aligned}\right\}\ \varepsilon > \mu, \\ \left.\begin{aligned} B_s(-\varepsilon + \mu) &= \frac{1}{\pi}\operatorname{Im} G_s(\varepsilon) \\ B_a(-\varepsilon + \mu) &= -\frac{i}{\pi}\operatorname{Re} G_a(\varepsilon) \end{aligned}\right\}\ \varepsilon < \mu, \right\} \tag{23.14}$$

or, alternatively,

$$\left.\begin{aligned} A_s(E) &= -\frac{1}{\pi}\operatorname{Im} G_s(\mu + E), \\ A_a(E) &= \frac{i}{\pi}\operatorname{Re} G_a(\mu + E), \\ B_s(E) &= \frac{1}{\pi}\operatorname{Im} G_s(\mu - E), \\ B_a(E) &= -\frac{i}{\pi}\operatorname{Re} G_a(\mu - E). \end{aligned}\right\} \tag{23.14'}$$

Substituting these expressions in the spectral equation, we have

$$G_s(\varepsilon) = -\frac{1}{\pi}\int_0^\infty dE\left\{\frac{\operatorname{Im} G_s(\mu+E)}{\varepsilon - E - \mu + i\delta} - \frac{\operatorname{Im} G_s(\mu - E)}{\varepsilon + E - \mu - i\delta}\right\},$$

$$G_a(\varepsilon) = \frac{i}{\pi}\int_0^\infty dE\left\{\frac{\operatorname{Re} G_a(\mu+E)}{\varepsilon - E - \mu + i\delta} - \frac{\operatorname{Re} G_a(\mu - E)}{\varepsilon + E - \mu - i\delta}\right\}.$$

These can be written in a more compact form, if we make the replacement $\mu + E \to E$ in the first term, and $\mu - E \to E$ in the second

$$\left.\begin{aligned} G_s(\varepsilon) &= -\frac{1}{\pi}\int_{-\infty}^\infty dE\,\frac{\operatorname{sign}(E-\mu)\operatorname{Im} G_s(E)}{\varepsilon - E + i\delta\operatorname{sign}(E-\mu)}, \\ G_a(\varepsilon) &= \frac{i}{\pi}\int_{-\infty}^\infty dE\,\frac{\operatorname{sign}(E-\mu)\operatorname{Re} G_a(E)}{\varepsilon - E + i\delta\operatorname{sign}(E-\mu)}. \end{aligned}\right\} \tag{23.15}$$

Separating the real and imaginary parts of eqns. (23.15), and using the relationship

$$\operatorname{Re}\,[a \pm i\delta]^{-1} = \mathrm{P}\,(1/a),$$

we finally obtain

$$\left.\begin{aligned} \operatorname{Re} G_s(\varepsilon) &= -\frac{1}{\pi}\mathrm{P}\int_{-\infty}^\infty dE\,\frac{\operatorname{sign}(E-\mu)\operatorname{Im} G_s(E)}{\varepsilon - E}, \\ \operatorname{Im} G_a(\varepsilon) &= \frac{1}{\pi}\mathrm{P}\int_{-\infty}^\infty dE\,\frac{\operatorname{sign}(E-\mu)\operatorname{Re} G_a(E)}{\varepsilon - E}. \end{aligned}\right\} \tag{23.16}$$

Using (23.11) and the fact that A and B are real, we find, for a spatially uniform system, that

$$\operatorname{Re} G(\boldsymbol{p}, \varepsilon) = -\frac{1}{\pi}\mathrm{P}\int_{-\infty}^\infty dE\,\frac{\operatorname{sign}(E-\mu)\operatorname{Im} G(\boldsymbol{p}, E)}{\varepsilon - E}. \tag{23.17}$$

It is easy enough to find the converse relation, expressing the imaginary part in terms of the real. To do this, we first prove that

$$\mathrm{P}\int_{-\infty}^\infty \frac{d\varepsilon}{(\varepsilon - E)(\varepsilon - E')} = -\pi^2\delta(E - E'). \tag{23.18}$$

When $E \neq E'$, direct computation gives a zero result; when $E = E'$, the left-hand side of (23.18) goes to infinity. To find the numerical coefficient of the δ-function, we evaluate the integral on the left-hand side in terms of $E - E' = \zeta$:

$$\operatorname*{Lim}_{N\to\infty} \mathrm{P} \int_{-N}^{N} d\zeta \int_{-\infty}^{\infty} d\varepsilon \frac{1}{\varepsilon(\varepsilon+\zeta)} = \operatorname*{Lim}_{N\to\infty} \mathrm{P} \int_{-\infty}^{\infty} \frac{d\varepsilon}{\varepsilon} \ln\left|\frac{N+\varepsilon}{N-\varepsilon}\right|.$$

Making the replacement $\varepsilon \to N\varepsilon$, we arrive at the integral

$$\mathrm{P} \int_{-\infty}^{\infty} \frac{d\varepsilon}{\varepsilon} \ln\left|\frac{1+\varepsilon}{1-\varepsilon}\right| = -\pi^2$$

which proves (23.18).

Multiplying both sides of (23.17) by $1/(\varepsilon - E)$, integrating over ε, and using (23.18), we obtain

$$\operatorname{Im} G(\boldsymbol{p}, \varepsilon) = \frac{1}{\pi} \operatorname{sign}(\varepsilon - \mu)\, \mathrm{P} \int_{-\infty}^{\infty} dE \, \frac{\operatorname{Re} G(\boldsymbol{p}, E)}{\varepsilon - E}. \qquad (23.19)$$

We will use these relations to find the rules for integrating around singularities, taking, as an example, the free-particle Green function. Suppose we know the real part of the Green function

$$\operatorname{Re} G_0(\boldsymbol{p}, \varepsilon) = \mathrm{P} \frac{1}{\varepsilon - \varepsilon_p}.$$

Substituting this expression into (23.19), we find

$$\operatorname{Im} G_0(\boldsymbol{p}, \varepsilon) = -\pi\delta(\varepsilon - \varepsilon_p) \operatorname{sign}(\varepsilon_p - \mu).$$

Thus we come down to the expression

$$G_0(\boldsymbol{p}, \varepsilon) = [\varepsilon - \varepsilon_p + i\delta \operatorname{sign}(\varepsilon_p - \varepsilon_F)]^{-1}$$

if we remember that the chemical potential coincides with ε_F.

The relations (23.14) for a spatially uniform system can be written in the form

$$\operatorname{Im} G(\boldsymbol{p}, \varepsilon) = \pi \begin{cases} -A(\boldsymbol{p}, \varepsilon - \mu) & \varepsilon > \mu \\ B(\boldsymbol{p}, \mu - \varepsilon) & \varepsilon < \mu. \end{cases}$$

The functions A and B are positive as well as real. The imaginary part of the Green function therefore changes sign at the point $\varepsilon = \mu$:

$$\left.\begin{aligned} \operatorname{Im} G(\boldsymbol{p}, \varepsilon) < 0 \qquad \varepsilon > \mu, \\ \operatorname{Im} G(\boldsymbol{p}, \varepsilon) > 0 \qquad \varepsilon < \mu. \end{aligned}\right\} \qquad (23.20)$$

In particular if we suppose that the function is continuous at this point, we can put

$$\operatorname{Im} G(p, \mu) = 0. \tag{23.21}$$

23.6. The two-particle Green function (20.1) can be investigated in the same way:

$$G(1, 2, 3, 4) = -\langle \Psi | T[\hat{\psi}_H(1)\, \hat{\psi}_H(2)\, \hat{\psi}_H^+(4)\, \hat{\psi}_H^+(3)] | \Psi \rangle .$$

It is sufficient for our purpose to consider the limit

$$\tilde{G}(1, 2) = \lim_{\substack{4\to 1,\ 3\to 2 \\ t_4 - t_1 \to +0 \\ t_3 - t_2 \to +0}} G(1, 2, 3, 4)$$

which is equal to

$$\langle \Psi | T[\hat{\psi}_H^+(1)\, \hat{\psi}_H(1),\, \hat{\psi}_H^+(2)\, \hat{\psi}_H(2)] | \Psi \rangle .$$

The subsequent calculations are similar to those discussed at the beginning of this section; the difference is that the operators $\hat{\psi}_H(1)$, $\hat{\psi}_H^+(2)$ are replaced by the combination $\hat{\psi}_H^+(1)\, \hat{\psi}_H(1)$ and $\hat{\psi}_H^+(2)\, \hat{\psi}_H(2)$. We introduce, as before, a complete set of intermediate states Ψ_n; now, however, these states correspond to the same number of particles N as the ground state. Introducing the function

$$\Phi_n(q_1, q_2, t) = \Phi_n(q_1, q_2) \exp(-i\Delta E_n t),$$

we can write†

$$\tilde{G}(1, 2) = \sum_n \begin{cases} \Phi_n^*(q_2, q_2)\, \Phi_n(q_1, q_1) \exp[-i\Delta E_n(t_1 - t_2)] & t_1 > t_2 \\ \Phi_n^*(q_1, q_1)\, \Phi_n(q_2, q_2) \exp[i\Delta E_n(t_1 - t_2)] & t_1 < t_2 \end{cases}$$

Introducing the Fourier transform of $\tilde{G}$ with respect to $t_1 - t_2$, we have

$$\tilde{G}(q_1, q_2, \varepsilon) = \sum_n \left[\frac{\Phi_n^*(q_2, q_2)\, \Phi_n(q_1, q_1)}{\varepsilon - \Delta E_n + i\delta} - \frac{\Phi_n^*(q_1, q_1)\, \Phi_n(q_2, q_2)}{\varepsilon + \Delta E_n - i\delta} \right].$$

† In this equation there is a + sign in front of the second line due to the permutations of pairs of operators.

Going over from summation to integration, we find

$$\tilde{G}(q_1, q_2, \varepsilon) = \int_0^\infty dE \left[\frac{A(q_1, q_2, E)}{\varepsilon - E + i\delta} - \frac{A(q_2, q_1, E)}{\varepsilon + E - i\delta} \right], \qquad (23.22)$$

where

$$A(q_1, q_2, E) = \sum_n \delta(E - \Delta E_n)\, \Phi_n^*(q_2, q_2)\, \Phi_n(q_1, q_1). \qquad (23.23)$$

For a spatially uniform system, $\Phi_n(q_1, q_1) = a_{n'p} \exp i(\boldsymbol{p} \cdot \boldsymbol{x}_1)$, where $\boldsymbol{p}$ is the total momentum of the excited state, and the n' are the other state subscripts. Then

$$\Phi_n^*(q_2, q_2)\, \Phi_n(q_1, q_1) = |a_{n', \boldsymbol{p}}|^2 \exp [i(\boldsymbol{p} \cdot \boldsymbol{x}_1 - \boldsymbol{x}_2)],$$

$$\Phi_n^*(q_1, q_1)\, \Phi_n(q_2, q_2) = |a_{n', -\boldsymbol{p}}|^2 \exp [-i(\boldsymbol{p} \cdot \boldsymbol{x}_1 - \boldsymbol{x}_2)].$$

We can show that $a_{n', \boldsymbol{p}} = a_{n', -\boldsymbol{p}}$ and, consequently, $|a_{n', \boldsymbol{p}}|^2 = |a_{n', -\boldsymbol{p}}|^2$. In fact, in this case, only the part of the operator $\hat{\psi}^+(q_1)\, \hat{\psi}(q_1)$ corresponding to the momentum $\boldsymbol{p}$ (namely $\int d\boldsymbol{x}_1 \exp [-i(\boldsymbol{p} \cdot \boldsymbol{x}_1)] \times \hat{\psi}^+(q_1)\, \hat{\psi}(q_1) = \hat{O}_{\boldsymbol{p}}$) contributes to the matrix element Φ_n. Owing to the Hermitian character of the operator $\hat{\psi}^+ \hat{\psi}$, $\hat{O}_{\boldsymbol{p}}^* = \hat{O}_{-\boldsymbol{p}}$, from which follows the assertion made above. Hence

$$\tilde{G}(\boldsymbol{p}, \varepsilon) = \int_0^\infty dE^2 \frac{A(\boldsymbol{p}, E)}{\varepsilon^2 - E^2 + i\delta}, \qquad (23.24)$$

where $A(\boldsymbol{p}, E)$ is a real, positive quantity.

It is immediately obvious that $A = -(1/\pi) \operatorname{Im} \tilde{G}$. Thus

$$\operatorname{Re} \tilde{G}(\boldsymbol{p}, \varepsilon) = -\frac{1}{\pi} \mathrm{P} \int_0^\infty \frac{dE^2 \operatorname{Im} \tilde{G}(\boldsymbol{p}, E)}{\varepsilon^2 - E^2}. \qquad (23.24')$$

23.7. Up to this point, the variables ε and E have been real quantities. In order to make a complete examination of the analytical properties of the Green function, it is essential to consider complex values of these quantities as well, i.e. to continue the Green function analytically into the complex energy plane.

We will transform eqn. (23.11): putting $E \to -E$ in the second term, and introducing the symbol

$$F(\boldsymbol{p}, E) = \begin{cases} A(\boldsymbol{p}, E) & E > 0 \\ B(\boldsymbol{p}, -E) & E < 0 \end{cases}$$

we obtain

$$G(\mathbf{p}, \varepsilon) = \int_{-\infty}^{\infty} dE \frac{F(\mathbf{p}, E)}{\varepsilon - \mu - E + i\delta \operatorname{sign}(E)}. \quad (23.25)$$

We will introduce a new complex variable

$$\zeta = E - i\delta \operatorname{sign}(E).$$

When $E > 0$, $\zeta = E - i\delta$; when $E < 0$, $\zeta = E + i\delta$.

The integration contour C, which corresponds to the real axis for the variable E, has the form shown in Fig. 47 in the complex ζ-plane. The explanation of this is that, when $\operatorname{Re} \zeta = E > 0$ (in the right-hand half-plane), we must have $\zeta = E - i\delta$, i.e. we must displace the contour below the real axis by an amount δ. The same argument applies to the left-hand half-plane.

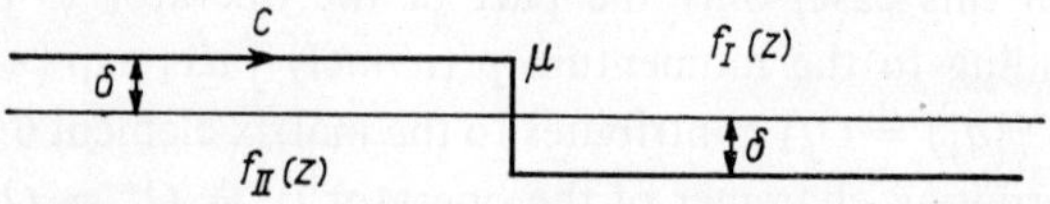

FIG. 47

Besides a complex variable of integration, we will also introduce a complex value for the quantity $\varepsilon - \mu$ which we will denote by z. The general complex plane for the quantities ζ and z is represented in Fig. 47. Substituting these new variables into (23.25), and denoting the quantity† $F[\zeta + i\delta \operatorname{sign}(\operatorname{Re}\zeta)]$ by $\tilde{F}(\zeta)$, we arrive at the basic equation:

$$G(\mathbf{p}, z) = \int_c \frac{d\zeta\, \tilde{F}(\mathbf{p}, \zeta)}{z - \zeta}. \quad (23.26)$$

This relation is similar to the familiar Cauchy integral [104], which defines two different, analytical functions on the two sides of the contour. We will denote the functions in the upper and lower half-planes by $f_I(z)$ and $f_{II}(z)$ respectively (see Fig. 47). Singularities, such as branch points, are possible on the contour itself.

† The quantity $\tilde{F}(\zeta)$ is real and positive on the contour. We note that $\operatorname{Re} \zeta = E$ and, therefore,

$$E = \zeta + i\delta \operatorname{sign} E = \zeta + i\delta \operatorname{sign}(\operatorname{Re}\zeta).$$

We will consider the connection between the function of the complex variable $G(z)$ and the Green function $G(\varepsilon)$. Since the region $\varepsilon > \mu$ lies above the contour, and $\varepsilon < \mu$ below it, we can write

$$G(\varepsilon) = \begin{cases} f_{\mathrm{I}}(z)|_{z\to\varepsilon-\mu>0} \\ f_{\mathrm{II}}(z)|_{z\to\varepsilon-\mu<0} \end{cases} \tag{23.27}$$

At the point $\varepsilon = \mu$, the function $G(\varepsilon)$ can (and must) have a singularity.

The difference between the functions f_I and f_{II} is that, when their arguments tend to some point on the boundary, the values obtained are the complex conjugates of each other. Suppose we consider, for example, the two points, $z_1 = -a$, $z_2 = -a + 2i\delta$ (Fig. 48), then we have

$$G(z_1) = \int_{-\infty}^{\infty} dE \frac{F(E)}{-a - E - i\delta}, \quad G(z_2) = \int_{-\infty}^{\infty} dE \frac{F(E)}{-a - E + i\delta}.$$

Remembering that the function $F(E)$ is real, we obtain $G(z_2) = G^*(z_1)$, or

$$\lim_{z_2 \to z_1} f_{\mathrm{II}}(z_2) = f_{\mathrm{I}}^*(z_1).$$

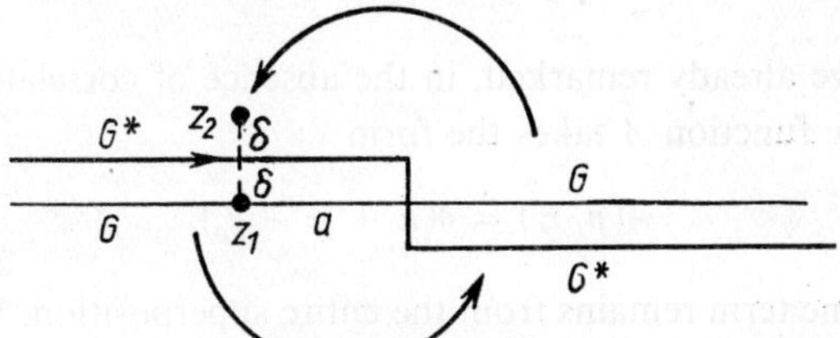

FIG. 48

Thus if we take the Green function $G(\varepsilon)$ with $\varepsilon > \mu$, and continue it analytically into the upper half-plane, we arrive at the function $f_{\mathrm{I}}(z)$. If, further, we approach the contour C in the region $\varepsilon < \mu$ from above, then we obtain the function $G^*(\varepsilon)$.

If we continue the Green function $G(\varepsilon)$, with $\varepsilon > \mu$ into the lower half-plane, i.e. into the second sheet, we obtain a function that has nothing in common with $f_{\mathrm{II}}(\varepsilon)$ and, in particular, is not analytical in this region. It must necessarily have a singularity, since no function exists which is everywhere analytical, different from zero, and decreasing towards infinity.† A similar analytical continuation of the function $G(\varepsilon)$ with $\varepsilon < \mu$ into the upper half-plane also leads to a function possessing singularities. All these singularities have an immediate physical significance.

In order to find these singularities, we can analytically continue the function Im G (i.e. the quantity A for $\varepsilon > \mu$; B for $\varepsilon < \mu$), as we did G. This depends on the fact that, when we continue the function G^* into the region concerned, we obtain an analytical function ($f_{\rm I}$ and $f_{\rm II}$). The functions G and $\text{Im}\, G = (G - G^*)/2i$ therefore have the same singularities in these regions.

The two-particle Green function can be discussed in a precisely similar way. The only difference is that the chemical potential μ must be put equal to zero (only states with an identical number of particles figure in the calculations).

23.8. We will return now to (23.11) and consider in detail its first term, which corresponds to the propagation of a particle

$$G(\boldsymbol{p}, \varepsilon) = \int_0^\infty dE \frac{A(\boldsymbol{p}, E)}{\varepsilon - \mu - E + i\delta}. \tag{23.28}$$

As we have already remarked, in the absence of correlation interactions the function A takes the form

$$A(\boldsymbol{p}, E) = \delta(E + \mu - \varepsilon_p),$$

i.e. only one term remains from the entire superposition. When the correlation interaction is taken into account, the quantity A does not reduce to a δ-function: it differs from zero for all values of E. It may turn out, however, that the function $A(\boldsymbol{p}, E)$ has a more or less sharply defined maximum near the point $E = E_p - \mu$, with a width $\Gamma_p \ll E_p$.

The fact that a function shows resonance behaviour means formally that, when the function is continued analytically into the complex plane of its argument, a pole appears which lies close to the real axis.

† This latter property stems from the asymptotic conditions (23.9).

This can be confirmed immediately. Suppose that the analytical continuation of the function $G(\mathbf{p}, E)$ into the lower half-plane leads to the appearance of a pole at the point $E_p - \mu - i\Gamma_p$. Then, near the pole, we can put

$$G = \frac{Z_p}{E - E_p + \mu + i\Gamma_p}$$

and

$$A(\mathbf{p}, E) \sim \operatorname{Im} G(\mathbf{p}, E) \sim \frac{Z_p \Gamma_p}{(E - E_p + \mu)^2 + \Gamma_p^2}, \tag{23.29}$$

where Z_p is some quantity independent of E. This expression holds for small Γ and on the real axis itself.

We will next consider what dependence of the Green function on the time difference $t_1 - t_2$ corresponds to this particular situation. It is evidently necessary to investigate the expression

$$G(\mathbf{p}, \tau) = \int_{-\infty}^{\infty} \frac{d\varepsilon}{2\pi} G(\mathbf{p}, \varepsilon) \exp(-i\varepsilon\tau)$$

$$= -i \int_{0}^{\infty} dE A(\mathbf{p}, E) \exp[-i(\mu + E - i\delta)\tau] \tag{23.30}$$

where $\tau = (t_1 - t_2) > 0$.

Here we have completed the contour in the lower half-plane (see Appendix C).

If we now substitute the expression for A corresponding to no correlation, we obtain

$$G(\mathbf{p}, \tau) \sim \exp(-i\varepsilon_p \tau), \tag{23.31}$$

which corresponds to thinking of G as the transition amplitude for a system of $N + 1$ particles from the state at time t_1 to the state at time t_2 (see § 10).

Considering the general case, we will suppose for simplicity that we are dealing with a single pole only, and will use the contour of integration shown in Fig. 49. The integrals along sections 3 and 3′ mutually compensate each other. After substituting (24.29), the contribution of the integration round the pole is

$$G_4(\mathbf{p}, \tau) = Z_p \exp(-iE_p\tau - \Gamma_p\tau). \tag{23.32}$$

We must finally estimate the contributions of 2, 2′ and 1. The integral along the line 2, 2′, corresponding to $E = E' - iz$ (where z is the distance between the real axis and the line), can be written as

$$G_{2,2'}(\boldsymbol{p}, \tau) = \int_0^\infty dE' \exp\left[-i(\mu + E')\tau - z\tau\right] A(\boldsymbol{p}, E')$$

By choosing a sufficiently large value of z, this integral can be made as small as desired.

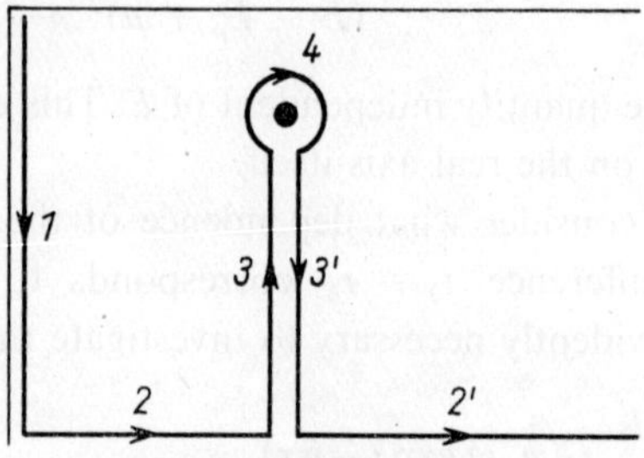

FIG. 49

For the section 1, we put $E = ix$ $(0 \geqq x > -\infty)$, and we can write

$$G_1(\boldsymbol{p}, \tau) \sim \int_{-\infty}^0 dx \exp(-i\mu\tau + x\tau) \frac{\Gamma_p}{(ix - E_p + \mu)^2 + \Gamma_p^2}.$$

The most important contribution to this integral is provided by values of x that satisfy the inequality

$$|x| \gtrsim \frac{1}{\tau} < \Gamma \ll E_p - \mu.$$

Whence

$$G_1(\boldsymbol{p}, \tau) \sim \left(\frac{\Gamma_p}{E_p - \mu}\right)^2 \frac{1}{\Gamma_p \tau}.$$

If $\exp(-\Gamma_p \tau)$ is not too small, this quantity is small compared with (23.32), and we can put $G(\boldsymbol{p}, \tau) = G_4(\boldsymbol{p}, \tau)$.

This result can be interpreted in the following way. Under our present assumptions, the function $A(\boldsymbol{p}, E)$ has a sharp maximum. This implies that we can replace the propagation of a particle described by the Green function and including interactions with the system by the propagation of a wave packet made up of free particles with a weight function for each particle of $A(\mathbf{p}, E)$. Such a wave packet necessarily moves in a non-stationary way and dies away with time: more quickly if the energy dispersion is greater.

23.9. Spectral representations can be written for other quantities related to the Green function, as well as for the Green function itself. In particular, one can be written for the mass operator [105].

We will start from the relation

$$G^{-1}(\mathbf{p}, \varepsilon) = \varepsilon - \varepsilon_p - M(\mathbf{p}, \varepsilon).$$

Changing to the complex plane $z = \varepsilon - \mu$, we consider the function

$$G^{-1}(\mathbf{p}, z) = z - (\varepsilon_p - \mu) - M(\mathbf{p}, z + \mu),$$

which is equal to $1/f_{\text{I}}(z)$ and $1/f_{\text{II}}(z)$ above and below C respectively (see Fig. 47). We will examine the analytical properties of the function G^{-1}. We know, first of all, that the functions G^{-1} and G have common singularities on the contour C; outside C, the function $G(\mathbf{p}, z)$ has no singularities, and decreases with distance from the origin as $|z|^{-1}$.

We next enquire whether G has any zeroes in the region, i.e. does the function $G^{-1}(\mathbf{p}, z)$ have a pole there. To find an answer, we consider the spectral representation of (23.25), where the function $F(\mathbf{p}, E)$ is real and positive. Continuing it into the complex plane, we can write

$$G(\mathbf{p}, z) = \int_{-\infty}^{\infty} dE \frac{F(\mathbf{p}, E)}{z - E + i\delta \operatorname{sign} E}.$$

In the complex z-plane (where $z = a + ib$), the imaginary part of $G(\mathbf{p}, z)$, equal to

$$\operatorname{Im} G(\mathbf{p}, z) = -b \int_{-\infty}^{\infty} dE \frac{F(\mathbf{p}, E)}{(a - E)^2 + b^2},$$

is different from zero. It is therefore evident that the function $G^{-1}(\mathbf{p}, z)$ has no singularities outside C. Similarly, the mass operator $M(\mathbf{p}, z + \mu)$ has no singularities.

For large values of ε, the mass operator can be computed in the lowest-order perturbation theory and, as is evident from the results of § 14, it tends to zero as ε^{-1} when $\varepsilon \to \infty$.

Let us consider the integral

$$I = \int_{-\infty}^{\infty} dE \frac{M(\boldsymbol{p}, E + \mu)}{\varepsilon - \mu - E + i\delta \operatorname{sign}(E)}$$

$$= \int_C d\zeta \frac{M(\boldsymbol{p}, \zeta + \mu + i\delta \operatorname{sign} \operatorname{Re} \zeta)}{z - \zeta}.$$

If we complete it by means of a large arc in the upper (for $\varepsilon > \mu$) or lower (for $\varepsilon < \mu$) half-planes, which is possible in view of the way that M decreases, and if we remember that the expression under the integral sign is analytic, we arrive at the following results. The contour integral obtained reduces to a residue at the point $E = \varepsilon - \mu + i\delta$ (for $\varepsilon > \mu$), or $E = \varepsilon - \mu - i\delta$ (for $\varepsilon < \mu$);

$$I = -2\pi i \operatorname{sign}(\varepsilon - \mu) M(\boldsymbol{p}, \varepsilon).$$

Equating the imaginary parts of both these expressions for I, we obtain finally

$$\operatorname{Re} M(\boldsymbol{p}, \varepsilon) = -\frac{\operatorname{sign}(\varepsilon - \mu)}{\pi} \mathrm{P} \int_{-\infty}^{\infty} dE \frac{\operatorname{Im} M(\boldsymbol{p}, E)}{\varepsilon - E}. \tag{23.33}$$

The quantity $\operatorname{Im} M = \operatorname{Im} G^{-1} = -\operatorname{Im} G/|G|^2$ changes sign at the point $\varepsilon = \mu$. If we introduce a quantity with a constant (positive) sign,

$$\operatorname{Im} \mathfrak{M}(\boldsymbol{p}, \varepsilon) = \operatorname{sign}(\varepsilon - \mu) \operatorname{Im} M(\boldsymbol{p}, \varepsilon),$$

where $\mathfrak{M}(\boldsymbol{p}, \varepsilon) = \operatorname{sign}(\varepsilon - \mu) M(\boldsymbol{p}, \varepsilon)$, then (23.33) assumes the same form as the spectral representation of the Green function itself

$$\operatorname{Re} \mathfrak{M}(\boldsymbol{p}, \varepsilon) = -\frac{1}{\pi} \mathrm{P} \int_{-\infty}^{\infty} dE \frac{\operatorname{sign}(E - \mu) \operatorname{Im} \mathfrak{M}(\boldsymbol{p}, E)}{\varepsilon - E}. \tag{23.34}$$

This relation is very convenient to use in many cases. It is usually simpler to calculate the imaginary part of the mass operator directly than the real part, since the δ-function in $\operatorname{Im} M$ reduces the number of variables of integration. It is then easy enough to calculate the real part from the imaginary part, using (23.34).

24. Quasi-Particles

24.1. The previous sections of this chapter have contained several references to descriptions of excited states of the system, in terms of the creation of some complex, of the propagation of a particle correlated with the system in terms of some wave packets, etc. The present section will deal with some general points concerning these questions, which are related to the concept of a quasi-particle (in the widest sense of this term).

We will begin by considering the spectral formula for a single-particle Green function. We will suppose that the analytical continuation of the Green function into the lower half-plane for $\varepsilon > \mu$ leads to a pole at the point $E_p - i\Gamma_p$. (We are considering, for simplicity, a spatially uniform system.) Then the exact Green function $G(\boldsymbol{p}, \varepsilon)$ can be put in the form

$$G(\boldsymbol{p}, \varepsilon) = \frac{Z_p}{\varepsilon - E_p + i\Gamma_p} + \varphi(\boldsymbol{p}, \varepsilon), \tag{24.1}$$

where we have explicitly separated out the singularity of this function. Here Z_p is some normalization factor; $\varphi(\boldsymbol{p}, \varepsilon)$ is a function regular at the point $\varepsilon = E_p - i\Gamma_p$.

We thus see that, for values of ε close to E_p, and values of $\boldsymbol{p}$ for which $\Gamma_p \ll E_p$, the exact expression for the Green function, $G(\boldsymbol{p}, \varepsilon)$, reduces to the first term of (24.1) and (to within a factor Z_p) can be represented by

$$G(\boldsymbol{p}, \varepsilon) \sim (\varepsilon - E_p + i\Gamma_p)^{-1}. \tag{24.2}$$

For the region considered, this leads us to the following conclusion. A system of interacting particles with arbitrary mutual correlations is equivalent to a system of a different type of objects whose dispersion law E_p is different from that for the true particles ε_p, and which have no correlation. In other words, so long as the necessary conditions are satisfied, a system of interacting particles can be replaced by a system with single-particle characteristics.

In fact if Γ_p is really small, the right-hand side of (24.2) cannot be distinguished from the expression for the Green function of a system in the Hartree–Fock approximation [see eqn. (10.8)]. This function contains an infinitely small imaginary component in the denominator; hence the requirement that Γ_p be small is a necessary condition for our conclusion to be true.†

Comparing (24.2) with (10.8), we see that E_p replaces ε_p, the dispersion law for a particle in the Hartree–Fock approximation. E_p may therefore be interpreted as the dispersion law for a new object. Introducing this object we can neglect the correlation interactions.

This object is called a quasi-particle (in the restricted sense of the word). The conclusion reached above can now be reformulated in the following fashion: a system of many interacting particles can be replaced by a perfect gas of quasi-particles.‡ This replacement can by no means always be made: it is limited by the requirement that Γ_p be small, and that ε be close to E_p. Consequently, the concept of a quasi-particle breaks down when these conditions are violated.

As an example of a system where the concept of a quasi-particle is very obvious, we can take a normal, classical, dilute gas, made up of neutral atoms. In the final analysis our system is composed of nuclei and electrons with a very large interactions (and correlations) between them. It is extremely easy to find the quasi-particle in this case (i.e. the object in terms of which system may be considered a perfect gas). It is simply the individual atom.

In this example, the quasi-particle is a complex formed from a small number of particles in the system. As the system evolves, this complex is not destroyed. In the general case a quasi-particle cannot be made to correspond to any specific particle in the system. Its constitution is affected by all the particles in the system: even if only a few of the particles enter into it at a given time, a short time later these are replaced by others. Quasi-particles are, so to speak, collective objects.

The system is far simpler to consider when it is reduced to a gas of quasi-particles. We can then take the customary physical

† The quantity $i\delta$ sign $(\varepsilon_p - \varepsilon_F)$ in (10.8) changes sign at the Fermi surface. Γ_p has similar properties (see § 24.3).

‡ The difference between a perfect gas and a system where correlation is ignored is not important here. We note that a quasi-particle transforms to an ordinary particle when the correlation interaction is excluded.

approach, using concepts and representations which formerly could only be applied to systems with weak interactions. Our use of quasi-particles is, in fact, simply an extension of the approach which we have previously been following, whereby we try to retain the single-particle character of the system by a suitable selection of particle properties. The transition to the Hartree–Fock approximation was the first step in this direction: by changing the dispersion law for the particles and introducing an additional external field, we excluded the self-consistent part of the particle interaction from consideration. The introduction of the quasi-particle represents the next step in the same direction. The correlation interactions, however, can only be excluded when very strict conditions on the smallness of the attenuation are satisfied.

If these conditions are disobeyed, the quasi-particle can be used to construct a suitable first approximation. In subsequent approximations we must introduce the interaction between the quasi-particles, which is necessarily less than the interaction between the real particles.

24.2. It is extremely important that the excited states of a system, under the given conditions, can also be treated in terms of quasi-particles. We can say that the transition of a system to an excited state results from the appearance in it of additional non-interacting quasi-particles. These quasi-particles may be of the same type as those from which the system itself is constructed, or they may be different.

The possibility of using this picture depends on the limitations mentioned in the previous section. These will be satisfied so long as the excitation energy is small.

If we consider, for simplicity, the excited states of a system of $N+1$ particles, we see that, when the necessary conditions are fulfilled, the function $A(\boldsymbol{p}, E)$ has a sharp maximum near the point $E_p - \mu$. This function represents the density of levels in the system near $E = \Delta' E_n = \Delta E_n - \mu$. Thus the excitation energy of the system coincides, in fact, with the quantity E_p, i.e. the excitation of the system is connected with the creation of a quasi-particle.

A similar situation occurs for a system of N particles. There are however, two ways of exciting the system in this case and, consequently, two different types of excitation spectrum [4]. Some are quasi-particle (particle-hole) states. They necessarily have an integral angular momentum and must be considered as bosons. The corresponding excitation spectra are of the Bose type, e.g. phonons, excitons, plasmons, magnons [47, 106–108].

We can also have the case of pair excitation with one momentum higher and the other lower than p_0—the effective Fermi surface for quasi-particles. The related Fermi excitation spectrum is qualitatively reminiscent of the excitation spectrum of a perfect gas, and the excitations correspond to the quasi-particles and "quasi-holes",† whose appearance leads to the excitation of a system of $N+1$ particles. From this point of view the Bose excitations can be treated as bound states of the quasi-particles and quasi-holes.

The excitation spectrum of a real many-body system includes, in the general case, both Fermi and Bose branches, the latter, itself, being made up of several subgroups (the phonons, excitons, etc., of solid state theory).

24.3. The fact that the correlation interaction between quasi-particles is small (within certain limits) makes it possible to speak of the individual characteristics of a quasi-particle, specifically, of its energy. To fix the ideas we shall speak of the quasi-particle proper—a single-particle, elementary excitation of the system.

In the two previous sections we have already, in essence, considered the problem of the energy of a quasi-particle from two different points of view. We have explained in § 22 that, so long as the attenuation is assumed to be small, the excitation energy of a system ΔE_n may be sought from the equation

$$\Delta E_n - \varepsilon_p - M(\boldsymbol{p}, \Delta E_n) = 0. \tag{24.3}$$

† A "quasi-hole" is an object whose properties are determined by the pole of the Green function for $\varepsilon < \mu$, and which transforms to an ordinary hole when the correlation interaction is excluded. The Fermi type includes the electron spectrum in metals [109], the polaron spectrum in ionic crystals [110], etc.

The solution of this equation has an imaginary part which corresponds to the attenuation of the function Φ_n

$$\Phi_n \sim \exp\left[-i(E_p - i\Gamma_p)\,t\right], \tag{24.4}$$

where $E_p - i\Gamma_p = \Delta E_n$ is the root of eqn. (24.3). The appearance of a non-zero attenuation can be explained by the fact that we have used an approximate equation for the function Φ_n, and have made assumptions concerning the stability of the wave function in the zero approximation. In actual fact the wave function we have selected is not completely stable. It is a superposition of several stable functions, each of which acquires its energy value as a result of correlation interactions: this leads to the observed non-stationary state. If the actual conditions can, in fact, be represented by an unstable wave function, then (24.4) describes a real attenuation for the given initial state of the system.

We can now answer a question that frequently crops up: why does the matrix element Φ_n, obtained from the solution of (24.3), decay with time, whilst the spectral function is deduced using an expression for Φ_n which contains no damping? The answer is that a precise expression for Φ_n is used in the latter case, corresponding to a true stationary state of the system; the equation for the function Φ_n, which contains only the mass operator, automatically corresponds to a non-stationary state.

If we wish to consider the evolution of a quasi-particle, we must specifically introduce an unstable wave function. Equations (24.3) and (24.4), were deduced assuming that the function $\Psi_{\mathrm{Int}\,n}(-\infty)$ corresponds to the presence of one particle in the background of the ground state, and describe the behaviour of this particle when the correlation interaction is included. The object which arises in this case is a quasi-particle. Similarly, when the correlation interaction is excluded, the quasi-particle transforms to an ordinary particle. Hence, the assumption we have made about the form of the function $\Psi_{\mathrm{Int}\,n}(-\infty)$ is obligatory. If we consider a true stationary state of the system, then the exclusion of correlation interactions leads to a state which contains, besides a particle, a certain number of pairs. This state contains, of course, more than one quasi-particle.† It follows immediately from the form of the wave function defined by (24.4), that E_p coincides with the energy of the quasi-particle. It is evident that eqn. (24.3) corresponds exact ly to the pole of the Green function. In looking for the roots of this equation, one only needs to watch that their imaginary part (the quantity $-\Gamma_p$) is negative, i.e. that there is damping.

These results come from arguments based on the spectral representation for the Green function. As was explained in § 23, if the analytical continuation of the Green function in the lower half-plane has a pole at the point $E_p - i\Gamma_p$, then there is a narrow packet of stationary states (width $\sim\Gamma_p$) present in the excited state of the system, and the system evolves according to (23.4). More accurately, if we suppose that the density of levels $A(\mathbf{p}, E)$ reduces to its pole term (23.29), i.e. that the excited state differs from the ground state only by the presence of a wave packet, then this packet (the quasi-particle) has an energy E_p and an attenuation Γ_p.

Thus the energy (the dispersion law) and the attenuation of a quasi-particle are directly connected with the single-particle Green function. In the uniform case both of these quantities are determined simply by the real and imaginary parts of the pole in the analytical continuation into the lower half-plane of the Fourier transform of the Green function.‡ In order to find them in the general case we must solve the equation obtained from (22.16) by substituting $\Phi \sim \exp(-iE_n t - \Gamma_n t)$, i.e.

$$(E_n - i\Gamma_n - \hat{T} - \hat{W})\,\Phi_n(q_1) - \int dq_2\, M(q_1, q_2, E_n - i\Gamma_n)\,\Phi_n(q_2) = 0, \qquad (24.5)$$

where $M(q_1, q_2, \varepsilon)$ is the Fourier transform of the mass operator.

We will consider the uniform case from this point of view in more detail. We have

$$E_p - i\Gamma_p - \varepsilon_p - M(\mathbf{p}, E_p - i\Gamma_p) = 0. \qquad (24.6)$$

† One cannot talk about an exact number of quasi-particles in a stationary state, since they do not correspond to an integral of motion.

‡ This refers to the case $\varepsilon > \mu$. When $\varepsilon < \mu$, the analytical continuation of $G(\mathbf{p}, \varepsilon)$ into the upper half-plane gives the characteristics of a quasi-hole, described by the quantity $\tilde{\Phi}_n(1)$, whose time-dependence is $\exp(iEt_1 - \Gamma t_1)$.

Assuming that $\Gamma_p \ll E_p$, and separating the real and imaginary parts, we arrive at a set of equations which give the following solutions:

$$\left.\begin{aligned} E_p &= \varepsilon_p + \operatorname{Re} M(\boldsymbol{p}, E_p), \\ \Gamma_p &= \frac{-\operatorname{Im} M(\boldsymbol{p}, E_p)}{1 - \partial \operatorname{Re} M(E_p)/\partial E_p}. \end{aligned}\right\} \qquad (24.7)$$

The imaginary part of the Green function can be written as

$$\operatorname{Im} G = \frac{\operatorname{Im} M}{|G|^2}.$$

Thus Γ and $\operatorname{Im} G$ go to zero simultaneously. As we have seen in § 23, $\operatorname{Im} G$ changes sign at the point $\varepsilon = \mu$, and, assuming it to be continuous, goes to zero at this point. The sign of Γ must be positive for $\varepsilon > \mu$, and negative for $\varepsilon < \mu$.

Let us consider the Green function near the point $\varepsilon = \mu$. We will suppose, for simplicity, that correlated pairs which lead to superconductivity cannot occur in the system.† The mass operator does not then have a singularity near $\varepsilon = \mu$. We may then expand the expression for G^{-1}

$$G^{-1}(\boldsymbol{p}, \varepsilon) = \varepsilon - \varepsilon_p - \operatorname{Re} M(\boldsymbol{p}, \varepsilon) - i \operatorname{Im} M(\boldsymbol{p}, \varepsilon)$$

in a series about the point $\varepsilon = \mu$ and $p = p_0$, where p_0 is the root of the equation

$$\mu = \varepsilon_p|_{p=p_0} + M(p_0, \mu) = E_p|_{p_0}. \qquad (24.8)$$

We find from (24.7) that

$$G^{-1}(\boldsymbol{p}, \varepsilon) = Z_p^{-1}[\varepsilon - \mu - v(p - p_0) - iZ_p \operatorname{Im} M], \qquad (24,9)$$

where

$$Z_p^{-1} = 1 - \left.\frac{\partial \operatorname{Re} M(\varepsilon, p_0)}{\partial \varepsilon}\right|_{\varepsilon=\mu}.$$

† The superconducting situation corresponds to a pole of the function M near the Fermi surface [111]. When there is no superconductivity, the mass operator can be expanded in the form:

$$M(\boldsymbol{p}, \varepsilon) = M_0(\boldsymbol{p}, \varepsilon) + i\alpha(\varepsilon - \mu)^2 \operatorname{sign}(\varepsilon - \mu) + \beta(\varepsilon - \mu)^3 \ln\left(\frac{|\varepsilon - \mu|}{\mu}\right),$$

where M_0 is a function regular at the point μ.

is the factor introduced above, and

$$v = Z_p \frac{\partial E_p}{\partial p}\bigg|_{p_0}$$

is the velocity of the quasi-particle at the Fermi surface. The quantity $Z_p \operatorname{Im} M$ is proportional to $(\varepsilon - \mu)^2 \operatorname{sign}(\varepsilon - \mu)$, i.e. close to the Fermi surface it is an infinitely small higher-order term [III]. We may therefore conclude that the quasi-particle representation holds near the Fermi surface. This is the reason why the quasi-particle concept can be used only for weakly excited states of a system.

24.4. If we compare (24.1) with the expression for the free-particle Green function, we see that the exact Green function is exactly the same as the free-particle Green function near the point $\varepsilon = \mu$. Since the quantity ε_F in (10.7) determines the occupation limit for the particles, it is obvious that the chemical potential μ represents the energy which determines the equivalent limit for the quasi-particles. We can thus say that the occupation of quasi-particle levels in the ground state of a system is the same as for a perfect gas. If we introduce the occupation number for quasi-particles, then

$$n = \theta(\mu - E_p). \qquad (24.10)$$

This only holds for small values of Γ_p, i.e. for small values of $E_p - \mu$.

It is natural to take p_0 in (24.8) as the limiting momentum of a quasi-particle. It coincides with the quantity p_0 introduced in § 24.2. When perturbation theory is applicable (when, in particular, there are no superconductivity effects), p_0 is the same as for a perfect gas, i.e.

$$p_0 = \left(\frac{6\pi^2 \varrho}{g}\right)^{1/3}, \qquad (24.11)$$

where ϱ is the number density of particles in the system [6, 112]. More accurately, one can say [113] that, if the equation $E_{p_0} = \mu$ has a real root (which is untrue for the superconducting situation [114]), then it coincides with (24.11).

The momentum distribution of the quasi-particles is thus a step-function, and can be written in the form

$$\varrho(\mathbf{p}) = \theta(p_0^2 - p^2). \tag{24.12}$$

It is worth noting that the momentum distribution of the real particles, which only coincides with (24.12) when there is no correlation interaction, also partially retains this discontinuity at the Fermi surface [115]. This can be proved by considering the general relation for $\varrho(\mathbf{p})$

$$\varrho(\mathbf{p}) = -i \int_C \frac{d\varepsilon}{2\pi} G(\mathbf{p}, \varepsilon).$$

If we substitute the expression for the Green function (24.1) in this, and remember that the damping Γ_p changes sign on passing through $\varepsilon = \mu$ (i.e. $p = p_0$), we may conclude that the pole of the Green function lies inside the contour C when $p < p_0$, and outside when $p > p_0$. Since the residue at this pole is equal to Z_{p_0}, we find

$$\varrho(p_0 - 0) - \varrho(p_0 + 0) = Z_{p_0}. \tag{24.13}$$

The quantity $\varrho(p)$ lies between zero and one, so we can write

$$|Z_p| \leqq 1,$$

i.e. the jump in the distribution is less than for a perfect gas. An approximate graph of $\varrho(p)$ is given in Fig. 50.

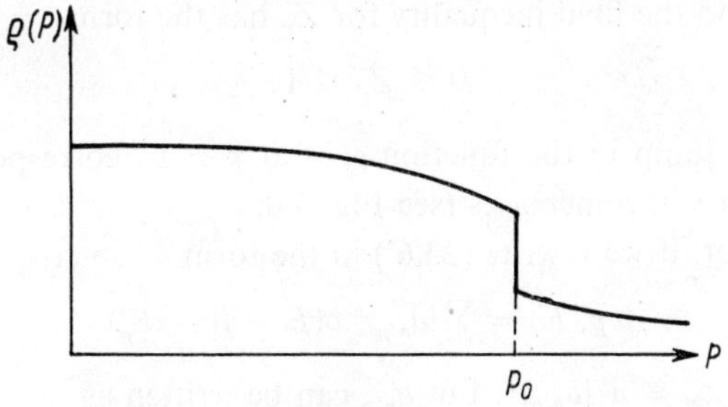

FIG. 50

There is a useful relation between the Fermi energy of the quasi-particles μ and the average energy, E/N, for a single particle in the ground state;

$$E/N = \mu. \tag{24.14}$$

This was introduced by Hugenholtz and Van Hove [112], and is applicable to self-condensed uniform systems which are in equilibrium in the absence of any external pressure (nuclear material).† We can prove this by means of the relationships $E = Nf(\varrho)$, $\partial E/\partial N = \mu$, $(\partial E/\partial \Omega)_N = -\varrho^2\, \partial(E/N)/\partial\varrho = -P$, where P is the external pressure. This gives

$$\mu = (E/N) + (P/\varrho)$$

whence (24.14) follows for $P = 0$. This equation can be used to test the consistency of the results obtained in calculating the single-particle energy E_p and the total energy of the system [112].

24.5. We will now explain the physical meaning of the quantity Z_p. The expression (24.1) for the Green function corresponds to the following choice of the weight functions A and B in (23.11) and (23.13):

$$A(\boldsymbol{p}, E) = Z_p \delta(E + \mu - E_p) + \cdots,$$

$$B(\boldsymbol{p}, E) = Z_p \delta(E - \mu + E_p) + \cdots,$$

where E_p is close to μ, as is essential if Γ_p is small. These expressions omit terms which are well-behaved and correspond to the non-singular term in (24.1).

The functions A and B have a positive sign; this implies that $Z_p > 0$, and the final inequality for Z_p has the form

$$0 \leqq Z_p \leqq 1. \tag{24.15}$$

Hence the jump in the function $\varrho(p)$ at $p = p_0$ corresponds to a decrease in ϱ as p increases (see Fig. 50).

Moreover, if we rewrite (23.6′) in the form

$$A(\boldsymbol{p}, E) = \sum |a_{n'p}|^2\, \delta(E + \mu - E_p),$$

we obtain $Z_p = n' |a_{n'p}|^2$. But $a_{n'p}$ can be written as

$$a'_{np} = \langle \Psi | \hat{a}_p | \Psi_n \rangle,$$

† This implies that the energy has a minimum at the given value of $\varrho(\partial E/\partial\varrho = 0)$.

where $\hat{a}_p$ is the annihilation operator for a particle in a state of momentum $\boldsymbol{p}$. Hence $a_{n'p}$ is the probability amplitude and Z_p is simply the probability of observing one quasi-particle in the state $\hat{a}_p^+\Psi$, when there is one normal particle added to the ground state. We remember that the state Ψ_n has a quasi-particle in the background of the ground state. We can analyse the function $B(\boldsymbol{p}, E)$ in the same way; it then appears that Z_p is also the probability of observing a quasi-hole in a state with one hole.

The quantity, analogous to Z_p, in relativistic field theory is called the renormalization constant. The renormalization of physical quantities—the properties of particles (mass) and their interactions—although normally associated with the divergences of field theory, is not, in essence, directly related to them. The aim of renormalization is to make the theoretical quantities correspond to those observed experimentally. In particular the mass of a free particle is an unobservable quantity in relativistic field theory, since there are no conditions under which the particle can be isolated from its self-field. Hence we cannot identify the observed mass, which must figure in calculations, with the mass of the free particle; we must use instead a quantity which takes into account all the effects due to interaction of the particle with its self-field. This constitutes the renormalization procedure, which in some special models, simultaneously eliminates the divergences.

In the non-relativistic many-body problem, which contains no divergences, there is no need to renormalize.† In this case the masses and charges of free particles (i.e. ones isolated from the system) are immediately observable quantities. The major task of microscopic theory is, in fact, to express the characteristics of the system in which we are interested in terms of the characteristics of free particles.

One should remember, of course, that when the interaction between particles is included, their properties change. The dispersion law for the particles and, in particular, their mass is changed.

† This does not refer to particles whose existence depends necessarily on interactions between the particles of a system—phonons, excitons, etc. Renormalization may become necessary for such particles, depending on the way in which the problem is formulated [115–117].

So is the law of interaction between the particles (see §§ 26 and 27)· It might seem convenient to carry on our discussions in terms of the changed properties of the particles, this being equivalent to renormalization. In the microscopic theory, however, it is eventually necessary to express these changed properties in terms of the characteristics of free particles. So far as phenomenological theory is concerned, a renormalization approach is therefore essential.

24.6. We will now consider the energy of the ground state of a system in terms of quasi-particles. Substituting (23.11) into (19.20) and applying the residue theorem, we obtain

$$E = \frac{1}{2}\,\mathrm{Tr}_{\sigma\tau}\int d^3\boldsymbol{p}\int_0^{\infty} dE\,B(\boldsymbol{p}, E)\,(\mu - E + p^2/2M). \qquad (24.16)$$

The quantity $\int_0^{\infty} dE\,B(\boldsymbol{p}, E)$ determines the momentum distribution in the ground state of the system (see § 23), and the function $B(\boldsymbol{p}, E)$ gives the density of excited states for a system of $N - 1$ particles near the point $E = \Delta E_n + \mu$. The function $B(\boldsymbol{p}, E)$ can be put in the form $(1/\pi)\,\mathrm{Im}\,G(\boldsymbol{p}, \mu - E)$. Hence,

$$E = \int d^3\boldsymbol{p}\int_{-\infty}^{\mu}\frac{dx}{2\pi}\,\mathrm{Im}\,G(\boldsymbol{p}, x)\left(x + \frac{p^2}{2M}\right).$$

Continuing Im G (or, simply, G) into the upper half-plane, we find a singularity at the point $E_p + i\Gamma_p$:†

$$G(\boldsymbol{p}, \varepsilon) = \frac{Z_p}{\varepsilon - E_p - i\Gamma_p}.$$

Substituting this expression into the previous equation, and carrying out a simple integration, we obtain, for small Γ,

$$E = \tfrac{1}{2}\,\mathrm{Tr}_{\sigma\tau}\int d^3\boldsymbol{p}\,\theta(\mu - E_p)\,Z_p(E_p + p^2/2M). \qquad (24.17)$$

† The quantities Z_p, E_p and Γ_p differ from those introduced above since the analytical continuation is made in another region and into another half plane. This difference can be neglected, however, near the Fermi surface where Γ_p is small. This is connected with the symmetry of particles and holes which is found near the Fermi surface.

This relation is comparable with the results of § 4. In the Hartree–Fock approximation the energy of a uniform system can be written as

$$E_0 = \tfrac{1}{2}\,\mathrm{Tr}_{\sigma\tau} \int d^3p\; \theta(\varepsilon_F - \varepsilon_p)\,(\varepsilon_p + p^2/2M).$$

It represents the sum of the particle energies ε_p minus half the interaction energy between them.

Hence the contribution of the quasi-particles to the true energy can be obtained by replacing ε_p by E_p and by introducing the probability of observing a quasi-particle Z_p. Thus a single-particle description of the system is also possible from the point of view of the energy. It is necessary in this case, however, to consider the self-consistent interaction between the quasi-particles.

25. The Equations for the Green Functions

25.1. We now consider the problem of how to calculate the single- and two-particle Green functions for a given interaction potential of the particles. Although this is simpler than finding an exact solution for the Schrödinger equation,† it is nevertheless highly complicated. We will first discuss why these complications arise.

We see, from the results in § 20, that single- and two-particle Green functions are completely determined if the vertex part Δ is known. Hence our problem reduces to determining Δ.

Even the most superficial examination of the perturbation theory diagrams which contribute to Δ shows that as n (the order of the perturbation theory) increases, the topological structure of the diagrams rapidly becomes complicated. This is immediately reflected in the mathematical structure of the theory. The equations determining the Green function (or the vertex part) cannot be written in terms of a finite set of integral equations. We either have to deal with an infinite set of coupled integral equations, each of which is connected with the neighbouring-order Green function (see § 20),

† The solution of the Schrödinger equation provides the information contained in all the Green functions. We particularly have in mind here the three-particle characteristics of the system, which cannot be described with single- or two-particle Green functions.

or with a finite set of equations in functional derivatives [5, 26]. These latter equations, which appear to be very compact, are mathematically far more complicated than integral equations. A discussion of functional methods in field theory is beyond the scope of this book.

We will, in the following, follow the simpler line of approach and make a straightforward analysis of the perturbation diagrams for the vertex part. This is possible because we are limiting ourselves to a restricted class of problems: the theories of dilute and dense systems. Such systems are characterized by the fact that it is relatively simple to select the appropriate subset from all the diagrams for the vertex part.

25.2. Thus we must turn to investigating the diagrams that enter into Δ. This quantity combines all vertex diagrams for which the external points are linked together. The topological structure of these diagrams is very complicated.

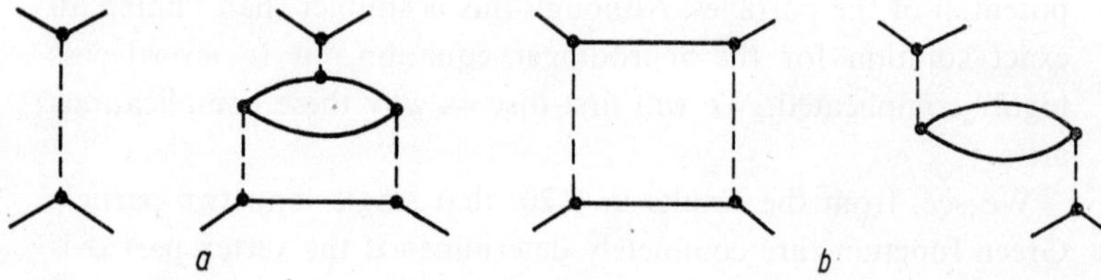

FIG. 51

We will introduce the concept of compact diagrams Δ_c in Δ. These diagrams cannot be crossed by a line which intersects only two full-drawn lines on the diagram. Figure 51a gives some examples of compact diagrams, Fig. 51b of non-compact diagrams.

If we combine the compact diagrams with one another, we can obtain all the diagrams entering into Δ. One might expect, in principle, that the topological complexity of the diagrams for Δ can be explained by the corresponding complexity of the diagrams entering into the compact part Δ_c. At the same time, the relation of Δ_c to Δ can be comparatively straightforward. This is what happens with the self-energy part of the S-matrix, where there is a simple relation between the functions Σ and M, and all the qualitative diversity of the Σ diagrams is reproduced in the compact part of M. If the situation is similar for the vertex part, then it would be reasonable to exclude, right from the start, the non-compact vertex part of Δ, replacing it by the simpler compact part Δ_c.

In actual fact the case here is different. The connection between the functions Δ and Δ_c is itself very complicated, and the topological complexity of the diagrams of Δ, is, to a considerable extent, due to this. (Though we must remember, of course, that the diagrams Δ_c are themselves highly complex.)

We will consider, for the purposes of illustration, several very simple combinations of the compact parts Δ_c (Fig. 52a). The diversity of the geometrical structures which are now obtained (see Fig. 52b) depends on the fact that the diagrams of Δ_c, have four external lines which can be "linked" with the external lines of other diagrams of Δ_c in a large number of ways. The diagram of the self-energy part, with two external lines, has only one mode of linkage. This represents the essential difference between the diagrams of the vertex and self-energy parts.

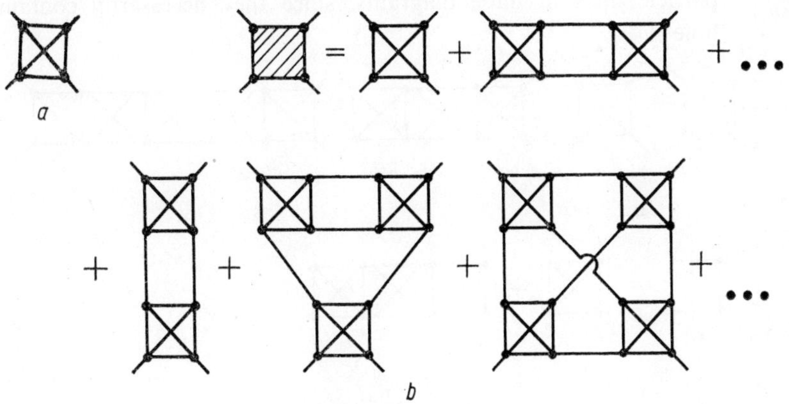

FIG. 52

Thus the transition to the compact vertex part is hardly possible in the general case. In those cases where we can select some simple subset from all the set of diagrams in Fig. 52, it is most convenient to establish the approximate equations for the non-compact vertex part of Δ.†

25.3. We will consider from this standpoint a dilute (non-resonant) many-body system. For such systems, the most important diagrams are those which have the least number of lines representing holes. The analysis carried out in § 16 referred directly only to the self-energy part; the results obtained there, however,

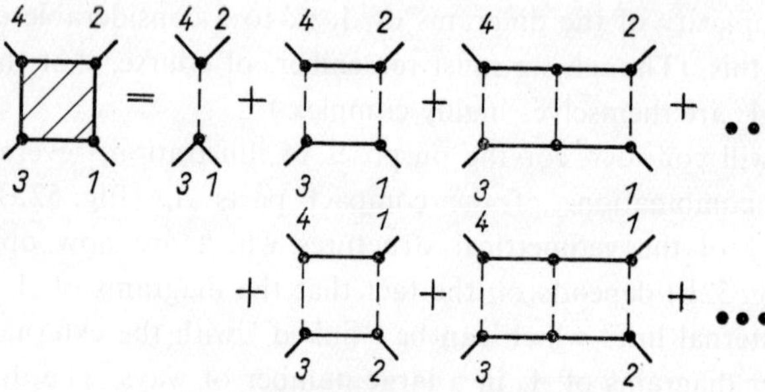

FIG. 53

also apply to the vertex part. The diagrams concerned are shown in Fig. 53. The self-energy insertions cannot be included from the particle lines in these diagrams, since they necessarily contain hole lines.

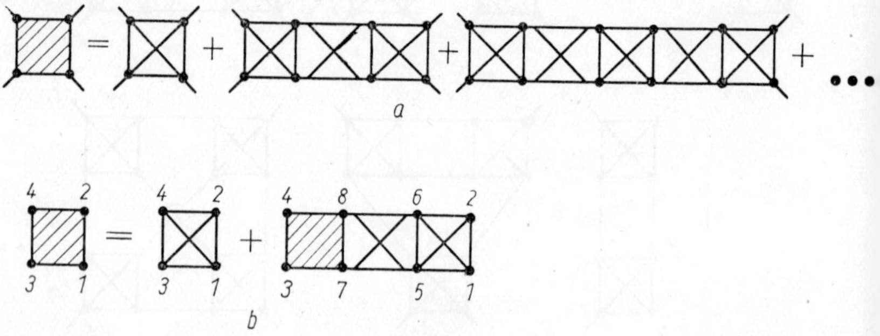

FIG. 54

† If we introduce the vertex part which is only compact relative either to vertical, or to horizontal sections, we can obtain simple integral equations similar to the Bethe–Salpeter equations. This does not, however, have any essential advantage over the approach outlined below for the problems which will be considered in the next few sections.

When Fig. 53 is compared with Fig. 52, it is evident that only the diagrams with a "horizontal" linkage of Δ_c (Fig. 54a) need be kept from all the diagrams in Fig. 52. The exchange diagrams can be included here if we introduce the two-particle Green function $G_0(1, 2, 3, 4)$, instead of the simple product of the Green functions $G_0(1, 3)$, $G_0(2, 4)$. The set of diagrams in Fig. 54a can be replaced by the diagram of Fig. 54b. This latter diagram can be reduced to the following integral equation using the Feynman rules:

$$\Delta(1, 2, 3, 4) = \Delta_c(1, 2, 3, 4) + \int d5\, d6\, d7\, d8\, \Delta_c(1, 2, 5, 6)\, G_0(5, 6, 7, 8)\, \Delta(7, 8, 3, 4).$$

For the case under consideration, we can replace the compact vertex part by its expression in the lowest-order perturbation theory

$$\Delta_c(1, 2, 3, 4) = -\frac{i}{4} V(1, 2)\, [\delta(1-3)\, \delta(2-4) - \delta(1-4)\, \delta(2-3)].$$

We then obtain the following equation:

$$\Delta(1, 2, 3, 4) = -\frac{i}{2} V(1, 2) \left\{ \frac{1}{2} [\delta(1-3)\, \delta(2-4) - \delta(1-4) \times \delta(2-3)] + \int d7\, d8\, G_0(1, 2, 7, 8)\, \Delta(7, 8, 3, 4) \right\}. \quad (25.1)$$

It is easy enough here to replace the two-particle Green function by an integral equation which is the analogue of the Dyson equation for a single-particle Green function. To this end we will consider the combination

$$\int d3\, d4\, \Delta(1, 2, 3, 4)\, G_0(3, 4, 1', 2') = \int d3\, d4\, \Delta_c(1, 2, 3, 4) [G_0(3, 4, 1', 2') + \int d5\, d6\, d7\, d8\, G_0(3, 4, 7, 8)\, \Delta(7, 8, 5, 6) G_0(5, 6, 1', 2')].$$

The quantity in the square brackets coincides with the two-particle Green function $G(3, 4, 1', 2')$, since the difference between the functions $\tilde{G}(1, 2, 3, 4)$ and $G_0(1, 2, 3, 4)$ can be ignored for a dilute system. Hence

$$\int \Delta(1, 2, 3, 4)\, G_0(3, 4, 1', 2')\, d3\, d4 = \int d3\, d4\, \Delta_c(1, 2, 3, 4)\, G(3, 4, 1', 2'). \quad (25.2)$$

If we substitute the relation we have obtained (Fig. 55) into eqn. (20.11), we arrive at the desired integral equation for the two-particle Green function

$$G(1, 2, 1', 2') = G_0(1, 2, 1', 2')$$
$$+ 2 \int d3\, d4\, d5\, d6\, G_0(1, 3)\, G_0(2, 4)\, \Delta_c(3, 4, 5, 6)\, G(5, 6, 1', 2').$$

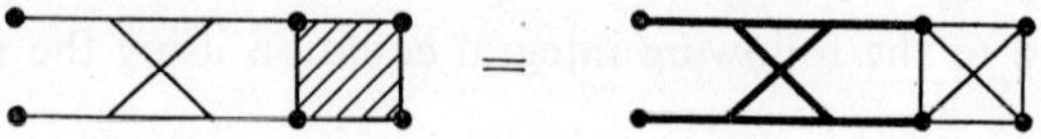

FIG. 55

Inserting the expression for Δ_c, we obtain

$$G(1, 2, 1', 2') = G_0(1, 2, 1', 2')$$
$$- i \int d3\, d4\, G_0(1, 3)\, G_0(2, 4)\, V(3, 4)\, G(3, 4, 1', 2') \qquad (25.3)$$

This equation is shown graphically in Fig. 56.

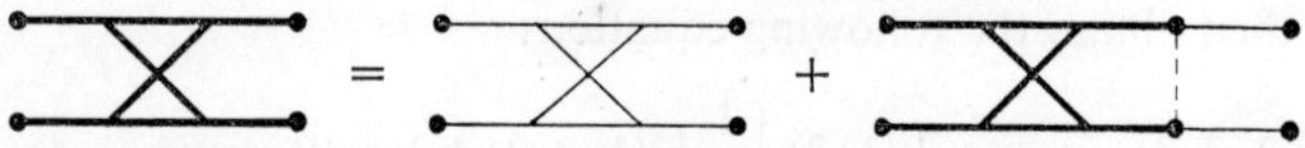

FIG. 56

25.4. Starting from eqn. (25.1), we can easily find an expression for the single-particle Green function. It is sufficient to restrict ourselves to the mass-operator definition. We can return to (20.18) with the proviso that the diagrams for M_1 and M_2 give a negligible contribution (including an additional number of holes) and that the function G must be replaced by G_0, then

$$M(1, 2) = -4i \int d3\, d4\, d5\, d6 \lim_{\substack{7 \to 3 \\ t_7 > t_3 > t_1}} V(1, 3)\, G_0(3, 4)\, G_0(1, 5)$$
$$\times\, \Delta(4, 5, 6, 2)\, G_0(6, 7). \qquad (25.4)$$

From eqn. (25.1), we can write

$$-iV(1, 3) \int d4\, d5\, G_0(3, 4)\, G_0(1, 5)\, \Delta(4, 5, 6, 2)$$

in the form

$$\Delta(3, 1, 6, 2) + \frac{i}{4} V(1, 3)\, [\delta(3 - 6)\, \delta(1 - 2) - \delta(3 - 2)\, \delta(1 - 6)].$$

Hence

$$M(1, 2) = 4\int d3\, d4 \lim_{\substack{5\to 3\\ t_5>t_3>t_1}} \Big\{\Delta(3, 1, 4, 2) + \frac{i}{4} V(1, 3)$$

$$\times [\delta(3-4)\,\delta(1-2) - \delta(3-2)\,\delta(1-4)]\Big\} G_0(4, 5). \quad (25.5)$$

This expression corresponds to closing the two "tails" of the vertex part. The second term in the curly brackets of (25.5) compensates the lowest-order term in the series expansion of Δ in perturbation theory.

We next turn to calculating the energy of a dilute system. It is most convenient for our purpose to use the general eqn. (19.19), after substituting Dyson's equation. This gives

$$E - E_0 = -\frac{i}{2}\int dq_1 \lim_{\substack{2\to 1\\ t_2>t_1}} \int d3\, M(1, 3)\, G_0(3, 2).$$

We have neglected the difference between G and G_0 in the last factor, and have replaced $i\partial/\partial t + \hat{T}$ by $i\partial/\partial t + \hat{T} + \hat{W}$ (the difference between these expressions is small). Substituting (25.5), we obtain

$$E - E_0 = -2i\int dq_1 \lim_{\substack{2\to 1\\ t_2>t_1}} \int_{6\to 4} d3\, d4\, d5$$

$$\times \Big\{\Delta(4, 1, 5, 3) + \frac{i}{4} V(1, 4)\, [\delta(4-5)\,\delta(1-3)$$

$$- \delta(4-3)\delta(1-5)]\Big\} G_0(5, 6)\, G_0(3, 2). \quad (25.6)$$

These relations will be investigated in the next section.

25.5. The most important diagrams for dense many-body systems are those which contain the maximum possible number of closed loops. The corresponding diagrams for the vertex part Δ are given in Fig. 57. Self-energy insertions in the continuous lines of the diagrams need not be taken into account. It is possible, moreover, to discard the exchange diagrams.

If Figs. 52 and 57 are compared, it is evident that the essential diagrams in this case are those with vertical linkage (Fig. 58), whilst the compact part Δ_c can also be replaced by the expression for it in lowest-order perturbation theory, i.e.

$$\Delta_c(1, 2, 3, 4) = -\frac{i}{4} V(1, 2)\,[\delta(1-3)\,\delta(2-4) - \delta(1-4)\,\delta(2-3)].$$

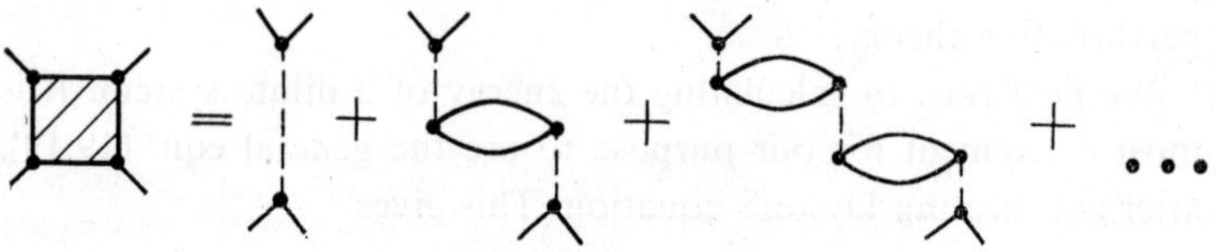

FIG. 57

It would seem that Δ can be looked for in the same form in dense systems

$$\Delta(1, 2, 3, 4) = -\frac{i}{4}\gamma(1, 2)\,[\delta(1-3)\,\delta(2-4) - \delta(1-4)\,\delta(2-3)]. \tag{25.7}$$

The following integral equation holds for γ [84]

$$\gamma(1, 2) = V(1, 2) - i\int d3\, d4\, V(1, 3)\, G_0(3, 4)\, G_0(4, 3)\, \gamma(4, 2). \tag{25.8}$$

Introducing the graphical form of the effective potential γ (Fig. 59a), this equation has an obvious interpretation (Fig. 59b). The two-particle Green function can be expressed easily enough in terms of γ. If (25.7) is substituted in (20.11), we get

$$G(1, 2, 1', 2') - \tilde{G}(1, 2, 1', 2') = -i\int d3\, d4\, G_0(1, 3)\, G_0(2, 4)\, \gamma(3, 4)\, G_0(3, 4, 1', 2'). \tag{25.9}$$

Here we have retained $\tilde{G}$ on the left-hand side: it takes into account the correlation of the particles with the system. Figure 60 illustrates the difference between dilute (a) and dense (b) systems. It is evident that, for the dense system, inclusion of the self-energy insertion in the free-particle term of (25.9) gives diagrams which are similar to those obtained when the correlation between the two particles is included.

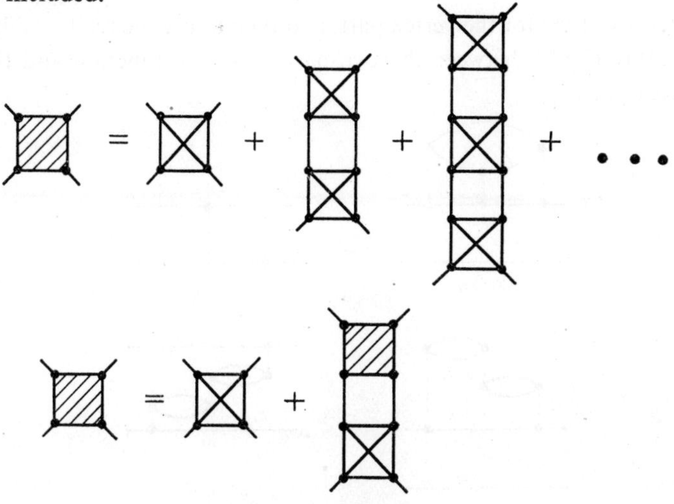

FIG. 58

Equation (25.9) can also be reduced to integral form [118]. It takes the rather unwieldy form:

$$G(1, 2, 1', 2') - \tilde{G}(1, 2, 1', 2') = \frac{i}{2}\int d3\, d4\, V(3, 4)\, [A(11'22'34)$$

$$+ A(22'11'34) - A(12'21'34) - A(21'12'34)], \qquad (25.10)$$

where

$$A(11'22'34) = [G(1, 3, 1', 3) - G_0(1, 1')\, G_0(3, 3)]\, G_0(2, 4)\, G_0(4, 2').$$

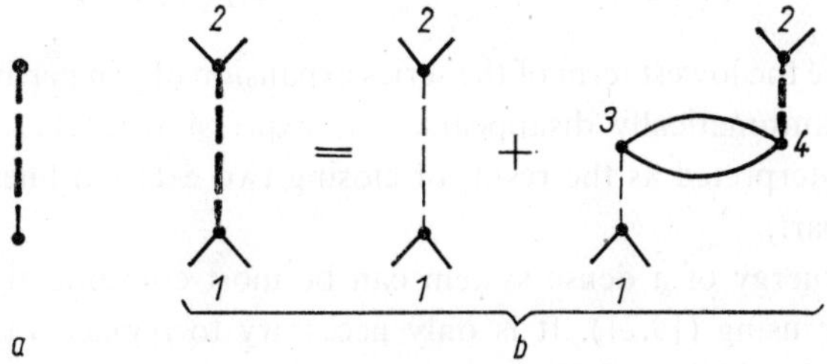

FIG. 59

This equation, like eqn. (25.3) for a dilute system, is of the general type we discussed in § 20.3. These equations are not very suitable for use, however; it is much more convenient to solve the equations (25.1) or (25.8) for the vertex part, and then to use the relationships (20.9) to (20.12) between the two-particle Green function and the vertex part.

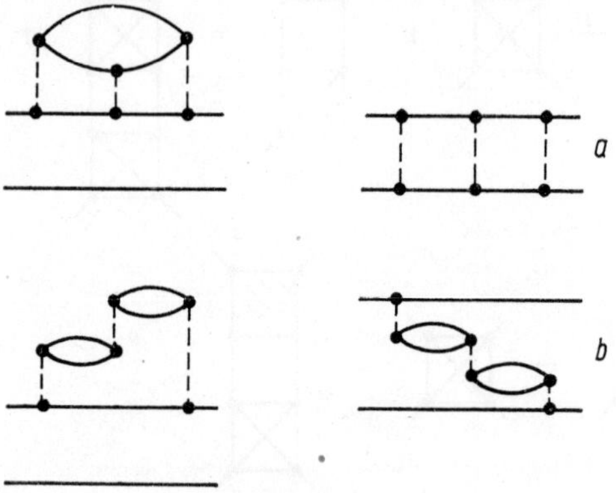

FIG. 60

25.6. We will now determine the single-particle Green function for a dense system. After substituting (25.7) into (25.4), the general expression (20.18) for the mass operator, in which the M_1 and M_2 terms have been ignored, takes the following form:

$$M(1,2) = -\int d3\, d4 \lim_{5\to 4} V(1,3)\, G_0(3,4)\, G_0(1,2)\, \gamma(4,2)\, G_0(4,5).$$

This expression for M is much simplified if we take eqn. (25.8) into account

$$M(1,2) = -i[\gamma(1,2) - V(1,2)]\, G_0(1,2). \qquad (25.11)$$

And here the lowest term of the series expansion of γ in perturbation theory automatically disappears. The expression (25.11) may be easily interpreted as the result of closing two external lines of the vertex part.

The energy of a dense system can be most conveniently calculated by using (19.21). It is only necessary to replace $V(1,2)$ by $\lambda V(1,2)$ in (25.8) and (25.11). We can neglect the last term, containing the operator $\hat{W}$, in (19.21). Replacing G by G_0, we obtain

$$E - E_0 = \frac{1}{2}\int_0^1 \frac{d\lambda}{\lambda}\int dq_1 \int d3\,[\gamma(1,3) - \lambda V(1,3)]\, G_0(1,3)\, G_0(3,1). \tag{25.12}$$

It is of considerable interest to obtain an integral equation for the matrix element $\Phi_n(1,2)$ in a dense system. We found in § 22.5 that ($G \approx G_0$):

$$\Phi_n(1,2) = \overline{\Phi}_n(1,2) + 4\int d3\,d4\,d5\,d6\, G_0(1,3)\, G_0(5,2) \times \Delta(3,4,5,6)\,\overline{\Phi}_n(6,4). \tag{25.13}$$

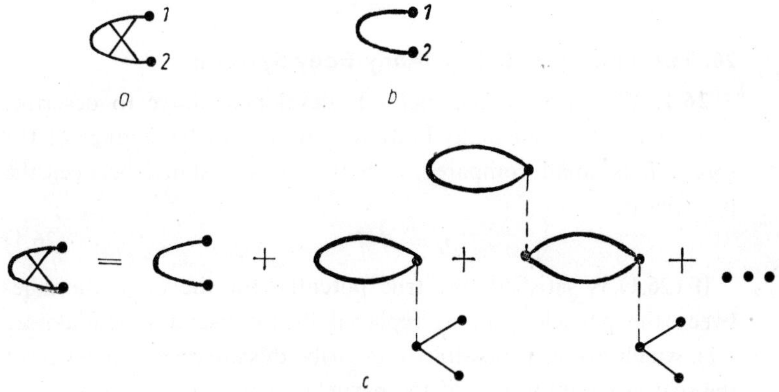

FIG. 61

Once the expressions for Δ have been substituted from (25.7), the integral on the right-hand side takes the form

$$-i\int d3\,d4\, G_0(1,3)\, G_0(3,2)\, \gamma(3,4)\, \overline{\Phi}_n(4,4).$$

The relationship

$$\int d4\, \gamma(3,4)\, \overline{\Phi}_n(4,4) = \int d4\, V(3,4)\, \Phi_n(4,4) \tag{25.14}$$

has the same significance as (25.2), and can be proved graphically. Introducing the graphical forms of the functions Φ (Fig. 61 a) and $\overline{\Phi}$ (Fig. 61 b), we can see that the determining diagrams for Φ_n in a dense system (see Fig. 61 c) are those which correspond to a maximum number of closed loops. The relationship (25.14), represented in Fig. 62, can then be verified immediately. We can use this relation to rewrite (25.13) as the following integral equation:

$$\Phi_n(1, 2) = \overline{\Phi}_n(1, 2) + i \int d3\, d4\, G_0(1, 3)\, G_0(3, 2)\, V(3, 4)\, \Phi_n(4, 4). \tag{25.15}$$

This equation plays an important part in determining the collective excited states of a system.

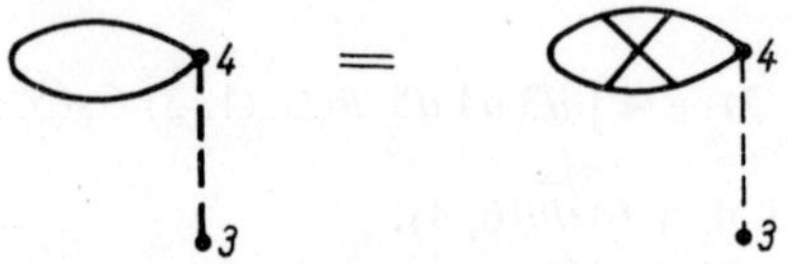

FIG. 62

26. The Theory of Dilute Many-Body Systems

26.1. We will use the methods developed above to describe, quantitatively, dilute many-body systems for which the range of the forces R is small compared with the mean distance between the particles d:

$$\eta = R/d \ll 1. \tag{26.1}$$

If (26.1) is satisfied, the true potential for the interaction between the particles can be replaced by the pseudopotential (see § 1), which makes it possible to describe this interaction in terms of the scattering amplitude of the particles. This is hardly surprising, since the interaction between particles in dilute systems can be treated as a consequence of interparticle scattering.

This section is concerned with deriving the pseudopotential within the framework of the general field approach: we will confirm the results obtained in §§ 1, 7 and 18. We will use the Green functions method to find the various properties of dilute many-body systems [83]. These results represent a series expansion of the physical quantities in term of the scattering length a, which, like the range of the forces R, must be considered small compared with the average distance between the particles:†

$$a/d \ll 1. \tag{26.2}$$

† The case where this ratio is not small (the resonance case) was partially considered in §§ 1 and 16.

The expansion in terms of this parameter (corresponding to what is called the gas approximation) in the first two orders describes only two-particle correlation between particles. Beginning with terms of order $(a/d)^3$, it is necessary to include the contribution of three-particle, and more complex, correlations. Estimates indicate [39] that the contribution of the three-particle correlations to the ground state energy of a dilute system is determined by

$$E \approx 0{\cdot}1(ap_0)^3 \frac{p_0^2}{M} N. \tag{26.3}$$

In the last few years Brueckner's method [2] has been widely used in the theory of nuclear matter. This method basically consists of allowing accurately for the two-particle correlations between particles, whilst completely ignoring three-particle, and more complex, correlations. The Hartree–Fock approximation describes the motion of an individual particle in the self-consistent field of the remaining particles; Brueckner's method describes accurately the interaction between each pair of particles moving in the self-consistent field of the remaining particles.

Brueckner's method can hardly be thought satisfactory. If the parameter a/d is not small compared with unity, Brueckner's method does not allow for the complex correlations, beginning with the three-particle, which contribute appreciably. If the parameter is small, then it is superfluous within the framework of this model to allow for all orders of a/d. One can say that, in the region where it applies (when three-particle correlation can be ignored), Brueckner's method coincides with the gas approximation.

The good agreement that Brueckner found between theory and experiment for a large number of nuclear characteristics can be put down to the fact that a/d is small. We can therefore avoid the unwieldy apparatus of Brueckner's method† as is shown by the results of §§ 7 and 18, which were calculated on the gas approximation including the first terms of the expansion in a/d. Brueckner's method is justified only in those cases where for some reason

† Brueckner and Gammel [2] indicated that they had to integrate numerically several hundred thousand Green functions of the corresponding equations.

the two-particle correlation is preferred, as, for example, in describing superfluidity of nuclear matter.

26.2. We now turn to the solution of the equations for the Green functions of a dilute many-body system. Since we only require (26.2) to be satisfied, the parameter for the Born approximation $\alpha' \sim MV_0R^2$ need not be small compared with unity.

We will start with the basic equation for this particular problem

$$\Delta(1, 2, 3, 4) = -\frac{i}{2} V(1, 2)\left\{\frac{i}{2}[\delta(1-3)\,\delta(2-4) - \delta(1-4) \times \delta(2-3)] + 2\int d5\, d6\, G_0(1, 5)\, G_0(2, 6)\, \Delta(5, 6, 3, 4)\right\}, \quad (26.4)$$

which can be obtained directly from (25.1).

We will limit our consideration to spatially uniform systems, and will transform (26.4) to momentum space, putting

$$\Delta(1, 2, 3, 4) = \int d^4p_1\, d^4p_2\, d^4p_3\, d^4p_4\, \Delta(p_1, \varrho_1; p_2, \varrho_2; p_3, \varrho_3; p_4, \varrho_4) \times \delta(p_1 + p_2 - p_3 - p_4) \exp[i(p_1 1) + i(p_2 2) - i(p_3 3) - i(p_4 4)].$$

After changing to the centre-of-mass system, and introducing the total momentum $P = p_1 + p_2 = p_3 + p_4$, and the relative momenta $k = (p_1 - p_2)/2$, $k' = (p_3 - p_4)/2$, we obtain the following equation:†

$$\Delta(P, k, k') = -\frac{i}{4}(2\pi)^4[v(\mathbf{k} - \mathbf{k}') - v(\mathbf{k} + \mathbf{k}')] + i\int d^4q\, v(\mathbf{k} - \mathbf{q})\, G_0(P/2 + q)\, G_0(P/2 - q)\, \Delta(P, q, k'), \qquad (26.5)$$

where

$$\Delta(P, k, k') \equiv \Delta(P/2 + k, P/2 - k, P/2 + k', P/2 - k').$$

Since there is no retardation, the Fourier transform of the potential, depends only on the spatial component of the momentum transfer vector. We will suppose that $\Delta(P, k, k')$ has the same property,

† This equation is analogous to the basic equation in Brueckner's method.

i.e. that apart from P it depends only on $\boldsymbol{k}$ and $\boldsymbol{k}'$. This supposition can be justified immediately: it makes it possible to integrate over ζ—the fourth component of the q-vector—on the right-hand side of (26.5):

$$I = \int \frac{d\zeta}{2\pi} G_0(P/2 + q)\, G_0(P/2 - q).$$

From the rules given in Appendix C, we can easily calculate that

$$I = -i\left[\frac{(1 - n_{\boldsymbol{q}+\boldsymbol{P}/2})(1 - n_{\boldsymbol{q}-\boldsymbol{P}/2})}{\varepsilon_{\boldsymbol{q}+\boldsymbol{P}/2} + \varepsilon_{\boldsymbol{q}-\boldsymbol{P}/2} - E + i\delta} - \frac{n_{\boldsymbol{q}+\boldsymbol{P}/2}\, n_{\boldsymbol{q}-\boldsymbol{P}/2}}{\varepsilon_{\boldsymbol{q}+\boldsymbol{P}/2} + \varepsilon_{\boldsymbol{q}-\boldsymbol{P}/2} - E - i\delta}\right],$$

where E is the total energy of the particles (the fourth component of the P-vector). Introducing the quantity

$$N(q) = 1 - n_{\boldsymbol{q}+\boldsymbol{P}/2} - n_{\boldsymbol{q}-\boldsymbol{P}/2},$$

we can write

$$I = i\frac{N(q)}{\varepsilon_{\boldsymbol{q}+\boldsymbol{P}/2} + \varepsilon_{\boldsymbol{q}-\boldsymbol{P}/2} - i\delta N(q)}. \tag{26.6}$$

Hence, the integral eqn. (26.5) takes the form

$$\Delta(P, \boldsymbol{k}, \boldsymbol{k}') = -\frac{i}{4}(2\pi)^4[\nu(\boldsymbol{k} - \boldsymbol{k}') - \nu(\boldsymbol{k} + \boldsymbol{k}')]$$
$$- \int d^3\boldsymbol{q}\, \frac{\nu(\boldsymbol{k} - \boldsymbol{q})\, N(q)}{\varepsilon_{\boldsymbol{q}+\boldsymbol{P}/2} + \varepsilon_{\boldsymbol{q}+\boldsymbol{P}/2} - E - i\delta N(q)} \Delta(P, \boldsymbol{q}, \boldsymbol{k}'). \tag{26.7}$$

Equation (26.7) indicates first of all, that assuming Δ to be independent of the fourth components of the vectors k and k' leads to no contradictions. This means physically that the inclusion of correlation between particles in a dilute system does not produce a retardation of their effective interaction. In the second place, the dependence of the effective interaction Δ on the total momentum of a pair is clearly evident. This represents the fact that Galilean invariance breaks down under these conditions. This is to be expected, however, since the particles in the medium interact, and since there is a privileged frame of reference in which the medium as a whole is at rest.

We note further that we have not written out the discrete subscripts explicitly, supposing that the Green function G_0 and ν are proportional to $\delta_{\varrho_1 \varrho_2}$.

The last point concerns the symmetry properties of $\Delta(P, \mathbf{k}, \mathbf{k}')$. From the antisymmetry of $\Delta(p_1; p_2; p_3; p_4)$ for interchanging p_1, p_2 or p_3, p_4, it follows that $\Delta(P, \mathbf{k}, \mathbf{k}')$ is antisymmetric under the substitutions $\mathbf{k} \to -\mathbf{k}$ and $\mathbf{k}' \to -\mathbf{k}'$. We will look for a solution of (26.7) in the form

$$\Delta(P, \mathbf{k}, \mathbf{k}') = \tfrac{1}{2}[U(P, \mathbf{k}, \mathbf{k}') - U(P, \mathbf{k}, -\mathbf{k}')]. \qquad (26.8)$$

Then we obtain for U

$$U(P, \mathbf{k}, \mathbf{k}') = -\frac{i}{2}(2\pi)^4 \, v(\mathbf{k} - \mathbf{k}') - \int d^3\mathbf{q} \, \frac{v(\mathbf{k} - \mathbf{q}) \, N(q) U(P, \mathbf{q}, \mathbf{k}')}{\varepsilon_{\mathbf{q}+P/2} + \varepsilon_{\mathbf{q}+P/2} - E - i\delta N(q)}. \qquad (26.9)$$

As can easily be shown, the solution of this equation has the property

$$U(P, -\mathbf{k}, \mathbf{k}') = U(P, \mathbf{k}, -\mathbf{k}').$$

Hence (26.8) gives a quantity which has all the required symmetry properties.

26.3. The direct solution of eqn. (26.9) is an extremely complicated task. The equation describes the mutual scattering of a pair of particles in the medium, i.e. allowing for the Pauli principle. The surroundings have comparatively little effect in a dilute system; it is therefore possible to express U (and also the effective potential Δ) in terms of the scattering parameters of a pair of isolated particles. In many cases these quantities (e.g. the phase of the scattering) are known from experiment, but the interaction potential is a derived quantity; the present approach is then of immediate physical value.

We will limit ourselves, for simplicity, to a square law for the particle dispersion (it is more convenient to consider the general case within the framework of pseudopotential theory). The change in the dispersion law due to self-consistent short-range interactions is unimportant when (26.2) is satisfied. Hence, if we wish to take our calculation up to $(a/R)^2$ inclusive, we can use a square law for the dispersion (see § 26.5).

We will consider the Schrödinger equation describing the scattering of a pair of particles in empty space. The results obtained represent an extension of those in § 1. Our initial equation has the form

$$(\nabla^2 + k^2)\, \psi_k(\boldsymbol{r}) = MV(\boldsymbol{r})\, \psi_k(\boldsymbol{r})$$

and, in momentum space,

$$(p^2 - k^2)\, \psi_k(\boldsymbol{p}) = -M \int d^3\boldsymbol{q}\, v(\boldsymbol{p} - \boldsymbol{q})\, \psi_k(\boldsymbol{q}).$$

Its solution can be expressed in the following way (see Appendix C):

$$\psi_k(\boldsymbol{p}) = (2\pi)^3\, \delta(\boldsymbol{p} - \boldsymbol{k}) - \frac{M}{p^2 - k^2 - i\delta} \int d^3\boldsymbol{q}\, v(\boldsymbol{p} - \boldsymbol{q})\, \psi_k(\boldsymbol{q}),$$

where the first term describes the incident wave and the second describes the scattered wave.

The quantity

$$a(\boldsymbol{p}, \boldsymbol{k}) = -\frac{M}{4\pi} \int d^3\boldsymbol{q}\, v(\boldsymbol{p} - \boldsymbol{q})\, \psi_k(\boldsymbol{q}) \tag{26.10}$$

can be called the scattering amplitude. More accurately, it is a generalization of the usual definition of the scattering amplitude to the case where we are interested not only in the asymptotic behaviour of the wave function at large distances, but also in the whole function $\psi_k(\boldsymbol{r})$. To confirm this, we will transform the equation

$$\psi_k(\boldsymbol{p}) = (2\pi)^3\, \delta(\boldsymbol{p} - \boldsymbol{k}) + \frac{4\pi}{p^2 - k^2 - i\delta}\, a(\boldsymbol{p}, \boldsymbol{k}) \tag{26.11}$$

to the coordinate representation. We then obtain the expression

$$\psi_k(\boldsymbol{r}) = \exp i(\boldsymbol{k} \cdot \boldsymbol{r}) + a(-i\nabla \cdot \boldsymbol{k}) \exp(ikr)/r.$$

For large r, we can replace $-i\nabla$ asymptotically by $k\boldsymbol{n}$, where $\boldsymbol{n}$ is the unit vector associated with $\boldsymbol{r}$, and we do, indeed, return to the usual definition of the scattering amplitude.

Our last task is to eliminate the interaction potential v from these equations and to replace it by the scattering amplitude a.

To this end, we consider (26.10). If we multiply it by $\hat{\psi}_k^*(\boldsymbol{p}')$, integrate over $\boldsymbol{k}$ and remember the condition for completeness

$$\int d^3\boldsymbol{k}\, \psi_k^*(\boldsymbol{p}')\, \psi_k(\boldsymbol{p}) = (2\pi)^3\, \delta(\boldsymbol{p} - \boldsymbol{p}'),$$

we find

$$\nu(\mathbf{p}-\mathbf{p}') = -\frac{4\pi}{M}\int d^3\mathbf{k}\, a(\mathbf{p},\mathbf{k})\, \psi_k^*(\mathbf{p}').$$

Substituting (26.11), we obtain the final relation between the scattering amplitude and the potential

$$\nu(\mathbf{p}-\mathbf{p}') = -\frac{4\pi}{M}a(\mathbf{p},\mathbf{p}') - \frac{(4\pi)^2}{M}\int \frac{d^3\mathbf{k}\, a^*(\mathbf{p}',\mathbf{k})\, a(\mathbf{p},\mathbf{k})}{p'^2 - k^2 + i\delta}. \tag{26.12}$$

It is convenient to get rid of the imaginary quantities on the right-hand side of this equation. We will therefore make the replacement

$$a(\mathbf{p},\mathbf{k}) = -\frac{l(\mathbf{p},\mathbf{k})}{1 + ikl(\mathbf{k},\mathbf{k})}. \tag{26.13}$$

The quantity l represents a generalization of the scattering length (see § 1 and 6) and is a real function. We will limit ourselves to the simplest case of s-scattering, when the quantity $l(\mathbf{p},\mathbf{k})$ is independent of angle.

Substituting (26.13) in (26.12), we are led to the following equations

$$\nu(\mathbf{p}-\mathbf{p}') = \frac{4\pi}{M}\cdot\frac{l(p,p')}{1 + p'^2 l^2(p',p')}$$
$$-\frac{(4\pi)^2}{M}\,\mathrm{P}\int d^3\mathbf{k}\frac{l(p',k)\, l(p,k)}{(p'^2-k^2)\,[1 + k^2 l^2(k,k)]}.$$

When (26.2) is satisfied, $l/d \ll 1$.

If we restrict ourselves to second-order terms in l, we finally find

$$\nu(\mathbf{p}-\mathbf{p}') = \frac{4\pi}{M}l(p,p') - \frac{(4\pi)^2}{M}\,\mathrm{P}\int d^3\mathbf{k}\frac{l(p',k)\, l(p,k)}{p'^2-k^2}. \tag{26.14}$$

26.4. Equation (26.14) enables us to eliminate the true interaction potential ν from all the calculations, and to replace it by the scattering length l.

We will substitute (26.14) in (26.9), taking $\varepsilon(p) = p^2/2M$. Restricting ourselves to terms of the order $(l/d)^2$, one iteration is sufficient. We obtain as a result

$$U(P, \boldsymbol{k}, \boldsymbol{k}') = -\frac{i}{2}(2\pi)^4 \frac{4\pi}{M}\left\{l(k, k') - 4\pi \int d^3\boldsymbol{q}\, l(k', q)\, l(k, q)\right.$$

$$\left.\times\left[\frac{N(q)}{q^2 + \frac{1}{4}P^2 - ME - i\delta N(q)} - \mathrm{P}\frac{1}{q^2 - (k')^2}\right]\right\}. \qquad (26.15)$$

When (26.1) is satisfied, the effective values of the momentum transfer are small compared with the momenta of the particles. The quantity l is, therefore, independent of k and k'. Defining the effective interaction potential $v_{\mathrm{eff}}(\boldsymbol{k} - \boldsymbol{k}')$ by the condition

$$U(P, \boldsymbol{k}, \boldsymbol{k}') = -\frac{i}{2}(2\pi)^4\, v_{\mathrm{eff}}(\boldsymbol{k} - \boldsymbol{k}'),$$

we have, to the lowest order in l,

$$v_{\mathrm{eff}} = \frac{4\pi}{M}\, l.$$

Whence, the effective potential

$$V_{\mathrm{eff}}(\boldsymbol{r}) = \frac{4\pi}{M}\, l\delta(\boldsymbol{r}). \qquad (26.16)$$

In this way we return to the pseudopotential introduced before.† The pseudopotential was introduced in terms of the two-body problem in § 1, and described the scattering of a pair of particles in empty space. The effects due to the presence of other particles in the system could only be obtained in the following stage of the calculation (when the perturbation theory was set up). The effective potential (26.15), on the other hand, already includes these effects (the second term). We will note several qualitative properties of the effective potential to the second-order in l. Replacing l by a constant in the second term of eqn. (26.15), and taking into account the identity

$$\int d^3\boldsymbol{q}\left(\mathrm{P}\frac{1}{q^2 - k'^2} - \mathrm{P}\frac{1}{q^2 - a^2}\right) \equiv 0, \qquad (26.17)$$

† The operator $1 + r(\partial/\partial r)$ has no effect in the lowest order in the expansion in l.

where a is any real number, it is evident that the quantity $U(P, \boldsymbol{k}, \boldsymbol{k}')$, also to second order in l, depends only on the total momentum P. We can thus write

$$V_{\text{eff}}(\boldsymbol{r}) = f(\hat{P})\,\delta(\boldsymbol{r}), \tag{26.18}$$

where $f(\hat{P})$ is a function of $\hat{P}$ that can be calculated explicitly.

26.5. The real interaction potential should also be replaced by l in expressions which correspond to the Hartree–Fock approximation. For this, we need an expression for the energy of particle ε_p.

The quantity ε_p is determined by the relation (see Chapter 2)

$$\varepsilon_p = (p^2/2M) + \frac{gp_0^3}{6\pi^2}\,\nu(0) - \int d^3p'\,\theta(p_0^2 - p'^2)\,\nu(\boldsymbol{p} - \boldsymbol{p}')$$

(we have replaced $\int d^3\xi V$ by $\nu(0)$ here). We must then substitute (26.14) for ν. From the identity (26.17), $\nu(\boldsymbol{p} - \boldsymbol{p}')$ does not depend on $\boldsymbol{p}$ and $\boldsymbol{p}'$ (for l = const.). We can therefore write

$$\varepsilon_p = (p^2/2M) + \frac{2(g-1)\,p_0^3}{3\pi M}\,l\left(1 - 4\pi l\,\mathrm{P}\int \frac{d^3\boldsymbol{k}}{a^2 - k^2}\right), \tag{26.19}$$

where a is an arbitrary number. We see that we retain a square law for the particle dispersion except for the addition of a constant term.

The expression (26.19) contains a divergent integral in k. The final expression, however, which includes correlation effects, is free of any divergence.

We next turn to the calculation of the mass operator. We must first transform (25.5) to the momentum representation

$$M(p) = \frac{4}{(2\pi)^4}\sum_{\varrho_3}\int d^4q\,G_0(q)\left\{\Delta\left(q + p, \frac{\boldsymbol{q} - \boldsymbol{p}}{2}, \frac{\boldsymbol{q} - \boldsymbol{p}}{2}\right)\right.$$
$$\left. + \frac{i}{4}(2\pi)^4[\nu(0) - \nu(\boldsymbol{q} - \boldsymbol{p})]\right\}\delta_{\varrho_1\varrho_2}\delta_{\varrho_1\varrho_3}\delta_{\varrho_2\varrho_4}.$$

From (26.8), (26.14) and (26.15), this can be rewritten as

$$M(p) = \frac{32(g-1)\,i}{M}\,l^2\int_C d^4q\,G_0(q)$$
$$\times\int d^4q'\left[\frac{N(q')}{q'^2 + (\boldsymbol{p} + \boldsymbol{q})^2/4 - M(\zeta + \varepsilon) - i\delta N(q')} - \mathrm{P}\frac{1}{q'^2 - a^2}\right],$$

where ζ and ε are the fourth components of q' and p. After a straightforward integration over ζ, we obtain

$$M(p) = M_0(p) - \frac{32(g-1)}{M} l^2 \int d^3\boldsymbol{q}\, d^3\boldsymbol{q}'$$

$$\times \left\{ N(q') \frac{N(q') - 1 + 2n_q}{2} \left[(q')^2 + \frac{(\boldsymbol{p}+\boldsymbol{q})^2}{4} - \varepsilon M - \frac{q^2}{2} \right.\right.$$

$$\left.\left. + i\delta(1 - 2n_q - N(q')) \right]^{-1} - n_q \mathrm{P} \frac{1}{q'^2 - a^2} \right\}, \qquad (26.20)$$

where

$$M_0(p) = -\frac{8}{3}(g-1)\frac{p_0^3}{M} l^2 \mathrm{P} \int \frac{d^3\boldsymbol{q}'}{q'^2 - a^2}.$$

accurately compensates the divergent term in (26.19).

Integrating over $\boldsymbol{q}$ and $\boldsymbol{q}'$, we are led to the following expression for the Green function:

$$G^{-1}(\boldsymbol{p}, \varepsilon) = \varepsilon - (p^2/2M) - \frac{2(g-1)\, p_0^3 l}{3\pi M} - [M(\boldsymbol{p}, \varepsilon) - M_0(\boldsymbol{p}, \varepsilon)]$$

(the quantity $M - M_0$ will be discussed below).

We will first of all calculate the energy and attenuation of the quasi-particles which make the foregoing expression equal to zero. Restricting our consideration to terms of the order l^2 we can substitute $p^2/2M$ for ε in $M(\boldsymbol{p}, \varepsilon)$. This gives (when $g = 2$)

$$E_p = \frac{p^2}{2M} + \frac{2}{3\pi M} p_0^3 l + \frac{2}{15\pi^2 M} p_0^4 l^2 (11 - 2 \ln 2)$$

$$- \frac{8}{15\pi^2 M} p_0^3 l^2 (7 \ln 2 - 1)(p - p_0) + \cdots \qquad (26.21)$$

The expression (26.21) holds near $p = p_0$. The damping in this region can be written as

$$\Gamma_p = \frac{1}{\pi M} p_0^4 l^2 (p - p_0)^2 \operatorname{sign}(p - p_0). \qquad (26.22)$$

This is in qualitative agreement with the general requirements which the damping must satisfy. It is a quantity which becomes small near the Fermi surface.

We next determine the chemical potential of the system, using the general equation $\mu = E_{p_0}$. From the relation

$$\mu = \frac{p_0^2}{2M}\left[1 + \frac{4}{3\pi}p_0 l + \frac{4}{15\pi^2}(11 - 2\ln 2)\,p_0^2 l^2\right], \qquad (26.23)$$

we can calculate the energy of the system immediately, using $\mu = (\partial E/\partial N)_\Omega$. To do this, we integrate (26.23) over N, remembering that $p_0^3/3\pi^2 = N/\Omega$. We obtain the result

$$E/N = \frac{3}{10}\cdot\frac{p_0^2}{M}\left[1 + \frac{10}{9\pi}p_0 l + \frac{4}{21\pi^2}(11 - 2\ln 2)\,p_0^2 l^2\right]. \qquad (26.24)$$

The general method, based on the pseudopotential (§ 18), leads to an analogous expression.†

In this case the theorem that μ and E/N are equal does not hold, since we are considering a gas which is not in a "self-condensed" state. We can easily confirm, by calculating $P = -(\partial E/\partial \Omega)_N$, that this is not equal to zero.

We will finally calculate the magnitude of the discontinuity in the momentum distribution of the particles Z_p. For this we need the expansion of the Green function near the points $p = p_0$ and $\varepsilon = \mu$. The quantity $M - M_0$ has the following form in this region:

$$M - M_0 = \frac{1}{\pi^2}\cdot\frac{p_0^4 l^2}{M}\left[\frac{2}{15}(11 - 2\ln 2) - \frac{8}{15}(7\ln 2 - 1) \times (p/p_0 - 1) - 4\ln 2\,\frac{\varepsilon - \mu}{p_0^2}\right].$$

Whence

$$Z_p = 1 - \frac{4}{\pi^2}(\ln 2)\,p_0^2 l^2. \qquad (26.25)$$

This is close to unity, so that the distribution $\varrho(p)$ is close to the Fermi distribution.

† There is some difference in the numerical coefficients, due to different values of the degeneracy factor g.

Summarizing our discussion, we note that the same results could be obtained in a shorter way using the pseudopotential. If we replace the true potential V by the pseudopotential right from the start, then the results obtained here are contained in the second-order self-energy part. We have dwelt especially on the method based on summing the perturbation theory diagrams, in order to illustrate a particular effective field theory method.

27. The Theory of Dense Many-Body Systems

27.1. We will now give a quantitative description of dense many-body systems, for which the average distance between particles d is small compared with the effective range of the forces R,

$$\eta = R/d \gg 1.$$

Systems with Coulomb interactions are considered as dense if

$$a_0/d \gg 1,$$

where a_0 is the Bohr radius of the particles.

We will suppose that the magnitude of the interaction parameter

$$\alpha \sim MV_0 d^2$$

satisfies the condition

$$\eta^{-1} > \alpha > \eta^{-3}.$$

The Born parameter

$$\alpha' \sim MV_0 R^2 \sim \alpha\eta^2$$

can be large in this case. If η is large, then the interaction parameter must be small for a system with Coulomb interactions. The basic equation in the theory of dense systems can be deduced from the results of §§ 16 and 25, and has the form

$$\gamma(1,2) = V(1,2) - i\int d3\, d4\, V(1,3)\, G_0(3,4)\, G_0(4,3)\, \gamma(4,2). \quad (27.1)$$

27.2. We now turn to the solution of eqn. (27.1), which determines the effective interaction potential between particles. We will suppose for simplicity that the potential is a function of the coordinates only. We will restrict ourselves to homogeneous systems, and transform (27.1) to the momentum representation

$$\gamma(k) = \nu(\boldsymbol{k})\,[1 - ig \textstyle\int d^4p\, G_0(p+k)\, G_0(p)\, \gamma(k)]$$

(g is the degeneracy factor which appears in the course of the summation over the discrete variables).

We can write explicitly

$$\gamma(k) = \nu(\boldsymbol{k})\,[1 + ig \textstyle\int d^4p\, G_0(p+k)\, G_0(p)\, \nu(\boldsymbol{k})]^{-1}. \qquad (27.2)$$

A calculation of the integral in this equation, according to the rules in Appendix C, gives

$$\Pi(k) = -ig \int d^4p\, G_0(p+k)\, G_0(p) = -g \int d^3\boldsymbol{p}$$

$$\times \left\{ \frac{n_p(1 - n_{p+k})}{-\omega + \varepsilon_{p+k} - \varepsilon_p - i\delta} + \frac{n_{p+k}(1 - n_p)}{\omega + \varepsilon_p - \varepsilon_{p+k} - i\delta} \right\},$$

where ω is the fourth component of the k-vector. The explicit dependence of the effective potential γ on this quantity leads to the appearance of retardation effects due to correlation between the particles.

Making the replacement $\boldsymbol{p} \to \boldsymbol{p} - \boldsymbol{k}$ and then $\boldsymbol{p} \to -\boldsymbol{p}$ in the second term, we arrive at the following final expression for $\Pi(k)$, which is called the polarization operator:

$$\Pi(k) = g \int d^3\boldsymbol{p}\, n_{p+k}(1 - n_p)$$

$$\times \left(\frac{1}{\varepsilon_{p+k} - \varepsilon_p - \omega - i\delta} + \frac{1}{\varepsilon_{p+k} - \varepsilon_p + \omega + i\delta} \right). \qquad (27.3)$$

This expression may be computed explicitly if we remember that k is effectively small compared with p_0. We can, moreover, assume a square law for the dispersion, since the exchange effects in the Hartree–Fock approximation play only a minor role.

Owing to the requirement $k \ll p_0$, the effective region of integration in (27.3) is narrow. Putting $p = p_0\,(1 + \alpha)$, where $\alpha > 0$, and allowing for the fact that $(\boldsymbol{p} + \boldsymbol{k})^2 < p_0^2$, we find

$$\alpha < -\frac{kx}{p_0} \quad \text{and} \quad x < 0,$$

where x is the cosine of the angle between the vectors $\boldsymbol{p}$ and $\boldsymbol{k}$. The integral (27.3) then becomes

$$\Pi(k) = -g\frac{Mp_0}{2\pi^2}f(\zeta^2), \tag{27.4}$$

where

$$f(\zeta^2) = \int_0^1 \frac{x^2\,dx}{x^2 - \zeta^2 - i\delta}\,; \quad \zeta \equiv \frac{M\omega}{p_0 k}. \tag{27.5}$$

In explicit form

$$f(\zeta^2) = 1 - \frac{\zeta}{2}\ln\left|\frac{1+\zeta}{1-\zeta}\right| - \frac{1}{2}\,i\pi|\zeta|\,\theta(1-|\zeta|). \tag{27.6}$$

This function goes from 1 to $-\infty$ as ζ^2 changes from 0 to 1, and increases for $\zeta^2 > 1$, having an asymptotic value, $f(\zeta^2) \approx -1/(3\zeta^2)$, for $\zeta^2 \to \infty$. The imaginary part of $f(\zeta)^2$ is non-zero only when $\zeta^2 < 1$.

The limiting values of the polarization operator are as follows:

when $\zeta \to 0$

$$\Pi(k) = -\frac{2gMp_0}{4\pi^2}, \tag{27.7}$$

when $\zeta \to \infty$

$$\Pi(k) = \frac{\varrho k^2}{\omega^2 M}, \tag{27.8}$$

where ϱ is the number density of the particles in the system.

27.3. The essential point to note about the expression for $\gamma(k)$,

$$\gamma(k) = \frac{v(\boldsymbol{k})}{1 - \Pi(k)\,v(\boldsymbol{k})}, \tag{27.9}$$

is that the denominator can go to zero for real values of ω and $\boldsymbol{k}$. This reflects the fact that a new (collective) type of excitation arises here (it will be considered in § 28).

It should be noted that the pole of $\gamma(k)$ presents a formal difficulty in the way of getting further information from the theory. In fact the expressions, e.g. for the mass operator or the energy, presuppose that quantities which include $\gamma(k)$ are integrated over ω and k. We therefore need additional information about the rules for integrating around the pole. This can be obtained from the spectral representation.

Before considering this further, we must investigate where, and under what conditions, the pole appears. Starting with the relation $1 = \Pi(k)\, v(k)$, where v is real, it is obvious that this equation only has real roots when $\Pi(k)$ is real. This requires, in turn, that

$$\zeta = M\omega/p_0 k > 1. \tag{27.10}$$

If we denote the pole of eqn. (27.9) by $\omega_0(k)$—it lies at the symmetrical points $\pm\omega_0(k)$—we have the following dispersion relation

$$1 = -\frac{gMp_0}{2\pi^2}\, v(\boldsymbol{k})\, f(\zeta_0^2), \tag{27.11}$$

where $\zeta_0 = M\omega_0/p_0 k$. Since the function $f(\zeta^2)$ is negative in the region of (27.10), eqn. (27.11) only has a solution for $v(\boldsymbol{k}) > 0$, i.e. when there is repulsion between the particles. If there are attractions between the particles, ω_0 may be imaginary. This indicates the instability of the system with respect to small variations in the density. The uniform solution $G_0(p)$ of the Hartree–Fock equation, used in the calculation of $\Pi(k)$, is unstable. It corresponds not to the minimum of the energy but to a stationary point. The systems spontaneously changes to an inhomogeneous state. More detailed considerations of this problem lie outside the framework of the present book.

We will first consider short-range forces. In the limit $k \to 0$, $v(\boldsymbol{k})$ tends to the finite value, $v(0) \sim V_0 R^3$. The coefficient of $f(\zeta_0^2)$ in eqn. (27.11) will therefore be of the order $\alpha\eta^3 \gg 1$, and, correspondingly, $f(\zeta_0^2) \ll 1$. This last inequality shows that ζ_0^2 is large. We therefore arrive at the equation $1 = gMp_0 v(0)/6\pi^2\zeta_0^2$, whence

$$\omega_0^2(k) = \varrho\,\frac{v(0)}{M}\,k^2. \tag{27.12}$$

This solution corresponds to what is known as zero sound [24, 25, 119].

In a system with Coulomb interactions, where $v(k) = 4\pi e^2/k^2$, eqn. (27.11) gives

$$\omega_0^2 \equiv \omega_L^2 = \frac{4\pi\varrho e^2}{M} \tag{27.12'}$$

as $k \to 0$, where ω_L is the plasma oscillation frequency, or the Langmuir frequency. It is important that ω_L is independent of k, when k is small (the optical spectrum). As k increases, terms proportional to $p_0^2 k^2/M^2$ begin to play a part in (27.11). When k is large, (27.11) gives $\omega_0 = p_0 k/M$. However, the equation itself only holds for small values of k.

We now turn to the question of establishing rules for integrating around the singularities of the function $\gamma(k)$. We will suppose that eqn. (27.11) has a real solution. We will use the spectral representation of the two-particle Green function, or, to be more accurate, of the function $\tilde{G}(1, 2) = G(1, 2, 1, 2)$. This representation has the form (see § 23.6)

$$\tilde{G}(\boldsymbol{p}, \varepsilon) = -\frac{1}{\pi}\int_0^\infty dE^2 \, \frac{\operatorname{Im} \tilde{G}(\boldsymbol{p}, E)}{\varepsilon^2 - E^2 + i\delta}.$$

The important point for us is that $\tilde{G}(\boldsymbol{p}, \varepsilon)$ depends only on the combination $\varepsilon^2 + i\delta$.

We will now relate the effective potential γ to the function $\tilde{G}$. Equating the arguments 1, 1′ and 2, 2′ in eqn. (25.9), we find

$$\tilde{G}(1, 2) - G(1, 1)\, G(2, 2) + G(1, 2)\, G(2, 1)$$

$$= -i \int d3\, d4\, G_0(1, 3)\, G_0(2, 4)\, \gamma(3, 4)\, G_0(3, 4, 1, 2).$$

Transforming to the momentum representation, we obtain

$$\tilde{G}(p) - \tilde{G}_0(p) = i\gamma(p)\, \Pi(-p)\, \Pi(p). \qquad (27.13)$$

We have here omitted terms with integrals of γ in which the effective value of the argument of γ is of the order p_0 or greater. $\tilde{G}_0(p)$ is the value of $\tilde{G}(p)$ when there is no correlation.

It follows from eqn. (27.13) that $\gamma(p)$ must indeed depend on the combination $\varepsilon^2 + i\delta$. We will consider the function $\gamma(k)$ near its pole with this in mind. Using eqn. (27.9) and the fact that $\gamma(k)$ is even in ω, we have

$$\gamma(\boldsymbol{k}) \approx -\frac{1}{(\partial \Pi/\partial \omega^2)_{\omega_0^2}\, [\omega^2 - \omega_0^2(\boldsymbol{k})]}.$$

It is obvious that we must replace ω^2 by $\omega^2 + i\delta$ in determining the required rules. The pole is then displaced from the real axis,

$\omega_0^2 \to \omega_0^2 - i\delta$. Hence, when $\omega_0 < 0$, the pole moves into the upper half-plane, when $\omega_0 > 0$, into the lower half-plane. The subsequent calculations are then unambiguous, as required.

27.4. We will investigate the effective interaction potential between the particles

$$\gamma(1, 2) = \int d^4k\, \gamma(k) \exp \{i[k(1 - 2)]\}.$$

This potential has the property of retardation and does not have a δ-type dependence on $t_1 - t_2$.

We can introduce an averaged effective potential, $V_{\text{eff}}(\boldsymbol{r}) \times \delta(t_1 - t_2)$, which does not contain retardation effects, and coincides with γ when averaged over time. Equating the integrals of the potentials γ and $V_{\text{eff}}\delta(\tau)$ over $\tau = t_1 - t_2$, we obtain

$$V_{\text{eff}}(\boldsymbol{r}) = \int d\tau\, \gamma(1, 2) = \int d^3\boldsymbol{k}\, \gamma(\boldsymbol{k}, 0) \exp i(\boldsymbol{k} \cdot \boldsymbol{r}). \quad (27.14)$$

In order to explain the physical significance of the quantity $V_{\text{eff}}(\boldsymbol{r})$, we will examine the potential set up by a certain static distribution of particles (an external source) with a density $\varrho(\boldsymbol{r})$. If correlation is excluded, this potential has the form

$$B(1) = \int d2\, V(1, 2)\, \varrho(2),$$

where $V(1, 2)$ is the true interaction potential between the particles. If the correlation is to be allowed for, we must make the replacement $V \to \gamma$. Remembering that ϱ is independent of time, we find

$$B(\boldsymbol{x}_1) = \int d^3\boldsymbol{x}_2\, V_{\text{eff}}(\boldsymbol{x}_1 - \boldsymbol{x}_2)\, \varrho(\boldsymbol{x}_2).$$

Thus V_{eff} determines the potential of the static sources.

If we substitute the value of the polarisation vector for $\zeta \to 0$ in (27.14), we find

$$V_{\text{eff}}(\boldsymbol{r}) = \int d^3\boldsymbol{k} \frac{\nu(\boldsymbol{k}) \exp i(\boldsymbol{k} \cdot \boldsymbol{r})}{1 + (gp_0M/2\pi^2)\, \nu(\boldsymbol{k})}. \quad (27.15)$$

Important physical consequences stem from this relation.

We will consider systems with long-range forces. Suppose the distance r is smaller than the range of the forces R, i.e. $k \gg 1/R$ in (27.15). Then, as was pointed out in § 16, the potential $\nu(\boldsymbol{k})$ can, in many cases of interest, be written as

$$\nu(\boldsymbol{k}) \sim V_0R/k^2.$$

The relationship (27.15) takes the form

$$V_{\text{eff}}(r) \sim V_0 R \frac{\exp(-k_0 r)}{r}, \tag{27.16}$$

where

$$k_0 \approx (V_0 R p_0 M)^{1/2} \sim (\alpha\eta^3)^{1/2} R^{-1}. \tag{27.17}$$

Since $\alpha\eta^3 \gg 1$, there is a screening of the initial interaction potential. The interaction forces between the particles become increasingly short-range, and change the range from R to

$$R_{\text{eff}} \sim (\alpha\eta^3)^{-1/2} R. \tag{27.18}$$

This property of dense systems is particularly clearly marked for Coulomb forces. In this case $\nu = 4\pi e^2/k^2$, and

$$V_{\text{eff}} = e^2 \frac{\exp(-k_0 r)}{r}, \tag{27.19}$$

where†

$$k_0 = 2(p_0/\pi a_0)^{1/2}.$$

The physical interpretation of the effect of Debye screening is especially obvious here. We will consider, for simplicity, a model of a uniform electron gas with a compensating positive background. It is evident that each of the electrons repels the others. Hence there is an excess of positive charge near an electron, which screens the electron charge. As a result the Coulomb forces become short-range.

If we return to our original expression (27.9), and write it in the form

$$\gamma(k) = \frac{4\pi e^2}{k^2[1 - (4\pi e^2/k^2)\,\Pi(\boldsymbol{k}, \omega)]},$$

then the quantity

$$\varepsilon(\omega, \boldsymbol{k}) = 1 - \frac{4\pi e^2}{k^2}\Pi(\boldsymbol{k}, \omega)$$

can be called the effective (longitudinal) dielectric constant of the medium. It depends on the wave vector $\boldsymbol{k}$ as well as the frequency ω, which indicates that there is both a frequency and a spatial

† There is a general equation for the Debye screening length which is based on an application of Ward's theorem and is not dependent on the limitations introduced here [26].

dispersion of the dielectric constant. The quantity $\varepsilon(\omega, \boldsymbol{k})$ can be used to express many of the physical properties of a system, and the ways in which it interacts with external agents. A "dielectric" formulation of many-body theory has been worked out in recent years [47, 120].

27.5. In calculating such quantities as the mass operator, the energy, etc., we come up against the special mathematical structure of the corresponding expressions (which is usually called logarithmic).

Suppose we are given an evidently convergent integral, which depends on some small parameter α,

$$I(\alpha) = \int_0^\infty dx f(x,\alpha)\,.$$

Suppose further that this integral becomes logarithmically divergent (at the lower limit), if we put $\alpha = 0$ in the expression under the integral sign; in other words $f(x, 0) \sim 1/x$ for small x. Then $I(\alpha)$ is a logarithmic function for $\alpha \to 0$; so long as the inequality $\ln(1/\alpha) \gg 1$ is satisfied, the calculation of $I(\alpha)$ can be considerably simplified.

We will assume, in fact, that for values of x less than some quantity x_0, which is independent of α, the function $f(x, \alpha)$ takes the form

$$f(x, \alpha) \approx \frac{\varphi(\alpha/x)}{x},$$

where the function φ is finite for a zero value of its argument (otherwise the integral $I(0)$ is not logarithmically divergent), and tends to zero when its argument tends to infinity (otherwise $I(\alpha)$ is not convergent). We will decompose $I(\alpha)$ into a sum of two integrals. The first, taken over the region from x_0 to ∞, tends to a constant limit

$$\int_{x_0}^\infty dx f(x, 0)$$

when $\alpha \to 0$. The second integral

$$\int_0^{x_0} \frac{dx}{x}\varphi(\alpha/x)$$

the form

$$\int_0^{x_0/\alpha} \frac{d\xi}{\xi} \varphi(1/\xi)$$

when $x = \alpha\xi$ is substituted. As α decreases, only the region near the upper limit is important in this convergent integral. Removing the quantity $\varphi(a/x_0) \approx \varphi(0)$ from under the integral sign in this region, we finally have

$$I(\alpha) = \varphi(0) \ln (x_0/\alpha) + O(\alpha^0). \qquad (27.20)$$

We thus obtain a simple answer which has logarithmic accuracy.

27.6. We will determine the mass operator for a dense system. We find, from (25.11), that

$$M(p) = -i \int d^4k[\gamma(k) - \nu(\boldsymbol{k})] \, G_0(p-k)$$

$$= -i \int d^4k \, G_0(p-k) \frac{\Pi(k) \, \nu^2(\boldsymbol{k})}{1 - \Pi(k) \, \nu(\boldsymbol{k})} \, .$$

We will restrict ourselves to the most important case of a Coulomb system. Then, introducing

$$\zeta = \frac{M\omega}{p_0 k} \, ; \quad \boldsymbol{k} = p_0 \boldsymbol{q}; \quad \boldsymbol{p} = p_0 \boldsymbol{x}$$

and

$$\varepsilon = \frac{p_0^2}{2M} y,$$

we obtain

$$M(p) = 8i \frac{e^2}{a_0} \int \frac{d^3\boldsymbol{q}}{q^3} \int_{-\infty}^{\infty} \frac{d\zeta}{2\pi} \frac{f(\zeta^2)}{1 + (k_0^2/2q^2p_0^2) f(\zeta^2)}$$

$$\times \{y - 2q\zeta - (\boldsymbol{x} - \boldsymbol{q})^2 + i\delta \operatorname{sign} [(\boldsymbol{x} - \boldsymbol{q})^2 - 1]\}^{-1}. \qquad (27.21)$$

We now turn to analysing the singularities of the expression under the integral sign in the complex plane. We must first deal with the pole of the Green function G_0 at the point

$$\zeta_1 = \frac{[y - (\boldsymbol{x} - \boldsymbol{q})^2]}{2q} + i\delta \operatorname{sign} [(\boldsymbol{x} - \boldsymbol{q})^2 - 1] \, .$$

We must also consider the poles of $\gamma(k)$. They are situated at the points

$$\zeta_2 = \begin{cases} \dfrac{M|\omega_0|}{p_0 k} - i\delta \\ -\dfrac{M|\omega_0|}{p_0 k} + i\delta \end{cases} \quad \frac{M|\omega_0|}{p_0 k} > 1.$$

Finally, the quantity $f(\zeta^2)$ itself has singularities. As is evident from its definition [see (27.5) and (27.6)], we have the branch points $\zeta_3^2 + i\delta = 1$, i.e.

$$\zeta_3 = \begin{cases} 1 - i\delta \\ -1 + i\delta \end{cases}$$

We thence arrive at the picture shown in Fig. 63.

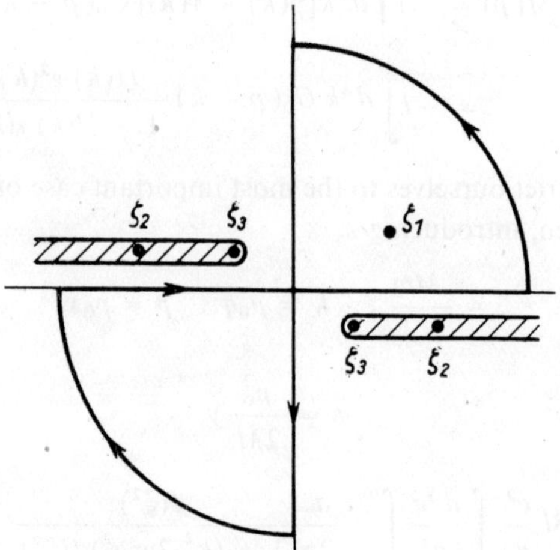

FIG. 63

The expression under the integral in (27.21) decreases reasonably rapidly as ζ increases, so the boundary of the integration can be displaced to the imaginary axis. We must now deal with the residues of the pole ζ_1, if it lies in the first or third quadrant, i.e. if the following conditions are satisfied

$$y > (x - q)^2 > 1$$

or

$$1 > (\mathbf{x} - \mathbf{q})^2 > y.$$

The integral round the closed contour C is equal to the residue in the first quadrant with a positive sign, or in the third quadrant with a negative sign. Hence the contribution of the residues to the integral over ζ in (27.21) is equal to

$$i\left\{\theta[y - (\mathbf{x} - \mathbf{q})^2] - \theta[1 - (\mathbf{x} - \mathbf{q})^2]\right\} \frac{f\{[y - (\mathbf{x} - \mathbf{q})^2]/4q^2\}}{1 + (k_0^2/2q^2p_0^2)f}.$$

When $y - x^2 > k_0/p_0$, the argument of the function f is large, and the function itself is proportional to q^2. Hence this is not a logarithmic situation and the contribution of the residues need not be taken into account. In the other limiting case ($y \approx 1$), the contribution of the residues is also extremely small owing to the narrow region of the integration over $\mathbf{q}$.

If we therefore omit the residues, we can replace the original integral by an integral along the imaginary axis, i.e. we make the replacement $\zeta \to i\zeta$. The function $f(\zeta^2)$ now takes the form

$$f(-\zeta^2) = 1 - \zeta \arctan(1/\zeta).$$

Thus the integral over ζ in (27.21) can be written as

$$I = i\int \frac{d\zeta}{2\pi} \frac{1 - \zeta \arctan(1/\zeta)}{1 + (k_0^2/2q^2p_0^2)\,[1 - \zeta \arctan(1/\zeta)]} \times [y - 2iq\zeta - (\mathbf{x} - \mathbf{q})^2]^{-1}.$$

We will consider the case $y - x^2 \gg k_0/p_0$. The small quantity here is the quantity k_0/p_0. If we neglect it, then, when $q \to 0$, $I \to$ const., and we have a typical logarithmic case. Then, from (27.20),

$$M(p) = -\frac{4e^2}{\pi^2 a_0}(y - x^2)^{-1} \times \int_{-\infty}^{\infty} \frac{d\zeta}{2\pi}[1 - \zeta \arctan(1/\zeta)] \ln(p_0/k_0)$$

or, working out the integral, and transforming to the usual notation

$$M(p) = -\frac{1}{2\pi^2}\frac{e^2}{a_0}\frac{p_0^2}{M}\frac{\ln(p_0/k_0)}{\varepsilon - p^2/2M}. \tag{27.22}$$

We will not dwell on further investigations of the mass operator.† We will only note that the difference between a quasi-particle and a true particle is, on the whole, unimportant in dense systems.

27.7. We will now calculate the energy of a dense system. Transforming (25.12) to the momentum representation, we find

$$E - E_0 = \Omega\frac{g^2}{2}\int_0^1\frac{d\lambda}{\lambda}\int d^4p\, d^4k\, [\gamma(k) - \lambda v(k)]\, G_0(p+k)\, G_0(p).$$

From our definition of the operator $\Pi(k)$, we can write

$$E - E_0 = -\frac{i\Omega}{2}g\int_0^1\frac{d\lambda}{\lambda}\int d^4k\,\frac{[\lambda v(k)\,\Pi(k)]^2}{1 - \lambda v(k)\,\Pi(k)}. \tag{27.23}$$

We have here replaced v by λv everywhere.

Replacing $\Pi(k)$ by the expression for it in (27.5) we transfer the integration to the imaginary axis. Here, unlike the example of § 27.6, no question about the residues arises, and we obtain

$$E - E_0 = \frac{\Omega g p_0}{2M}\int d^3\boldsymbol{k}\, k\int_{-\infty}^{\infty}\frac{d\zeta}{2\pi}\,[Q - \ln(1+Q)], \tag{27.23'}$$

where

$$Q = g\frac{Mp_0 v(k)}{2\pi^2}\,[1 - \zeta\arctan(1/\zeta)].$$

This relation, for the particular case of Coulomb forces, was first derived by Gell–Mann and Brueckner [123].

We will also consider this case, putting $g = 2$ and $v(\boldsymbol{k}) = 4\pi e^2/k^2$. The quantity Q has the form

$$Q = \frac{k_0^2}{2k^2}\,[1 - \arctan(1/\zeta)]. \tag{27.24}$$

The function $Q - [\ln(1+Q)]/Q^2$ coincides with the function $f(x)$ which we discussed in § 16. It tends to a constant for small Q, and to Q^{-1} for large Q.

† These can be found in [121, 122].

The substitution of (27.24) into (27.23′) leads to a typical logarithmic situation: as k_0^2 tends to zero, the logarithmic integral $\int dk/k$ appears. An application of (27.20) gives†

$$E - E_0 = -N\frac{e^2}{\pi^2 a_0}(1 - \ln 2)\ln(a_0 p_0), \qquad (27.25)$$

where N is the number of particles.

Special emphasis must be given to the anomalously small numerical coefficient in this equation—only three-hundredths. We might suppose that the smallness of the correlation effects is also a property of systems which are not very dense and for which the parameter η is only slightly greater than unity.

For uniform Coulomb systems, there is a logarithmic term in the expression for $E - E_0$. The general equation has the form [123]

$$E - E_0 = -N[0{\cdot}0311 \ln(a_0 \varrho^{1/3}) + 0{\cdot}0628)]\frac{e^2}{a_0}. \qquad (27.26)$$

The negative sign for the correlation correction to the energy agrees with results from the variational principle (see § 7).

Various interpolation formulae for $E - E_0$ have been given in the literature, covering a wide range of densities [124]. Most of these are based on Wigner's investigation of a dilute electron gas (see § 6) [49].

We must next consider the limits of validity of these relations. We have already remarked on the limitations on the degree of condensation of the system. We must now look at the other side of this question.

The results we have obtained refer only to an idealized uniform model, corresponding to an infinite uniformly distributed electron gas with a compensating positive background. Such systems do not exist in nature.‡ It is therefore important to discover to what

† The integral $\int_{-\infty}^{\infty} d\zeta[1 - \zeta \arctan(1/\zeta)]^2$ which appears in this is most simply calculated using the parametric equation (27.5), replacing ζ^2 by $-\zeta^2$.

‡ In a plasma, where there is a uniform distribution, the correlation of the heavy particles is of considerable importance.

extent an inhomogeneity in the particle distribution can affect our results. Beginning from our discussion of § 16, we can say that the essential condition for (27.25) to be applicable is that the following inequality holds:

$$x_0 k_0 \gg 1, \tag{27.27}$$

where x_0 is the scale length of the disturbance and k_0 is the Debye momentum.

This condition breaks down, for example, for a normal atom (see § 16), where $x_0 \sim a_0 Z^{-1/3}$ and $k_0 \sim a_0^{-1} Z^{1/3}$. It is interesting to consider a dense substance from this point of view. As the density increases, the particle distribution becomes more and more uniform so that one might expect (27.27) to be satisfied for sufficiently high pressures. The results of § 6 indicate that the distribution for the limiting momentum $p_0(x)$ in the range of pressures II and III is given by

$$p_0(x) \sim \varrho^{1/3}\left[1 + \frac{Z^{2/3}}{\varrho^{1/3} a_0} f\left(\frac{x\varrho^{1/3}}{Z^{1/3}}\right) + \cdots\right],$$

so that

$$x_0 \sim p_0/\nabla p_0 \sim a_0 Z^{-1/3} \gg k_0^{-1}.$$

There is, however, still another characteristic scale length for a disturbance in a compressed substance. This is determined by the Laplacian of the Fermi momentum. From the Thomas–Fermi equation (5.13) and the requirement that the substance must be neutral, it follows that the parameter $x_0 k_0$ is of the order of unity, except near the nuclei [see (5.14)]. Our expressions, however, actually contain quantities which are averaged over a distance determined by the range of the correlation $1/k_0$. This region, as has been pointed out in § 17.6, contains many neutral cells. Hence the effective value of the parameter $x_0 k_0$, which is equal in order of magnitude to $(\varrho^{1/3} a_0/Z^{1/3})^{1/4}$, is small for pressures much greater than the lower bound of region II. Under these conditions we can use a "quasi-uniform" expression

$$E - E_0 = -\frac{e^2}{\pi^2 a_0}(1 - \ln 2)\int d^3\mathbf{x}\, \varrho(\mathbf{x}) \ln\left[a_0 \varrho^{1/3}(\mathbf{x})\right]. \tag{27.25'}$$

The correlation contribution to the pressure is of particular interest. Differentiating (27.25) in terms of Ω with N constant $[p_0 \sim (N/\Omega)^{1/3}]$, we find [125]

$$P - P_0 = -\frac{1 - \ln 2}{3\pi^2} \cdot \frac{e^2}{a_0} \varrho . \tag{27.28}$$

This is small compared with the quantum and exchange corrections to the pressure.

27.8. We will apply the expression for the correlation energy of a dense system (27.23), to systems with short-range forces[81]. If we wish to include spin and isotopic spin dependent forces, this expression requires some modification. Using eqns. (27.1) and (27.2), and remembering that the polarization operator is diagonal in the discrete suffixes for filled shells, we can consider the matrix structure of the true and effective interaction potentials to be the same. This applies to a case of interest to us—the Serber forces. The operator $\hat{S} = \frac{1}{2}(1 - \hat{\mathscr{P}}_\sigma \hat{\mathscr{P}}_\tau)$ has the property $\hat{S}^2 = \hat{S}$. Hence, making the replacements $\gamma \to \gamma\hat{S}$, $V \to V\hat{S}$, we arrive at a relationship of the type (27.2), in which $\hat{S}$ only appears on the lefthand side and in the numerator on the right; it is excluded from the denominator on the right. This simplifies the calculations considerably and results in the quantity Q of (27.23′) retaining its form (g must be equal to 4 for nuclear matter). The only change is in the coefficient of the expression for $E - E_0$, in which we must substitute $\mathrm{Tr}_{\sigma\tau}\hat{S}^2 = 3/2$, instead of g.

We therefore obtain the following expression for the correlation energy of a dense system containing two different types of particles with short-range forces (nuclear matter):

$$\left.\begin{aligned} E - E_0 &= \frac{3}{4}\frac{\Omega p_0}{M}\int d^3k\, k \int_{-\infty}^{\infty} \frac{d\zeta}{2\pi}\left[Q - \ln(1 + Q)\right], \\ Q &= \frac{2Mp_0}{\pi^2}\, v(k)\left[1 - \zeta \arctan(1/\zeta)\right]. \end{aligned}\right\} \tag{27.29}$$

We will analyse the contributions by the different values of k to the integral over k, assuming that the condition $\alpha\eta < 1$ holds. We introduce the Debye momentum for this system, as defined by (27.17)

$$k_0 \sim (V_0 R p_0 M)^{1/2} \ll p_0 .$$

In the region from $k = 0$ to $k = 1/R$, $\nu(k) \sim V_0 R^3$, and $Q \sim \alpha < 1$. Thus the contribution of this region to the energy is relatively small, of the order $\Omega p_0/MR^4$.

In the region $k > 1/R$, the Fourier transform of the potential has the following form for the cases of greatest interest:

$$\nu(k) = \frac{C}{k^2}; \quad C \sim V_0 R.$$

We will work out the values of C for the most interesting potentials. The Yukawa potential $V(r) = V_0 \exp(-r/R)/(r/R)$ corresponds to

$$\nu(k) = \frac{4\pi V_0 R^3}{1 + (kR)^2}$$

and $C = 4\pi V_0 R$. For the potential well, $V(r) = -V_0\,(r < R)$, $V(r) = 0\,(r > R)$, and we have

$$\nu(k) = \frac{4\pi V_0 R^3}{kR} \cdot \frac{\partial}{\partial(kR)} \left(\frac{\sin(kR)}{kR} \right).$$

For large kR, ν becomes $4\pi V_0 R^3 \cos(kR)/(kR)^2$. This does not have the required form; however, its square† can be written as $\nu^2(k) = 8\pi^2 V_0^2 R^2/k^4$. We have here used $\cos^2 x \to 1/2$ when $x \to \infty$. We can therefore put $C = 2\sqrt{(2)}\,\pi V_0 R$. In the range $1/R < k < k_0$, Q has the value $Q \sim Mp_0 c/k^2 \sim k_0^2/k^2 \gg 1$. The logarithm in (27.29) can be neglected, and we arrive at an estimate for $E - E_0$ in this region:

$$E - E_0 \sim \frac{\Omega p_0 k_0^4}{M}.$$

This contribution can also be considered logarithmically small. Finally, in the region $k_0 < k < p_0$ (we will not consider values of k larger than p_0), we have

$$Q \sim k_0^2/k^2 \ll 1.$$

The function $Q - \ln(1 + Q)$ can be expanded as a series in Q; if we restrict ourselves to the first term $Q^2/2$, we obtain the

† It is evident from (27.29) that only even powers of ν are important—more specifically the second.

expression

$$E - E_0 = -\frac{3}{8}\frac{\Omega p_0}{M}\cdot\frac{2M^2p_0^2}{\pi^6}\int_{k_0}^{p_0}\frac{dk}{k}\int_{-\infty}^{\infty}\frac{d\zeta}{2\pi}\,[1-\zeta\arctan(1/\zeta)]^2.$$

We find on integration

$$E - E_0 = -N\frac{3(1-\ln 2)}{128\pi^4}C^2M\ln\left(\frac{p_0}{MV_0R}\right). \qquad (27.30)$$

For the potential well,

$$E - E_0 = -N\frac{3}{16\pi^2}(1-\ln 2)\,MV_0^2R^2\ln\left(\frac{p_0}{MV_0R}\right). \qquad (27.31)$$

These expressions refer immediately only to systems with repulsive forces. This appears formally in the fact that, when $v(k) < 0$, $1 + Q$ in (27.29) can go to zero for momenta less than the Debye momentum. Similarly, these expressions contain an imaginary component which corresponds to an indentation round the singularity. This reflects the "instability" of the ground state in the form used for calculating the Green function G_0 (see the end of § 16). An estimate indicates, however, that the contribution of the imaginary part to (27.30) and (27.31) is small. We may suppose that an alteration of the ground state would have little effect on the results.

27.9. We now turn to a basic consideration of the model for nuclear matter discussed in §§ 7 and 18 [46].† We remember that the nucleons in this model are subject to direct correlations due purely to the short-range repulsive forces $V_{(c)}$. The long-range attractive forces $V_{(a)}$ only appear as a general potential well and a modified dispersion law for the nucleons.

Going over to operator language, we can say that we are considering the replacement of the exact Hamiltonian of the system

$$\hat{H} = \hat{H}_F + \hat{H}_{(c)} + \hat{H}_{(a)}$$

($\hat{H}_{(a,c)}$ are the Hamiltonians corresponding to the forces $V_{(a)}$ and $V_{(c)}$) by the approximate Hamiltonian

$$\hat{H} = \hat{H}_{0(a)} + \hat{H}_{(c)},$$

† Related questions are also investigated in [122].

where $\hat{H}_{0(a)}$ is the Hamiltonian in the Hartree–Fock approximation, including only the attractive forces.

If the repulsive Hamiltonian $\hat{H}_{(c)}$ is absent, the possibility of making this replacement can be verified directly. In this case, we simply have to show that the correlation effects (due to the attractive forces only) are small. If the values for V_0, R and p_0 (see § 1) are substituted in (27.31), we obtain the following estimate:†

$$\frac{E - E_0}{N} \approx -1\,\text{MeV}. \tag{27.32}$$

This is small even in comparison with the binding energy of nuclear matter.

We must now show that the presence of short-range forces $\hat{H}_{(c)}$ does not prevent us replacing the Hamiltonian $\hat{H}_F + \hat{H}_{(a)}$ by $\hat{H}_{0(a)}$. The problem is equivalent to determining the interference between the correlations of the forces $V_{(c)}$ and $V_{(a)}$. This cannot be large. In fact the correlation effects due to these and other forces are small. The effect of Pauli's principle is important for $V_{(a)}$, due to the large value of the condensation parameter η. The correlation effects of the repulsive forces are small because the scattering length c is small compared with the distance between the particles. The interference effect must be governed, to a great extent, by a product of these small parameters. This is confirmed by the calculations [47]: the energy correction is of the order -1 MeV.

This model of nuclear matter is therefore accurate (so far as the energy is concerned) to within a few MeV. The effects of three-particle correlations [see (26.3)] are of this order of magnitude, but have the opposite sign.

We now return to the general problem of the stability of nuclear configurations (see § 7). We will consider the equation of state for high-density nuclear matter, supposing initially that there are no forces of the hard-core type. The energy of such a system is determined firstly by eqn. (7.31), corresponding to the Hartree–

† In this case $\alpha\eta$ is not small compared with unity. This does not, however change the order of magnitude of the estimate.

Fock approximation, and secondly by the correlation term (27.31), which makes a negligible contribution to the pressure.

The results obtained in § 7 go beyond the Hartree–Fock approximation and provide not only a sufficient, but also a necessary criterion for nuclear matter to be unstable. The deduction that kinetic energy plays a relatively small part in highly condensed nuclear matter is fully confirmed. One frequently finds a contrary assertion in the literature, supported by a reference to systems with Coulomb interactions. This analogy is incorrect, however, since, owing to the electrical neutrality of Coulomb systems as a whole, the basic part of the energy corresponding to the self-consistent interaction disappears.

The hard-core potential radically changes the stability problem. The equation of state for nuclear matter at high compressions including this potential has been investigated more than once [77, 126]. However, the idea of impenetrable spheres is only true for a density which is not too high.

This is true of all questions concerning high-density nuclear matter. There can be no doubt that at sufficiently high compressions, significant many-body forces appear which depend on the state of the interacting nucleons. This, needless to say, greatly complicates the problem.

28. Applications to the Theory of Collective Oscillations

28.1. We will now consider an example which illustrates the application of the Green function method to the problem of describing the excited states of a system.

We will take a system of particles with Coulomb interactions and suppose that it is dense. Such a system is characterized by a small interaction ($\alpha \sim \eta^{-2}$). Hence, when it is excited with no change in the number of particles, we can always reach states which correspond to the creation of a pair—a particle and a hole—with negligible interaction between the pair.

It is a property of these excited states that the excitation affects only a small number of particles in the system—in this case, one. We may therefore call them individual-particle states. The spectrum of individual-particle excited states in a system can be found in the following way. From the results of § 21, we can put

$$\Delta E_n = \varepsilon_\nu - \varepsilon_\mu .$$

Here, a particle acquires energy and gets into a state ε_ν, at the same time leaving a vacancy in the state ε_μ. In a spatially uniform system the states are characterized by the momentum of the particles. If $\boldsymbol{p} + \boldsymbol{k}$ is the momentum of state ν, and $\boldsymbol{p}$ of state μ, then

$$\Delta E_n = \frac{(\boldsymbol{p} + \boldsymbol{k})^2 - p^2}{2M} .$$

The quantity k represents the momentum acquired by the system in the excitation. Since the term $p^2/2M$ has an upper limit set by the Fermi energy, $p_0^2/2M$, the spectrum of the individual-particle excitations has an upper and lower bound of $(p_0 k/M) + (k^2/2M)$ and $(-p_0 k/M) + (k^2/2M)$ respectively. This region is shown in Fig. 64.

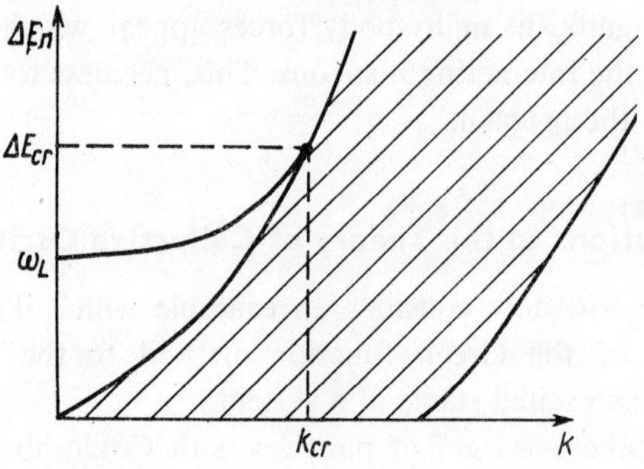

FIG. 64

The energy spectrum of a system with weak interactions is by no means limited to levels corresponding to individual-particle excitations of the system. There are also what are known as the collective levels. The corresponding state of the system is characterized by the simultaneous excitation of a large number of particles. These collective excitations appear classically as waves which propagate throughout the system; they are therefore referred to as collective oscillations.

Collective excitations have been studied in connection with longitudinal waves in plasmas, X-ray absorption in metals, zero sound in dense systems, giant resonances in photo-nuclear reactions, etc.

A collective excitation is an additional type of excitation which does not appear in the spectrum of a perfect gas. The cause is the interaction, or, more accurately, the correlation between the particles, which makes it possible for collective coordinated motions of the particles to occur. If the interaction disappears, a collective excited state goes over to an individual-particle excited state (the mechanism of this transformation will be looked at below).

28.2. The transition of a system to a collective excited state can be interpreted as the creation of a quasi-particle,representing a bound state of a particle and a hole.† The wave function of such a state can be represented as a superposition of wave functions corresponding to individual excitations of the different particles. This serves as another illustration of the collective nature of these states.‡

By no means all systems have collective energy levels. Of the various conditions that must be satisfied if these levels are to appear, there is first of all the requirement that repulsive forces must predominate among the particles in the system. This guarantees that the oscillation frequencies are not purely imaginary (see § 27.3).

Furthermore, the interaction forces between particles in a dilute system cannot lead to the appearance of coordinated multi-particle motion (only pair correlation between particles is important in this case). Even if such motion did appear, it would decay very rapidly into unrelated individual motions. Hence the second requirement for the existence of collective levels is that the system be dense, i.e. there must be a comparatively large number of particles within the range of each particle.

† More accurately, we should speak of the state of a particle–hole system with discrete energy levels.

‡ We will be talking about collective excitations corresponding to a single bound pair. In principle more involved complexes can also be formed.

The wave function for a collective excited state of a system is a superposition of the following type:

$$\Psi_n(N) = \sum_p [\alpha_{p,k}\hat{a}^+_{p+k}\hat{b}^+_p + \alpha'_{p,k}\hat{b}_{p+k}\hat{a}_p]\,\Psi, \qquad (28.1)$$

where α and α' are certain coefficients [101]. If we now consider the way in which these coefficients depend on $\boldsymbol{k}$—the total momentum of the quasi-particle—then it appears that, as $\boldsymbol{k}$ increases, the relative value of one of the coefficients α_{pk} increases at the expense of the others. For some particular value, $k = k_{cr}$ (see Fig. 64), only this coefficient remains, and the quasi-particle changes into an individual excitation. This means that we only need investigate relatively long wavelength excitations for which

$$k < k_{cr}. \qquad (28.2)$$

In a non-homogeneous system the quantity k is itself indeterminate, owing to the momentum exchange between the quasi-particle and the inhomogeneities in the system. As a measure of this exchange, we can use the inverse of the scale-length of the disturbances $1/x_0$, where x_0 is the distance within which the characteristics of the system change appreciably. Hence, when the condition

$$k_{cr} - k > 1/x_0 \qquad (28.3)$$

is broken, there is a finite probability that a quasi-particle will transform to an individual excitation. There are also other reasons why the quasi-particle might be an unstable formation. It is essential that the damping be small compared with ΔE_n.†

28.3. We now provide a quantitative description of collective oscillations. Since the quasi-particle is formed from a particle–hole pair, we must consider the quantity

$$\Phi_n(1,2) = \langle\Psi|T(\hat{\psi}_H(1)\,\hat{\psi}^+_H(2)|\Psi_n\rangle$$

which, for a dense system, is determined by eqn. (25.15)

$$\Phi_n(1,2) = \overline{\Phi}_n(1,2) - i\int d3\, d4 G_0(1,3)\, G_0(3,2)\, V(3,4)\, \Phi_n(4,4). \qquad (28.4)$$

† When we speak of damping, we are thinking not of a true stationary state of a system, but of a state that arises as the result of a definite excitation mechanism, and which leads to the creation of a quasi-particle (see §§ 22 and 24).

The function $\bar{\Phi}_n(1, 2)$ describes the propagation of a particle and a hole which are not correlated with one another, and it is not related to their bound state. We can therefore restrict our consideration to a homogeneous integral equation.

The function $\Phi_n(1, 2)$ has the following time dependence:

$$\Phi_n(1, 2)|_{t_2 - t_1 \to 0} = \exp(-i\Delta E_n t_1)\, \Phi_n(q_1, q_2).$$

If we substitute this into (28.4), and remember that $V(3, 4) = V(q_3, q_4)\, \delta(t_3 - t_4)$, we arrive at the following integral equation for the function $\Phi_n(q_1, q_2)$:

$$\Phi_n(q_1, q_2) = -i \int dq_3\, dq_4 \int \frac{d\varepsilon}{2\pi}\, G_0(q_1, q_3, \varepsilon)\, G_0(q_3, q_2, \varepsilon + \Delta E_n)$$
$$\times V(q_3, q_4)\, \Phi_n(q_4, q_4). \tag{28.5}$$

If we put $q_1 = q_2$, we obtain the equation

$$\Phi_n(q_1, q_1) = -i \int dq_3\, dq_4 V(q_3, q_4) \int \frac{d\varepsilon}{2\pi}\, G_0(q_1, q_3, \varepsilon)$$
$$\times G_0(q_3, q_1, \varepsilon + \Delta E_n)\, \Phi_n(q_4, q_4), \tag{28.6}$$

which can be used to find the excitation spectrum ΔE_n when specific boundary conditions are imposed.

These equations simplify for a spatially uniform system. We will consider the function $\Phi_n(q_1, q_2)$ in the form†

$$\Phi_n(q_1, q_2) = \int d^3\boldsymbol{p}\, \varphi(\boldsymbol{p}, \boldsymbol{k}) \exp\{i[(\boldsymbol{p} + \boldsymbol{k} \cdot \boldsymbol{x}_1) - (\boldsymbol{p} \cdot \boldsymbol{x}_2)]\}\, \delta_{\sigma_1 \sigma_2}. \tag{28.7}$$

Here $\boldsymbol{p} + \boldsymbol{k}$ is the momentum of the particle; $\boldsymbol{p}$ is the momentum of the hole; $\boldsymbol{k}$ is the total momentum of the quasi-particle. Substituting (28.7) in (28.5) gives

$$\varphi(\boldsymbol{p}, \boldsymbol{k}) = -2iv(\boldsymbol{k}) \int \frac{d\varepsilon}{2\pi}\, G_0(\boldsymbol{p}, \varepsilon)\, G_0(\boldsymbol{p} + \boldsymbol{k}, \varepsilon + \Delta E_n)$$
$$\times \int d^3\boldsymbol{p}' \varphi(\boldsymbol{p}', \boldsymbol{k}). \tag{28.8}$$

† Since the spins of particles and holes are the same, this indicates that the total spin of a quasi-particle will be zero.

Integrating both sides of this equation in terms of $\boldsymbol{p}$, and using the definition of the polarization operator, we obtain

$$[1 - \Pi(\boldsymbol{k}, \Delta E_n)\, \nu(\boldsymbol{k})] \int d^3\boldsymbol{p}\, \varphi(\boldsymbol{p}, \boldsymbol{k}) = 0. \qquad (28.9)$$

Thus the required value for the excitation energy is, simultaneously, both the pole of the effective potential and the pole of the two-particle Green function.

The roots of the equation $1 = \Pi\nu$ have been considered in a previous section, where ΔE_n was denoted by $\omega_0(\boldsymbol{k})$. Figure 64 shows the way in which ΔE_n varies with k. For regions of k near k_{cr}, we have used a more accurate expression for the polarization operator which includes terms in $k^2/2M$. This expression has the following form [101]:

$$\Pi(\boldsymbol{k}, \omega) = \frac{M^3}{4\pi^2 k^3}\left\{\Delta_1\Delta_2 \ln\frac{\Delta_1}{\Delta_2} + \Delta_3\Delta_4 \ln\frac{\Delta_3}{\Delta_4} - \frac{2k^3 p_0}{M^2}\right\}, \qquad (28.10)$$

where

$$\Delta_{1,2} = \omega \mp \frac{kp_0}{M} - \frac{k^2}{2M},$$

$$\Delta_{3,4} = \omega \pm \frac{kp_0}{M} + \frac{k^2}{2M}.$$

The position of the critical point k_{cr} is determined by the intersection of the ΔE_n curve, corresponding to eqn. (28.9), with the curve corresponding to the upper bound of the individual-particle excitation region. We have here the following equation:

$$\left(\frac{k_{cr}}{p_0}\right)^2 = \frac{1}{\pi a_0 p_0}\left[\left(2 + \frac{k_{cr}}{p_0}\right)\ln\left(1 + \frac{2p_0}{k_{cr}}\right) - 2\right],$$

$$\frac{\Delta E_{cr}}{\omega_L} = \left(\frac{3\pi a_0 p_0}{4}\right)^{1/2}\left(\frac{k_{cr}}{p_0} + \frac{k_{cr}^2}{2p_0^2}\right).$$

These relationships are correct so long as k_{cr} and k are small compared with the Fermi momentum p_0. As p_0 increases, so does k_{cr}, but not so rapidly; for relatively low densities, k_{cr} is of the order of the Debye momentum.

The quantity ΔE_{cr} does not in general exceed the Langmuir frequency ω_L by any appreciable amount. Collective levels therefore appear in a comparatively narrow region of the spectrum.

For the particular case of a uniform system, the damping is exact zero.† Hence the problem of finding the true excited energy levels and the behaviour of the quasi-particle is already solved in this case.

When interaction is excluded, a collective excitation goes over to a single-particle excitation. It is of interest to investigate the mechanism of this transition. We refer to Fig. 65, and consider a quasi-particle with a definite momentum k (point 1). If we decrease

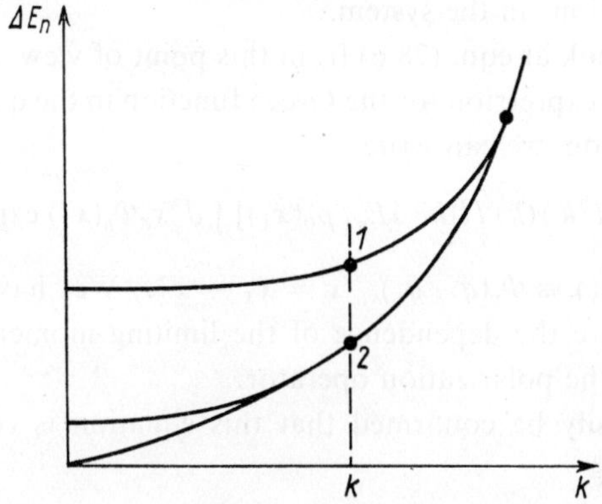

FIG. 65

the electron charge e, the entire collective branch is displaced downwards, as is evident from the expressions for ω_L and ΔE_{cr}; the energy of the quasi-particle is decreased correspondingly. For some value of e, the collective curve passes through point 2, which lies on the upper bound of the individual-particle excitation region. There is then no difference between collective and individual excitations. For small values of e, only the individual-particle excitation remains. The formation of quasi-particles, when interaction is turned on, occurs in a similar way.

28.4. For collective oscillations of inhomogeneous systems, we should begin with the general equations (28.5) and (28.6). This

† This conclusion is true (for a dense system) to within small terms describing the decay of a quasi-particle into two individual particle excitations.

would lead, however, to very tedious calculations [84]. We will restrict our consideration to systems which are only weakly inhomogeneous [100].

If the condition

$$x_0 k_0 \gg 1$$

is satisfied, where x_0 is a characteristic scale-length for the disturbance, then a "quasi-uniform" formulation of the problem is possible, i.e. we can neglect the gradients of quantities characterizing the distributions in the system.

We will look at eqn. (28.6) from this point of view. If we substitute into the expression for the Green function in the quasi-classical approximation, we can write

$$\Phi_n(\boldsymbol{x}_1) = \int d^3\boldsymbol{k}\, \nu(\boldsymbol{k})\, \Pi[\boldsymbol{k}, \Delta E_n, p_0(\boldsymbol{x}_1)] \int d^3\boldsymbol{x}_2 \Phi_n(\boldsymbol{x}_2) \exp\left[-i(\boldsymbol{k}\cdot\boldsymbol{x})\right],$$

where $\Phi_n(\boldsymbol{x}_1) \equiv \Phi_n(q_1, q_1)$, $\boldsymbol{x} = \boldsymbol{x}_1 - \boldsymbol{x}_2$. We have explicitly indicated here the dependence of the limiting momentum on the point $\boldsymbol{x}_1$ in the polarization operator.

It can easily be confirmed that this equation is equivalent to the following:

$$\Phi_n(\boldsymbol{x}_1) = \nu(-i\nabla_1)\, \Pi(-i\nabla_1, \Delta E_n, p_0(x_1))\, \Phi_n(\boldsymbol{x}_1), \qquad (28.11)$$

where we have formally replaced $\boldsymbol{k}$ by the operator $-i\nabla_1$, which acts only on the function $\Phi_n(\boldsymbol{x}_1)$.

If we remember that the function $p_0(x)$ changes only slightly, we can replace this equation by the following:

$$[\nabla^2 + k^2(x)]\, \Phi_n(\boldsymbol{x}) = 0, \qquad (28.12)$$

where the quantity $k(x)$ is the spatially variable wave vector of the quasi-particle, and can be obtained from the equation

$$1 = \nu[k(x)]\, \Pi[k(x), \Delta E_n, p_0(x)]. \qquad (28.13)$$

It depends on the excitation energy ΔE_n, which is determined by eqn. (28.12) with the appropriate boundary conditions.

As proof of this, we notice that the polarization operator is an even function of k, i.e. depends, in effect, only on k^2. If, therefore, we substitute the solution of (28.12) into (28.11) instead of Φ_n, and use (28.13), we arrive at the required identity.

Equation (28.12) can be thought of as the Schrödinger equation for a quasi-particle.† We can solve it by the normal methods of quantum mechanics. The function $k^2(x)$, like $p_0(x)$, has only a slow spatial variation. We can therefore seek a solution of (28.12) in the quasi-classical approximation. In particular,

$$\Phi_n(x) = \frac{1}{\sqrt{k(x)}} \exp\left[i \int dx\, k(x)\right], \tag{28.14}$$

for an s-wave and a spherically symmetric $p_0(x)$. We can find the spectrum by using the equality

$$\oint k(x)\, dx = 2\pi(n + \tfrac{1}{2}).$$

28.5. We will consider the problem of collective oscillations for a heavy atom [100]. This has been discussed in the literature in connection with the energy losses of fast particles and the plasma oscillations of the residual ions in metals [127].

The essential difference between an atom and an extended medium is that the former is limited in space and, hence, the wave number k is quantized. The region of the atom where the density requirement is satisfied has a radius of the order $Z^{-1/3}a_0$. It is in this region that the excitation first appears. The smallest value of k is of the order $Z^{1/3}/a_0$. Hence the limiting case for long wavelengths does not appear in an atom—or in any other finite system. This radically restricts the existence of possible collective levels [see (28.2)].

We will not be interested in how much a quasi-particle is attenuated, and can therefore put the emphasis on finding the number of collective levels satisfying requirement (28.2). This requirement is very stringent, so it may be that no collective levels will be found.

The investigation of an uncompressed atom is made more difficult by the fact that the condition $x_0 k_0 \gg 1$ is not satisfied. We will therefore consider a simple model, replacing the true density

† The function Φ_n is not, however, the wave function of the quasi-particle.

distribution in an atom by a square distribution, and choosing the radius of the distribution r_0 by the minimum energy requirement. This gives

$$r_0 \approx 2{\cdot}5 a_0 Z^{-1/3} \quad \text{and} \quad p_0 \approx \frac{0{\cdot}8 Z^{2/3}}{a_0}.$$

We must put $k^2(x) = \text{const.}$ in (28.12), which holds inside the atom in this model. Its solution can then be written as

$$\Phi_n(x) \sim \frac{1}{\sqrt{kr}} J_{l+1/2}(kr)\, Y_{lm}(\theta, \varphi),$$

where J and Y are the Bessel function and the spherical harmonic function, respectively. The quantity k can be found from the boundary conditions. In the most interesting case of a quasi-particle localized in an atom, there must be no flow of matter through the surface of the atom. It has been shown [100] that this condition can be written in the form

$$\left(\frac{\partial}{\partial r} - \frac{\partial}{\partial r'}\right) \Phi_n(r, r')|_{r=r'=r_0} = 0.$$

If this relation and the dispersion relation are considered together as functions of k and ΔE_n, we come to the following results.

For all cases, except $l = 1$, the value of ΔE_n lies higher than ΔE_{cr}; all the corresponding levels are essentially single particle. There is just one admissable solution for $l = 1$:

$$\Delta E_n = 1{\cdot}41 \omega_L \approx 20Z \text{ eV}$$

Hence if collective oscillations of an atom exist, we should expect that there will most probably be only one level.

If we impose the stricter boundary condition ($\Phi_n(r)|_{r=r_0} = 0$), there is no solution for which ΔE_n is less than ΔE_{cr}.

The existence of a collective level can radically affect atomic reactions in the energy region of a few hundred electron volts. The resultant picture is closer to the Bohr concept of nuclear reactions. A resonance electron (an X-ray quantum) can hand on its energy immediately to a large number of atomic electrons, transferring these latter to a comparatively long-lived excited state (a "compound" atom). This state can decay by various routes—both elastic and inelastic. In particular it is possible for a group of several electrons to be ejected.

The available experimental data on multiple ionization of rare-gas atoms agree qualitatively with this picture. The ionization cross-section has an additional, anomalously high (for a degree of ionization $k = 2$ to 6) maximum. As Z increases, the cross-section increases rapidly for constant k, which unmistakably points to a collective process. Finally, the ratio of the corresponding cross-sections for different atoms increases rapidly with k, thus indicating a preferential ejection of electron groups [128].

28.6. The results of the previous section are purely preliminary since an extremely simple atomic model was used and, even more important, the calculations were made in the quasi-uniform approximation. We will now formulate a way of calculating the polarization operator that depends only on the assumption that the parameter $1/x_0 p_0$ is small. The relative magnitudes of x_0 and $1/k_0$ may be arbitrary.

We will begin from the general equation (28.6), writing it in the form:

$$\Phi(\mathbf{x}_1) = \int d^3\mathbf{x}_2 Q(\mathbf{x}_1, \mathbf{x}_2)\,\Phi(\mathbf{x}_2). \qquad (28.15)$$

Here $\Phi(\mathbf{x}_1) = \Phi_n(q_1', q_1)$ and Q coincides with the integral kernel of (28.6). Using the operator form of the Green function (10.9), we have

$$Q = 2i \lim_{x_1' \to x_1} \int \frac{d\varepsilon}{2\pi} \int d^3\mathbf{x}_3 [\hat{G}_0(\varepsilon + \Delta E_n)\,\delta(\mathbf{x}_1 - \mathbf{x}_3)]$$
$$\times [\hat{G}_0(\varepsilon)\,\delta(\mathbf{x}_3 - \mathbf{x}_1')]V(\mathbf{x}_3 - \mathbf{x}_2)],$$

where

$$\hat{G}_0(\varepsilon) = [\varepsilon - \hat{T} - \hat{W} + i\delta \operatorname{sign}(\hat{T} + \hat{W} - \varepsilon_F)]^{-1}.$$

Integrating over $\mathbf{x}_3$, and expanding the second δ-function into a Fourier integral, we obtain

$$Q = 2i \int d^4p\, \langle \hat{G}_0(\varepsilon + \Delta E_n)\, V(\mathbf{x}_1 - \mathbf{x}_2)\, \hat{G}_0(\varepsilon) \rangle_p .$$

Finally, expanding the Green function G_0 in a Fourier integral in terms of $\hat{T} + \hat{W}$ gives

$$Q = \int_{-\infty}^{\infty} d\tau_1\, d\tau_2 \varphi(\tau_1, \tau_2) \int d^3\mathbf{p}\, \langle V(\mathbf{x}'(\tau_1) - \mathbf{x}_2)$$
$$\times \exp{[i(\tau_1 + \tau_2)(\hat{T} + \hat{W})\rangle_p}, \qquad (28.16)$$

where

$$\varphi(\tau_1, \tau_2) = \frac{1}{\pi(\tau_1 + \tau_2)} [\theta(\tau_2) e^{i\tau_2 \Delta E_n} - \theta(-\tau_1)\, e^{-i\tau_1 \Delta E_n}]$$

and

$$\hat{x}(\tau) = e^{i(\hat{T}+\hat{W})\tau}\, x_1 e^{-i(\hat{T}+\hat{W})\tau}$$

is the Heisenberg coordinate operator.

If we now allow for the fact that we are dealing with a dense system, we can first neglect the exchange effects, and write

$$\hat{T} + \hat{W} = \frac{\hat{p}^2 - p_0^2(x_1)}{2M} + \varepsilon_F.$$

Moreover, the basic contribution to the momentum integral of (28.16), as in the uniform case, comes from a narrow region near the Fermi surface. We can therefore make the replacement $p \to p_0(1 + \alpha)\, n$, and

$$\int d^3p \to p_0^3/(2\pi)^3 \int_{-\infty}^{\infty} d\alpha \int d\Omega_n.$$

The result of the action of the exponent in (28.16) on $\exp i(p \cdot x_1)$ can be written as (see Appendix B):

$$\hat{K}(\tau_1 + \tau_2) \exp\left(i(\tau_1 + \tau_2) \frac{p^2 - p_0^2}{2M} \right) \exp i(p \cdot x_1),$$

where $\hat{K}$ is a function of the commutators of the operators $\hat{p}^2$ and p_0^2, which reduces to unity for $\tau_1 + \tau_2 = 0$. It may easily be shown that the difference between p and p_0 need only be taken into account in the second factor of the last expression, which becomes $\exp [i(\tau_1 + \tau_2)\, p_0 \alpha / M]$. The subsequent integration over α leads to $\delta(\tau_1 + \tau_2)$ as a result of which the operator $\hat{K}$ disappears completely from the calculations.

We must finally remember that the operator $V(\hat{x}(\tau_1) - x_2)$ becomes $V(x(\tau_1) - x_2)$ due to the quasi-classical requirements, where $x(\tau)$ is the classical coordinate of the particle. It is defined by the equation

$$M\ddot{x}(\tau) = \nabla p_0^2 / 2M$$

with the initial conditions

$$M\dot{x}(0) = p_0(x_1) n\,; \quad x(0) = x_1.$$

We finally obtain

$$Q(\boldsymbol{x}_1, \boldsymbol{x}_2) = \frac{p_0(\boldsymbol{x}_1)}{\pi^2 a_0} \left\{ \frac{1}{|\boldsymbol{x}_1 - \boldsymbol{x}_2|} + i\Delta E_n \int_{-\infty}^{0} d\tau \, \overline{\frac{\exp(-i\tau\Delta E_n)}{|\boldsymbol{x}(\tau) - \boldsymbol{x}_2|}} \right\}. \tag{28.17}$$

The bar here signifies an averaging in the direction of the initial velocity $\boldsymbol{n}$. The fact that the polarization operator (like the mass operator—see § 23.9) satisfies a dispersion relation can be expressed explicitly, if we rewrite (28.17) in the form

$$Q(\boldsymbol{x}_1, \boldsymbol{x}_2) = -\frac{p_0(\boldsymbol{x}_1)}{2\pi^3 a_0} \int_{-\infty}^{\infty} \frac{d\omega' \omega'^2}{\omega^2 - \omega'^2 + i\delta} \overline{\left(\frac{1}{|\boldsymbol{x}(\tau) - \boldsymbol{x}_2|} \right)_{\omega'}}, \tag{28.18}$$

where the subscript ω' indicates the Fourier transform in terms of τ.

28.7. The relations obtained in the previous section show that, if we want to determine the polarization operator for an inhomogeneous quasi-classical medium, it is sufficient to find the classical trajectory of a particle in the self-consistent field. We note that the quasi-uniform approach corresponds to replacing the true particle trajectory by a free-particle trajectory with the same initial conditions.

The integration over the initial velocity angles, and over τ is distinctly laborious, and represents a serious technical difficulty. We will now show that this obstacle can be overcome: at least so far as determining the spectrum of the collective excitations is concerned. To this end, we introduce a new function Ψ, defined by $\Phi(\boldsymbol{x}) = \nabla^2 \Psi(\boldsymbol{x})$. Neither of these functions blow up at infinity and are regular at the origin. We can then write (28.15) as

$$\nabla^2 \Psi(\boldsymbol{x}_1) = \int d^3 x_2 \nabla^2_{x_2} Q(\boldsymbol{x}_1, \boldsymbol{x}_2)\, \Psi(\boldsymbol{x}_2),$$

where, from (28.17),

$$\nabla^2_{x_2} Q(\boldsymbol{x}_1, \boldsymbol{x}_2) = -\frac{4 p_0(\boldsymbol{x}_1)}{\pi a_0} \{ \delta(\boldsymbol{x}_1 - \boldsymbol{x}_2) + i\Delta E_n \int_{-\infty}^{0} d\tau \exp(-i\tau\Delta E_n)\, \overline{\delta(\boldsymbol{x}(\tau) - \boldsymbol{x}_2)} \}.$$

The average in this equation represents the integral

$$\frac{1}{4\pi}\int_0^{2\pi} d\varphi \int_{-1}^{1} d\cos\theta,$$

where φ and θ are the angles defining the direction of the initial velocity of the particle. We also have

$$\delta(\boldsymbol{x}(\tau) - \boldsymbol{x}_2) = \sum_i \frac{1}{|D_i|}\delta(\tau - \tau_i)\,\delta(\varphi - \varphi_i)\,\delta(\cos\theta - \cos\theta_i),$$

where τ_i, φ_i and $\cos\theta_i$ are the solutions of the equation $\boldsymbol{x}(\tau) = \boldsymbol{x}_2$, and D_i is the Jacobian

$$\left.\frac{\partial(x_1(\tau), x_2(\tau), x_3(\tau))}{\partial(\tau, \varphi, \cos\theta)}\right|_{\tau=\tau_i,\, \varphi=\varphi_i,\, \cos\theta=\cos\theta_i}.$$

The quantities entering into this have a simple physical connotation: τ_1 is the time taken by the particle to go from $\boldsymbol{x}_1$ to $\boldsymbol{x}_2$, φ_i and θ_i are the corresponding initial velocity angles. The suffix i represents the index of a trajectory which connects the points concerned, and corresponds to an initial velocity p_0/M. There is not, generally speaking, only one such trajectory: under certain conditions, there can even be a continuous set. This last possibility will not be considered here. The derivatives $\partial\boldsymbol{x}/\partial\tau|_{\tau=\tau_i}$ in the Jacobian are obviously equal to the velocity components of a particle at the point $\boldsymbol{x}_2$.

We finally obtain the following simple equation

$$\left(\nabla^2 + \frac{4p_0(\boldsymbol{x}_1)}{\pi a_0}\right)\Psi(\boldsymbol{x}_1) = -\frac{i\Delta E_n p_0(\boldsymbol{x}_1)}{\pi^2 a_0}\sum_i \int d\boldsymbol{x}_2$$

$$\times \frac{\exp(-i\tau_i \Delta E_n)}{|D_i|}\Psi(\boldsymbol{x}_2), \qquad (28.19)$$

which, together with the boundary conditions mentioned above, determines the spectrum and the damping of the collective excitations.

It must be noted that the damping determined by this last equation (corresponding to the mechanism at the end of § 28.2) has the same order of magnitude as the frequency itself. Since there are, as yet, no direct observations of collective oscillations in atoms, the possibility of their existence remains open. We can hope that further calculations with eqn. (28.19), using a more realistic atomic model, may throw light on this interesting problem.

Chapter 5

The Method of Green Functions in Quantum Statistics

29. General Equations

29.1. This chapter will be devoted to some questions concerning the quantum theory of many-body systems which are kept in a thermostat at some temperature T, differing from zero, and which are in statistical equilibrium. We are assuming, that is, that the system has a sufficiently large number of particles. We can suppose that the total number of particles in the system is not strictly fixed, but can undergo thermodynamical fluctuations by the exchange of particles between the system and its surroundings. This approach, which fixes the average number of particles, is particularly suitable for describing the statistical properties of a system made up of a large number of particles [144].

The basic quantum-mechanical statement on this approach is the following. An equilibrium many-body system at a temperature T is not in a state with a definite wave function, but in a mixed state, the weights for which are given by Gibbs' equation and are temperature dependent. More accurately, the state of a system in a thermostat is described by a density matrix

$$\sum_n P_n \Psi_n^*(Q') \Psi_n(Q),$$

where the summation is taken over the complete set of stationary states, Ψ_n, and the weights are determined by the relation

$$P_n = \exp [\beta(\Omega + \mu N_n - E_n)]. \tag{29.1}$$

Here $\beta \equiv 1/kT$ (k is Boltzmann's constant), Ω is the thermodynamic potential of the system, μ is its chemical potential, and E_n is the energy of a state with N_n particles.

As the temperature tends to zero ($\beta \to \infty$), only one term, corresponding to the lowest value of E_n, remains; we then return to the ground state of the system as considered in the previous chapters.†

Using these quantities, we can write down the mean value of any other quantity connected with the system. Suppose is it characterized by some operator $\hat{\alpha}$, then

$$\langle \hat{\alpha} \rangle = \sum_n P_n \int dQ \Psi_n^*(Q)\, \hat{\alpha} \Psi_n(Q).$$

The averaging here has a double significance (quantum-mechanical and statistical). If we introduce the operators for the Hamiltonian $\hat{H}$ and the total number of particles $\hat{N}$ which satisfy the following obvious relationships

$$(\hat{H} - E_n)\,\Psi_n = 0, \quad (\hat{N} - N_n)\,\Psi_n = 0,$$

then the expression for $\langle \hat{\alpha} \rangle$ can be rewritten in the more convenient form

$$\langle \hat{\alpha} \rangle = \mathrm{Tr}\,\{\exp\,[\beta(\Omega + \mu \hat{N} - \hat{H})]\,\hat{\alpha}\}, \tag{29.2}$$

which is indepedent of the representation chosen. (See Appendix A for the properties of the trace and the methods of calculating it.)

The thermodynamic potential Ω can be found from the normalization condition, which can be obtained by substituting unity for the operator $\hat{\alpha}$ in (29.2). This gives

$$\Omega = -\frac{1}{\beta} \ln \mathrm{Tr} \exp\,[\beta(\mu \hat{N} - \hat{H})]. \tag{29.3}$$

It is convenient to introduce the operator

$$\zeta = \exp\,[\beta(\mu \hat{N} - \hat{H})] \tag{29.4}$$

† This is true only if the state is non-degenerate. Otherwise one obtains a sum of expressions (equally weighted) each of which refers to one of the substates. This is the reason why one cannot consider quantum statistics to contain quantum mechanics as a limiting case.

which is called the statistical operator, and the quantity

$$Z = \mathrm{Tr}\hat{\zeta} \tag{29.5}$$

which is called the partition function. Then (29.2) and (29.3) can be rewritten in the form

$$\langle\hat{\alpha}\rangle = \mathrm{Tr}\hat{\zeta}\hat{\alpha}/\mathrm{Tr}\hat{\zeta}, \tag{29.6}$$

$$\Omega = -\frac{1}{\beta}\ln Z. \tag{29.7}$$

From this latter relationship, we can determine all the thermodynamical properties of the system†

$$\left.\begin{aligned} P = -\Omega/V;\quad S = -(\partial\Omega/\partial T)_{\mu},\\ N = -(\partial\Omega/\partial\mu)_{T,V};\quad E = \Omega + \mu N + TS,\end{aligned}\right\} \tag{29.8}$$

where P, S, N and E are the average values of the pressure, entropy, number of particles and energy of the system, respectively [144].

29.2. Besides the general thermodynamic properties of a system, considerable interest attaches to more specific ones, the distribution functions of physical quantities, the spectrum and damping of the quasi-particles, etc. As for the zero-temperature case, the corresponding information is contained in the Green functions. In subsequent sections we will examine in detail the properties of the Green functions in quantum statistics.

In many cases the probability distribution of a dynamical variable, i.e. the probability of observing particular values, can be computed in the following direct way.

Suppose we are interested in observing a value α of the quantity described by the operator $\hat{\alpha}$. The corresponding probability density is given by

$$W(\alpha) = \frac{\mathrm{Tr}[\hat{\zeta}\delta(\alpha - \hat{\alpha})]}{\mathrm{Tr}\hat{\zeta}}. \tag{29.9}$$

† Equation (29.7) gives Ω as a function of the volume V, the temperature T and the chemical potential μ. This latter can be expressed in terms of T, V and the average number of particles N, using the third relationship of (29.8). Throughout this chapter we will use V to denote the volume.

To confirm this, we take the trace in terms of the set of functions which are the eigenfunctions of the operator $\hat{\alpha}$. Then

$$W(\alpha) = \frac{\sum_n \langle n|\zeta|n\rangle \, \delta(\alpha - \alpha_n)}{\sum_n \langle n|\zeta|n\rangle} . \tag{29.10}$$

The average value of any function of $\hat{\alpha}$ is given on the one hand by

$$\langle f(\hat{\alpha})\rangle = \int d\alpha f(\alpha) \, W(\alpha)$$

and, on the other, by

$$\langle f(\hat{\alpha})\rangle = \frac{\mathrm{Tr}[\zeta f(\hat{\alpha})]}{\mathrm{Tr}\zeta} = \frac{\sum_n \langle n|\zeta|n\rangle \, f(\alpha_n)}{\sum_n \langle n|\zeta|n\rangle} .$$

The fact that these two expressions coincide when (29.10) is applied, confirms the correctness of (29.9). It is also easy to confirm that the distribution of (29.9) has the necessary normalization.

It is convenient to consider, instead of (29.9), the Fourier transform of the function $W(\alpha)$, which we will denote by $A(q)$

$$A(q) = \int d\alpha \exp(-iq\alpha) \, W(\alpha) = \frac{\mathrm{Tr}[\xi \exp(-iq\hat{\alpha})]}{\mathrm{Tr}\xi} . \tag{29.11}$$

If we now calculate the derivative

$$\frac{\partial \ln A(q)}{\partial q} = -i \frac{\mathrm{Tr}[\zeta \exp(-iq\hat{\alpha}) \, \hat{\alpha}]}{\mathrm{Tr}[\xi \exp(-iq\hat{\alpha})]} ,$$

then it obviously coincides (to within some factor) with the average value of the operator $\hat{\alpha}$ over the states of a fictitious system, whose statistical operator is the quantity $\xi \exp(-iq\hat{\alpha})$. We will denote averages of this type by the symbol $\langle ... \rangle_\alpha$. Remembering that $A(0) = 1$, we find [129]

$$A(q) = \exp\left\{ -i \int_0^q dq \langle \hat{\alpha} \rangle_\alpha \right\} . \tag{29.12}$$

We will consider the distribution of the number of particles in the system as an example. We can then find out when fluctuations in the number of particles can be ignored. Putting $\hat{\alpha} = \hat{N}$, and allowing for the commutation of the operators $\hat{N}$ and $\hat{H}$, we can write

$$\xi \exp(-iq\hat{\alpha}) = \exp\left\{\beta\left[\left(\mu - \frac{iq}{\beta}\right)\hat{N} - \hat{H}\right]\right\}.$$

Hence (29.3) gives

$$W(N) = \frac{1}{2\pi}\int dq \exp\left\{iq\,N - \beta\left[\Omega\left(\mu - \frac{iq}{\beta}\right) - \Omega(\mu)\right]\right\}. \quad (29.13)$$

It all reduces to calculating the thermodynamic potential for complex values of the argument, or, which is the same thing, to the analytical continuation of Ω into the complex μ-plane [129].

Since $q/\beta \sim 1/N\beta$ is small compared with μ, if we expand the expression in square brackets in (29.13) as a series in iq/β, we obtain

$$\Omega\left(\mu - \frac{iq}{\beta}\right) - \Omega(\mu) = \frac{iq}{\beta}\langle N\rangle + \frac{q^2}{2\beta^2}\frac{\partial\langle N\rangle}{\partial\mu} + \cdots$$

Substituting this in (29.13), we get the Gaussian distribution

$$W(N) = (2\pi(\overline{\Delta N})^2)^{-1/2}\exp\left\{-\frac{(N - \langle N\rangle)^2}{2(\overline{\Delta N})^2}\right\}, \quad (29.14)$$

where

$$(\overline{\Delta N})^2 = \frac{1}{\beta}\frac{\partial\langle N\rangle}{\partial\mu}$$

is the mean square fluctuation of the number of particles in the system. This is proportional to the total number of particles, whence it follows that, as this latter increases, the relative fluctuations in N decrease.

We can consider fluctuations in the total energy of the system in a similar way. We will use (29.12) for this, remembering that

$$\xi \exp(-iq\hat{H}) = \exp\left\{\beta\left[\mu\hat{N} - \left(1 + \frac{iq}{\beta}\right)\hat{H}\right]\right\}.$$

Whence

$$\langle\hat{H}\rangle_E = E\left(\beta \to \beta + iq, \mu \to \mu\frac{\beta}{\beta + iq}\right).$$

For small q

$$\langle \hat{H} \rangle_E = \langle E \rangle - iq\left(-\frac{\partial \langle E \rangle}{\partial \beta} + \frac{\mu}{\beta} \frac{\partial \langle E \rangle}{\partial \mu} \right).$$

Hence

$$W(E) = [2\pi(\overline{\Delta E})^2]^{-1/2} \exp\left[-\frac{(E - \langle E \rangle)^2}{2(\overline{\Delta E})^2} \right], \qquad (29.15)$$

where

$$(\overline{\Delta E})^2 = \frac{\mu}{\beta} \cdot \frac{\partial \langle E \rangle}{\partial \mu} - \frac{\partial \langle E \rangle}{\partial \beta}.$$

In this, as in the preceding calculations, the other arguments are kept constant during the differentiation.

29.3. We will now consider the single-particle Green function in quantum statistics.

This quantity, as before, is defined as the average value of the time-ordered operator product. The averaging now, however, is not over the ground state of the system, but over the statistical ensemble:

$$G(1, 2) = -i \frac{\mathrm{Tr}\, \{\xi [T(\hat{\psi}_H(1)\, \hat{\psi}_H^+(2))]\}}{\mathrm{Tr}\xi}. \qquad (29.16)$$

The field operators are taken in the Heisenberg representation.

We can obtain a spectral representation for the Green function as for the case of zero temperature [130]. To this end, we will rewrite (29.16) in the form

$$G(1, 2) = -i \sum_n \exp\left[\beta\,(\Omega + \mu N_n - E_n)\right] \langle \Psi_n | T[\hat{\psi}_H(1)\, \hat{\psi}_H^+(2)] | \Psi_n \rangle$$

and introduce a complete, intermediate set of functions Ψ_m. We now have to deal with matrix elements of the field operators of the type $\langle \Psi_n | \hat{\psi}_H | \Psi_m \rangle$, $\langle \Psi_m | \hat{\psi}_H^+ | \Psi_n \rangle$. If we include the time dependence explicitly, and take into account the properties of the operators $\hat{\psi}_H$ and $\hat{\psi}_H^+$, we can write

$$\langle \Psi_n | \hat{\psi}_H(1) | \Psi_m \rangle = \langle \Psi_n | \hat{\psi}(q_1) | \Psi_m \rangle \exp\left[-i(E_m - E_n)\, t_1\right],$$

$$\langle \Psi_m | \hat{\psi}_H^+(1) | \Psi_n \rangle = \langle \Psi_m | \hat{\psi}^+(q_1) | \Psi_n \rangle \exp\left[i(E_m - E_n)\, t_1\right],$$

where the number of particles in the states Ψ_m and Ψ_n, for which there is a non-zero matrix element, are connected by $N_m = N_n + 1$.

When $t_1 > t_2$,

$$G(1,2) = -i \sum_{n,m} \exp\left[\beta(\Omega + \mu N_n - E_n)\right] \times \exp\left[-i(E_m - E_n)(t_1 - t_2)\right] \langle \Psi_n|\hat{\psi}(q_1)|\Psi_m\rangle \langle \Psi_m|\hat{\psi}^+(q_2)|\Psi_n\rangle;$$

when $t_1 < t_2$,

$$G(1,2) = i \sum_{n,m} \exp\left[\beta(\Omega + \mu N_n - E_n)\right] \exp\left[i(E_m - E_n)(t_1 - t_2)\right] \times \langle \Psi_m|\hat{\psi}(q_1)|\Psi_n\rangle \langle \Psi_n|\hat{\psi}^+(q_2)|\Psi_m\rangle.$$

It is convenient to interchange the indices $n \rightleftarrows m$ in this last relation. Allowing for the connection between N_m and N_n, this gives

$$G(1,2) = -i \sum_{n,m} \exp\left[\beta(\Omega + \mu N_n - E_n)\right] \times \exp\left[-i(E_m - E_n)(t_1 - t_2)\right] \langle \Psi_n|\hat{\psi}(q_1)|\Psi_m\rangle \langle \Psi_m|\hat{\psi}^+(q_2)|\Psi_n\rangle \times \begin{cases} 1 & t_1 > t_2 \\ -\exp\left[\beta(E_n - E_m + \mu)\right] & t_1 < t_2. \end{cases} \quad (29.17)$$

If we take the Fourier transform with respect to $t_1 - t_2$, replace the summation over n and m by an integration over the difference $E_m - E_n = E$, and sum over the remaining variables, we arrive at the final spectral function:

$$G(q_1, q_2, \varepsilon) = \int_{-\infty}^{\infty} dE\, A(q_1, q_2, E) \times \left\{\frac{1}{\varepsilon - E + i\delta} + \frac{\exp\left[-\beta(E - \mu)\right]}{\varepsilon - E - i\delta}\right\}. \quad (29.18)$$

For a spatially uniform system we have simply

$$G(\mathbf{p}, \varepsilon) = \int_{-\infty}^{\infty} dE\, a(\mathbf{p}, E) \left\{\frac{1}{\varepsilon - E + i\delta} + \frac{\exp\left[-\beta(E - \mu)\right]}{\varepsilon - E - i\delta}\right\}, \quad (29.19)$$

where $a(\mathbf{p}, E) \geqq 0$.

We will separate a factor $\{1 + \exp[\beta(\mu - E)]\}^{-1}$ from $a(\mathbf{p}, E)$, and using

$$\frac{e^x}{1 + e^x} = \frac{1}{1 + e^{-x}},$$

we can rewrite (29.19) in the following form:

$$G(\mathbf{p}, \varepsilon) = \int_{-\infty}^{\infty} dE\, a'(\mathbf{p}, E) \left\{ \frac{1}{(\varepsilon - E + i\delta)\{1 + \exp[-\beta(E-\mu)]\}} + \frac{1}{(\varepsilon - E - i\delta)\{1 + \exp[\beta(E-\mu)]\}} \right\}. \tag{29.20}$$

It is particularly obvious from this relation how the limiting transition to the zero-temperature case takes place when $\beta \to \infty$.

Equation (29.20) provides, moreover, a simple physical interpretation of the Green function as representing the simultaneous propagation of free particles and holes with different energies E. As distinct from the case $T = 0$, the corresponding propagation functions here have additional factors

$$n = \{1 + \exp[\beta(E-\mu)]\}^{-1}; \quad 1 - n = \{1 + \exp[-\beta(E-\mu)]\}^{-1}$$

which represent the well-known, average occupation numbers for the particles and holes in the Fermi distribution. Hence the weights, which the particles and the holes in the Green function have, are determined not only by the dynamics of the system (the factor $a'(\mathbf{p}, E)$), but also by the occupation numbers of the free particles and holes with the same energy E.

29.4. We will examine some of the consequences of the spectral function. We will first establish the relationship between the real and imaginary parts of the Green function for real values of the energy ε.

It is easily seen from (29.19) that

$$\operatorname{Re} G(\mathbf{p}, \varepsilon) = \mathrm{P} \int_{-\infty}^{\infty} \frac{dE\, a(\mathbf{p}, E)}{\varepsilon - E} \{1 + \exp[-\beta(E-\mu)]\},$$

$$\operatorname{Im} G(\mathbf{p}, \varepsilon) = -\pi a(\mathbf{p}, \varepsilon)\{1 - \exp[-\beta(\varepsilon-\mu)]\}.$$

This latter equation shows that the sign of $\operatorname{Im} G(\mathbf{p}, \varepsilon)$ is the opposite of the sign of $\varepsilon - \mu$, and that both of these quantities go to zero together. Using the simple formula

$$\coth \frac{x}{2} = \frac{e^x + 1}{e^x - 1},$$

we arrive at the required equation [130]

$$\operatorname{Re} G(\boldsymbol{p}, \varepsilon) = -\frac{1}{\pi} \mathrm{P} \int_{-\infty}^{\infty} \frac{dE}{\varepsilon - E} \coth\left[\frac{\beta(E-\mu)}{2}\right] \operatorname{Im} G(\boldsymbol{p}, E), \tag{29.21}$$

which can be used to find the rules for integrating around singularities in the Green function. It is used for this in superconductivity theory [114].

When $T \neq 0$, the Green function does not have the same analytical properties in the complex region as the Green function in quantum mechanics (see § 23). This is due to the impossibility of representing the Green function as a Cauchy-type integral. As an illustration, we will write $G = \operatorname{Re} G + i \operatorname{Im} G$ in the following way, using (29.21):

$$G(\boldsymbol{p}, \varepsilon) = \frac{1}{\pi} \int_{-\infty}^{\infty} dE \operatorname{Im} G(\boldsymbol{p}, E) \left\{ -\coth \frac{\beta(E-\mu)}{2} \cdot \frac{1}{\varepsilon - E} + i\pi\delta(\varepsilon - E) \right\}.$$

When $T \neq 0$ ($\beta \neq \infty$), the terms in the curly brackets cannot be combined as the kernel of a Cauchy integral. Combination is possible for the function

$$G'(\boldsymbol{p}, \varepsilon) = \operatorname{Re} G(\boldsymbol{p}, \varepsilon) - i \coth \frac{\beta(\varepsilon-\mu)}{2} \operatorname{Im} G(\boldsymbol{p}, \varepsilon),$$

which has the same analytical properties as the Green function in quantum mechanics. In other words its analytical continuation into the upper ($\varepsilon > \mu$), and lower ($\varepsilon < \mu$), half-planes of the complex variable ε leads to analytical functions devoid of singularities. However, the analytical continuation of $G'(\boldsymbol{p}, \varepsilon)$ in the upper ($\varepsilon > \mu$), and lower ($\varepsilon < \mu$), half-plane leads unavoidably to functions with singularities. From the point of view of locating these singularities and their residues, it is immaterial whether we continue the function G' or $\operatorname{Im} G' = -\coth [\beta(\varepsilon - \mu)/2] \operatorname{Im} G$. The singularities of the factor $\coth [\beta(\varepsilon - \mu)/2]$, which differentiates the functions $\operatorname{Im} G$ and $\operatorname{Im} G'$, are at the points $\varepsilon = \mu + 2\pi n i$, where n is a whole number. These singularities do not affect the problem of finding the properties of the quasi-particles, since they represent, so to speak, the "zero" energy of a quasi-particle. Hence there is no essential difference between the functions G and G'.

The physical meaning attached to the analytical continuation of G in § 23, can be repeated for G'. From the location of the pole of the analytical continuation of G' (or G), we can determine the dispersion law and the attenuation of a quasi-particle (or elementary excitation) for a non-zero temperature. The corresponding physical information about the system is the same as for a zero temperature. It is only necessary to emphasize that the quasi-particle properties found are necessarily temperature dependent, since this is a property of the weight function $a(\boldsymbol{p}, E)$. Hence it is impossible to make a purely mechanical interpretation from the point of view of the energy spectrum of the total Hamiltonian. We must speak of some effective, statistically averaged, excitation spectrum describing slowly decaying motions in the system.

We often require in quantum statistics what are known as the advanced and retarded Green functions [5, 131, 132], such as, for example,

$$G_a(1, 2) = i\theta(t_2 - t_1)\,\mathrm{Tr}\,\{\zeta[\hat{\psi}_H(1), \hat{\psi}_H^+(2)]_+\}/\mathrm{Tr}\zeta$$

$$G_r(1, 2) = i\theta(t_1 - t_2)\,\mathrm{Tr}\,\{\zeta[\hat{\psi}_H(1), \hat{\psi}_H^+(2)]_+\}/\mathrm{Tr}\zeta.$$

These functions, unlike G, have the same analytical properties as the Green functions in quantum mechanics, and are closely related to the function G'.

30. The Hartree–Fock Approximation in Quantum Statistics

30.1. In order to make practical use of the general relationships introduced in previous sections, we must solve the dynamical problem including interactions between particles. This is no easier (except for very high temperatures) than in quantum mechanics. Hence, as at zero temperature, we must begin by choosing a zero approximation; it is convenient to choose the Hartree–Fock approximation. We can obtain this approximation by looking for the best single-particle approximation, i.e. we must require that the norm of the difference between the exact and the single-particle Hamiltonians, averaged over the state of the system concerned, has a minimum value. In actual fact all the states of a system figure in quantum statistics, not only one. We must therefore enquire what we mean by the "best approximation".

One might, in principle, demand the best description for each of the states figuring in the statistical equations (see § 29.1). Each of these states would then have its corresponding self-consistent field. This approach, however, might destroy the factorizability of the higher-order density matrix (see Appendix A), and prevent that mathematical simplicity which is essential for a zero approximation.

Hence, in quantum statistics, the Hartree–Fock approximation usually means a cruder approximation: a single, statistically averaged, self-consistent field for all configurations Ψ_n. This corresponds to demanding that the statistical average over the square of the difference of the exact Hamiltonian and the single-particle Hamiltonian is a minimum:

$$\langle(\hat{H} - \hat{H}_0)^2\rangle_0 = \min, \tag{30.1}$$

where the symbol $\langle\ldots\rangle_0$ denotes averaging in the single-particle approximation

$$\left.\begin{aligned} \langle\ldots\rangle_0 &= \frac{\mathrm{Tr}\,[\hat{\zeta}_0(\ldots)]}{\mathrm{Tr}\,\hat{\zeta}_0}, \\ \hat{\zeta}_0 &= \exp\,[\beta(\mu\hat{N} - \hat{H}_0)]. \end{aligned}\right\} \tag{30.2}$$

As in Chapter 2, we can take $\hat{H}_0$ in the form

$$\hat{H}_0 = \int dq\, \hat{\psi}^+(q)\,(\hat{T} + \hat{W})\,\hat{\psi}(q) + C.$$

Then the operator $\hat{\zeta}_0$ can be written as

$$\hat{\zeta}_0 = \exp\,\{-\beta C + \beta \sum_\nu (\mu - \varepsilon_\nu)\,\hat{A}_\nu^+ \hat{A}_\nu\},$$

where ε_ν is the energy of the single-particle state

$$(\hat{T} + \hat{W})\,\chi_\nu(q) = \varepsilon_\nu \chi_\nu(q).$$

We will not, as yet, change to the hole formulation.

We now introduce rules for averaging, using the operator ζ_0. We have

$$\langle \hat{A}_\mu^+ \hat{A}_\nu \rangle_0 = \delta_{\mu\nu} \langle \hat{A}_\nu^+ \hat{A}_\nu \rangle_0 = -\frac{\delta_{\mu\nu}}{\beta} \frac{\partial \operatorname{Tr}\zeta_0'/\partial \varepsilon_\nu}{\operatorname{Tr}\zeta_0'},$$

where

$$\zeta_0' \equiv \zeta_0 \exp(\beta C).$$

Similarly,

$$\langle \hat{A}_\mu^+ \hat{A}_\nu^+ \hat{A}_\sigma \hat{A}_\tau \rangle_0 = (\delta_{\nu\sigma}\delta_{\mu\tau} - \delta_{\nu\tau}\delta_{\mu\sigma}) \frac{1}{\beta^2} \frac{\partial^2 \operatorname{Tr}\zeta_0'/\partial \varepsilon_\mu \, \partial \varepsilon_\nu}{\operatorname{Tr}\zeta_0'},$$

etc. It reduces to calculating the partition function $Z_0 = \operatorname{Tr}\zeta_0$. Writing it in the form

$$Z_0 = \sum_n \exp[\beta(\mu N_n - E_n)]$$

and putting $N_n = \sum_\nu n_\nu$, $E_n = \sum_\nu \varepsilon_\nu n_\nu + C$, where n_ν is the occupation number for the level ν, we have

$$Z_0 = \exp(-\beta C) \sum_n \prod_\nu \{\exp[\beta(\mu - \varepsilon_\nu)]\}^{n_\nu}$$

$$= \exp(-\beta C) \prod_\nu \{1 + \exp[\beta(\mu - \varepsilon_\nu)]\}. \qquad (30.3)$$

We have used here the fact that, if we change the occupation number n_ν at each level in an arbitrary way, i.e. put n_ν equal to 0 and 1, we can work our way through all states n.

We thus arrive at the expressions

$$\langle \hat{A}_\mu^+ \hat{A}_\nu \rangle_0 = \delta_{\mu\nu} \bar{n}_\nu,$$

$$\langle \hat{A}_\mu^+ \hat{A}_\nu^+ \hat{A}_\sigma \hat{A}_\tau \rangle_0 = (\delta_{\nu\sigma}\delta_{\mu\tau} - \delta_{\nu\tau}\delta_{\mu\sigma}) \bar{n}_\mu \bar{n}_\nu,$$

etc., where we have introduced the average occupation number of a state

$$\bar{n}_\nu = \{1 + \exp[\beta(\varepsilon_\nu - \mu)]\}^{-1}. \qquad (30.4)$$

Comparing these expressions with the results in Appendix A and § 4, we see that, if we introduce the occupation operator $\hat{\varrho}$ into the statistics,

$$\hat{\varrho}\chi_\nu(q) = \bar{n}_\nu \chi_\nu(q),$$

where, explicitly,

$$\hat{\varrho} = \{1 + \exp[\beta(\hat{T} + \hat{W} - \mu)]\}^{-1}, \tag{30.5}$$

then all the rules for averaging, which were used in quantum mechanics, remain in force here (the factor βC is eliminated in the averaging). In particular, for single-particle and two-particle operators, we have

$$\langle\hat{\alpha}_1\rangle_0 = \mathrm{Tr}\,\hat{a}_1\,\hat{\varrho},$$

$$\langle\hat{\alpha}_2\rangle_0 = \tfrac{1}{2}\,\mathrm{Tr}\,[\hat{a}_2(1 - \hat{\mathscr{P}}_{12})\,\hat{\varrho}_1\hat{\varrho}_2].$$

Varying (30.1) with respect to $\hat{H}_0$ leads to the equations

$$\hat{W} = \mathrm{Tr}_{q'}[\hat{V}(q, q')\,(1 - \hat{\mathscr{P}}_{qq'})\,\hat{\varrho}_{q'}], \tag{30.6}$$

$$C = -\tfrac{1}{2}\,\mathrm{Tr}\,[(\hat{V}q, q')\,(1 - \hat{\mathscr{P}}_{qq'})\,\hat{\varrho}_q\hat{\varrho}_{q'}], \tag{30.7}$$

which completely determine the Hamiltonian in the Hartree–Fock approximation. The matrix elements of the operator $\hat{\varrho}$, as before, give the density matrix of the system (in the Hartree–Fock approximation)

$$R_0(q, q') = \sum_\nu \bar{n}_\nu \chi_\nu^*(q')\,\chi_\nu(q). \tag{30.8}$$

We can transform to the distribution function $f(\boldsymbol{x}, \boldsymbol{p})$ in the usual way (see § 4). The relationships (4.29) to (4.34) of Chapter 2 also remain in force, so long as we alter the occupation operator appropriately.

It is not difficult to write an expression for the Green function in the Hartree–Fock approximation. It is simply necessary to replace the quantity n_ν in the equations of § 10 by

$$\{\exp[\beta(\varepsilon_\nu - \mu)] + 1\}^{-1}.$$

In particular, for a uniform system,

$$G_0(\boldsymbol{p}, \varepsilon) = \frac{1 - n_p}{\varepsilon - \varepsilon_p + i\delta} + \frac{n_p}{\varepsilon - \varepsilon_p - i\delta}. \tag{30.9}$$

Here n_p is the average occupation number of the level, equal to

$$\{\exp[\beta(\varepsilon_p - \mu)] + 1\}^{-1}.$$

The chemical potential μ can be found from the normalization condition

$$g \int d^3p \, n_p = \varrho.$$

It is evident that (30.9) agrees with the general spectral function (29.21), and corresponds to the choice

$$\operatorname{Im} G(p, \varepsilon) = \pi(2n_p - 1)\, \delta(\varepsilon - \varepsilon_p).$$

We now return to the expression for the average occupation number. This depends only on the energy, and all substates corresponding to a given energy value are occupied to the same extent. This represents a link between a statistical system and a system at absolute zero with a closed shell. As a result of this, subsequent construction of a perturbation theory is not complicated by degeneracy of the levels in the zero approximation.

30.2. One of the main questions we must study is the calculation of the energy and the other thermodynamical quantities in the Hartree–Fock approximation. The average energy of a system can be defined first of all as the average value of the Hamiltonian $\hat{H}_0$ (the dynamical definition of the energy)

$$E_0 = \langle \hat{H}_0 \rangle_0$$

and, secondly, purely thermodynamically,

$$E_0 = \left(1 + \beta \frac{\partial}{\partial \beta} - \mu \frac{\partial}{\partial \mu}\right) \Omega_0 .$$

These definitions do not coincide in the Hartree–Fock approximation. This is connected with the fact that the effective Hamiltonian of the approximation $\hat{H}_0$, as distinct from the exact Hamiltonian $\hat{H}$, is itself dependent on β and μ.

We will consider the thermodynamical definition of the energy, substituting the quantity $-(1/\beta) \ln \operatorname{Tr} \{\exp [\beta(\mu \hat{N} - \hat{H}_0)]\}$ for Ω_0. A straightforward calculation gives

$$E_0 = \langle \exp(\beta \hat{H}_0) \left(\frac{\mu}{\beta} \frac{\partial}{\partial \mu} - \frac{\partial}{\partial \beta}\right) \exp(-\beta \hat{H}_0) \rangle_0 .$$

If the Hamiltonian $\hat{H}_0$ is independent of β and μ, we return to the dynamical definition. In the present case, however, we must in addition differentiate $\hat{H}_0$ with respect to β and μ.

The disagreement between the two definitions of the energy can be explained by the simplification introduced in the self-consistent procedure. If this procedure had been carried out for each of the configurations Ψ_n, then the effective Hamiltonian $\hat{H}_0$ (different for the different configurations) would be independent of β and μ.

Many thermodynamical relations lose their significance in the Hartree–Fock approximation for this reason; notably, the differential relations of (29.8). Suppose we consider, for example, the derivative

$$(\partial\Omega/\partial\mu)_T = -N - \frac{1}{\beta}\langle \exp(\beta\hat{H}_0)\frac{\partial}{\partial\mu}\exp(-\beta\hat{H}_0)\rangle_0 .$$

Because $\hat{H}_0$ depends on μ, the second term on the right-hand side of these equality is different from zero.† It is therefore impossible in this approximation to use thermodynamical definitions of physical quantities which are based on equations of the type (29.8). One must thus either rely on the dynamical definitions, particularly

$$E_0 = \langle \hat{H}_0\rangle_0 \quad \text{and} \quad N = \langle \hat{N}\rangle_0,$$

or use the thermodynamical relationships, formally supposing $\hat{W}$ and C to be independent of μ and β. This means that the necessary differentiations with respect to μ and β should be carried out as far as the self-consistent procedure which leads to the dependence of $\hat{W}$ and C on μ and β.

Suppose we consider the pressure P from this standpoint. This quantity can be defined in a purely dynamical way as the force per unit area acting on the system; such an approach has been taken in some papers [70]. The other possibility is to use a purely thermodynamical relationship

$$P = -\Omega_0/V,$$

† Numerous general relationships of the type $\partial P/\partial\mu = \varrho$ break down for this reason in quantum mechanics.

which can only be derived if $\hat{W}$ and C are assumed to be independent of μ and β.

A direct calculation of the thermodynamic potential Ω_0 leads to†

$$\Omega_0 = -\int_{-\infty}^{\mu} d\mu \operatorname{Tr} \hat{\varrho} + C, \tag{30.10}$$

which can be obtained using the equality

$$\ln \{1 + \exp [\beta(\mu - \varepsilon_\nu)]\} = \beta \int_{-\infty}^{\mu} d\mu \, n_\nu .$$

The quantity $C = -\frac{1}{2} \operatorname{Tr} \hat{W}\hat{\varrho}$, as in § 4, can be put in a simpler form if we eliminate the exchange operator $\hat{A}$ [51].

30.3. The theoretical apparatus for slightly inhomogeneous systems can be simplified by going over to the quasi-classical approximation. As for the zero-temperature case, it reduces to neglecting the commutators of terms in the Hamiltonian $\hat{T} + \hat{W}$.

Let us consider the limits of applicability of the quasi-classical approximation. One can say, in general, that the non-zero temperature makes the quasi-classical condition more applicable. This is because, in a "heated" medium, the kinetic energy and the mean momentum of the particles is higher (other things being equal) than for a cold system. The particles correspondingly have a smaller de Broglie wavelength.

The appropriate parameter ξ^2 (see § 5) has different expressions for a degenerate gas ($\beta\varrho^{2/3}/M \gg 1$) and for high temperatures ($\beta\varrho^{2/3}/M \ll 1$), where Boltzmann statistics hold. In the first case the estimates made in § 5 may be used

$$\xi^2 \sim \varrho^{-1/3}, \tag{30.11}$$

in the second [70]

$$\xi^2 \sim \varrho\beta^2. \tag{30.12}$$

The dependence of the quantum parameter ξ^2 on the density, unlike its dependence on the temperature, is non-monotonic: for small densities—corresponding to the Boltzmann case—ξ^2 in-

† The integration is carried out for fixed $\hat{W}$ and C.

creases with the density, reaching a maximum at the edge of the degeneracy zone, and then falling off. The approximate form of lines of constant ξ in the (T, ϱ) coordinates is given in Fig. 66.

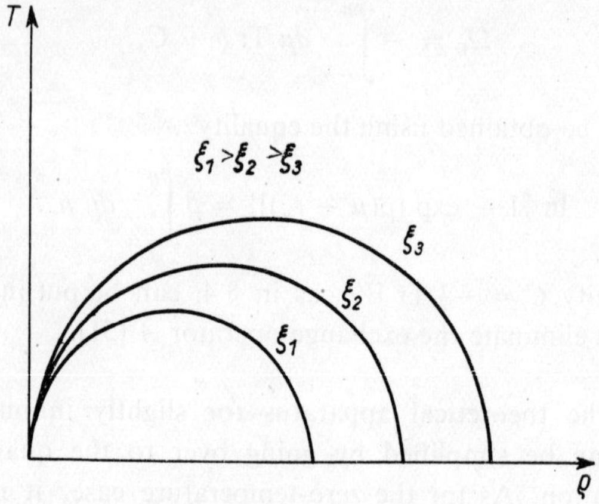

FIG. 66

The relative role of exchange effects for systems with Coulomb interactions is determined by the parameter

$$2MA/p_0^2 \sim e^2\varrho/p_0^2, \tag{30.13}$$

where, as before,†

$$p_0^2 = 2M(\mu - B - U). \tag{30.14}$$

This parameter is of the same order of magnitude as ξ^2 for all ϱ and T. Hence, for a non-zero temperature too, we cannot justify the Thomas–Fermi–Dirac method; the quantum and exchange effects must be considered on an equal footing.

If we ignore the exchange effects, we can write the Hartree equation in the following operator form ($\lambda = \beta/2M$):

$$\left.\begin{aligned} f(x, p) &= \langle\{\exp[\lambda(\hat{p}^2 - p_0^2(x))] + 1\}^{-1}\rangle_p, \\ \varrho(x) &= g \int d^3p\, f(x, p), \end{aligned}\right\} \tag{30.15}$$

and

$$\nabla^2 p_0^2 = \frac{8\pi}{a_0}[\varrho(x) - \sigma(x)].$$

† In this case p_0^2 cannot be interpreted as the square of the Fermi momentum, but simply acts as some characteristic quantity with the same dimensions.

If the parameter ξ^2 is sufficiently small, then, ignoring the fact that the operators $\hat{p}^2$ and $p_0^2(x)$ do not commute, we find the following expression for the distribution function:

$$f(x, p) = \{\exp [\lambda(p^2 - p_0^2(x))] + 1\}^{-1}. \tag{30.16}$$

The density has the following form as a function of p_0^2:

$$\varrho = (2\pi^2\lambda^{3/2})^{-1} I_{1/2}(\lambda p_0^2). \tag{30.17}$$

We are introducing here the Fermi–Dirac function [133]

$$I_n(x) = \int_0^\infty \frac{y^n\, dy}{\exp (y - x) + 1}, \tag{30.18}$$

which satisfies the relation

$$I_n' = nI_{n-1} \tag{30.19}$$

and has the asymptotic value for $x \to \infty$

$$I_n(x) = \frac{x^{n+1}}{n+1} + \frac{\pi^2}{6} nx^{n-1} + \cdots \tag{30.20}$$

and for $x \to -\infty$

$$I_n(x) = \Gamma(n+1)\, e^x. \tag{30.21}$$

We must look at the question of how we can go to the limit $\beta \to \infty$ and $\beta \to 0$ in (30.17). In the first case $\mu \to \varepsilon_F$, $\lambda p_0^2 \to \infty$, and using (30.20), we get back immediately to the usual expression for ϱ. The situation for $\beta \to 0$ is more complicated. In this case the chemical potential μ is a rapidly varying function of β. We will suppose that, as $\beta \to 0$, the quantity $\lambda p_0^2 \sim \lambda\mu \to -\infty$. Then (30.17) takes the form

$$\varrho \sim \lambda^{-3/2} e^{\lambda\mu},$$

whence it is evident that, as $\lambda \to 0$, $\lambda\mu \to -\infty$ as postulated, and, consequently, $\mu \to -\infty$. Hence, in the high-temperature limit, we must use the asymptotic value (30.21), which reduces to the Boltzmann equation for the density distribution.

30.4. We will consider the equation of state for a very dense "hot" substance, restricting ourselves to a region where the electron distribution is practically uniform (for the more general case, see [70], which we follow here). The distribution may be considered uniform either when there is a high density, or when there is a high temperature. Qualitatively, the boundary of the uniform region coincides with the curves of Fig. 66. Within this region, interaction contributes very little to the equation of state. It follows from the results of § 30.2 that, if the term C is neglected,

$$P = \int_{-\infty}^{\mu} d\mu\, \bar{\varrho}.$$

Using (30.17) and (30.19), we obtain

$$P = \tfrac{2}{3}\mathscr{E}_k = (6\pi M\lambda^{5/2})^{-1}\, I_{3/2}(\lambda p_0^2). \tag{30.22}$$

This relation, together with (30.17) gives a parametric representation of $P(\varrho, T)$. As $\beta \to \infty$, we return to the relations of § 6; as $\beta \to 0$, we obtain the usual equation $P = \varrho kT$. An estimate of the errors due to inhomogeneities in the distribution leads to the result

$$\delta P \approx \theta Z^{2/3} \varrho^{4/3},$$

where the coefficient θ was defined in § 6. We must emphasize that δP is not explicity dependent on the temperature.

We now introduce some results concerning the contribution of exchange and quantum effects to the pressure. It seems that the quantum and exchange corrections to the pressure are always negative for all temperatures and densities, and their ratio does not exceed $\frac{1}{3}$. For a degenerate gas, this ratio becomes $\frac{2}{9}$ (see § 6); it achieves its maximum value of $\frac{1}{3}$ at high temperatures.

So far as the expressions for the quantum and exchange corrections themselves are concerned, we will restrict ourselves to the region where the distribution is uniform. The ratio of the correction to the actual pressure in this region is ($i = 1, 2$):

$$\frac{\delta_i P}{P} = -\frac{1}{2\sqrt{2}} \left(\frac{e^2\beta}{a_0}\right)^{1/2} \frac{I_{1/2}^2(x)}{\pi I_{3/2}(x)\, I'_{1/2}(x)} \frac{d}{dx} \left(\frac{\int_{-\infty}^{x} \Psi_i(x)\, dx}{I_{1/2}(x)} \right) \tag{30.23}$$

Here $x = \lambda p_0^2$ is determined from (30.17): $\Psi_1 = I_{1/2}I''_{1/2} + (I'_{1/2})^2$, $\Psi_2 = 6(I'_{1/2})^2$; the suffix 1 refers to the quantum correction, and 2 to the exchange correction. For low temperatures,

$$\frac{\delta_1 P}{P} = -\frac{5}{18\pi(3\pi^2)^{1/3}\varrho^{1/3}a_0}$$

which corresponds to the results of § 6. In the high-temperature region

$$\frac{\delta_1 P}{P} = -\frac{\pi}{6}\varrho\left(\frac{e^2\beta}{a_0}\right)^2 a_0^3.$$

The corresponding exchange corrections can be then obtained by a simple recalculation.

We can use these results to decide on the region within which the quasi-classical equation of state for a substance is applicable.

31. Thermodynamical Perturbation Theory

31.1. It is by no means always possible to restrict ourselves to the Hartree–Fock approximation in solving statistical problems. We must, therefore, as in quantum mechanics, consider methods of including the correlation interactions between particles.

If we look at the expression for the Green function (29.16), it is fairly obvious that the interaction between the particles enters into this expression in two ways. On the one hand, the interaction affects the operators $\hat{\psi}_H$ and $\hat{\psi}_H^+$, on the other, the operator ζ, which determines the statistical averaging, is dependent on it. We will formulate later a single method for including both aspects of the interaction.

We will only consider the statistical operator ζ in this section. Besides a discussion the systematic approach developed first by Matsubara [134], there is also the possibility of constructing the thermodynamics of a system of interacting particles.

We will consider the expression for the statistical operator

$$\zeta = \exp\left[-\beta(\hat{H} - \mu\hat{N})\right], \tag{31.1}$$

where the operators $\hat{H}$ and $\hat{N}$ are taken in the second-quantization and Schrödinger representations, and are of the same form as those in § 3. It is convenient to subdivide the field operators $\hat{\psi}(q)$ and $\hat{\psi}^+(q)$ into creation and annihilation parts in statistics in a way different from that which was used in § 8. We put

$$\left.\begin{aligned} \hat{\psi}(q) &= \sum_\nu \chi_\nu(q) \,[(1 - \sqrt{n_\nu})\, \hat{a}_\nu + \sqrt{(n_\nu)}\, \hat{b}_\nu^+], \\ \hat{\psi}^+(q) &= \sum_\nu \chi_\nu^*(q) \,[(1 - \sqrt{n_\nu})\, \hat{a}_\nu^+ + \sqrt{(n_\nu)}\, \hat{b}_\nu], \end{aligned}\right\} \tag{31.2}$$

where $n_\nu = \{\exp[\beta(\varepsilon_\nu - \mu)] + 1\}^{-1}$. The operators $\hat{a}_\nu$, $\hat{a}_\nu^+$ and $\hat{b}_\nu$, $\hat{b}_\nu^+$ will be called the particle and hole operators. If we consider the ground state of a system with the particles distributed over the levels according to $n_\nu = \theta(\varepsilon_F - \varepsilon_\nu)$, a free level below the Fermi surface is considered as a hole, one above the Fermi surface as a particle. When the temperature differs from zero, the average particle occupation is given by the expression for $n_\nu(\beta)$ introduced above. We find in this case that there is no clear division between the regions containing particles and holes. There is, in principle, a possibility of observing a particle, or a hole, at each level. The factors $\sqrt{n_\nu}$ and $1 - \sqrt{n_\nu}$ reflect this: they transform to $\theta(\varepsilon_F - \varepsilon_\nu)$ and $\theta(\varepsilon_\nu - \varepsilon_F)$ for $\beta \to \infty$.†

We will employ the normal product of the field operators as before, but using the new division of the operators into creation and annihilation parts. It can easily be seen that

$$\hat{\psi}^+(q')\,\hat{\psi}(q) = N[\hat{\psi}^+(q')\,\hat{\psi}(q)] + R(q, q'),$$

where $R(q, q')$ is the density matrix (30.8). We can generalize (8.10) for the N-product of four field operators in the same way. The choice of the coefficients $\sqrt{n_\nu}$ and $1 - \sqrt{n_\nu}$ is here shown to be justified.

We can now obtain expressions for the Hamiltonian and for the particle number operator

$$\hat{H} = \hat{H}_0 + \hat{H}' \tag{31.3}$$

$$\hat{H}_0 = E_0 + \int dq N[\hat{\psi}^+(q)\,(\hat{T} + \hat{W})\,\hat{\psi}(q)], \tag{31.4}$$

$$\hat{H}' = \tfrac{1}{2} \int dq\, dq'\, N[\hat{\psi}^+(q)\,\hat{\psi}^+(q')\,\hat{V}\hat{\psi}(q')\,\hat{\psi}(q)], \tag{31.5}$$

$$\hat{N} = N + \int dq\, N[\hat{\psi}^+(q)\,\hat{\psi}(q)]. \tag{31.6}$$

† It is necessary to introduce the root of n_ν because the operator $\hat{n}_\nu$ is a bilinear combination of the ψ and ψ^+.

Here E_0 is the average energy in the Hartree–Fock approximation; N is the average number of particles in the system. These relationships enable us to separate the total Hamiltonian of the system into a zero-approximation Hamiltonian $\hat{H}_0$ and the perturbation Hamiltonian $\hat{H}'$.

We will rewrite (31.1) in the form

$$\hat{\zeta} = \exp\left[-\beta(\hat{H}_0 - \mu\hat{N} + \hat{H}')\right]. \tag{31.1'}$$

If we formally differentiate both parts of (31.1') with respect to β, which we consider as a variable, then we obtain what is known as Bloch's equation

$$-\frac{\partial \hat{\zeta}}{\partial \beta} = (\hat{H}_0 - \mu\hat{N} + \hat{H}')\,\hat{\zeta}, \tag{31.7}$$

which is analogous to Schrödinger's equation with the replacement $it \to \beta$. Matsubara also started from this analogy.

31.2. The aim of the account which now follows is to construct a special interaction representation in terms of the temperature, so that we can introduce the standard diagram technique for calculating the trace of the operator $\hat{\zeta}$.

If we ignore the correlation interaction, this operator has the form

$$\hat{\zeta}_0(\beta) = \exp\left[-\beta(\hat{H}_0 - \mu\hat{N})\right]. \tag{31.8}$$

We will represent $\hat{\zeta}(\beta)$ by

$$\hat{\zeta}(\beta) = \hat{\zeta}_0(\beta)\,\hat{S}(\beta), \tag{31.9}$$

where $\hat{S}(\beta)$ is some unknown operator. Then we obtain for $\hat{S}$ the following equation†

$$-\frac{\partial \hat{S}(\beta)}{\partial \beta} = \hat{H}'(\beta)\,\hat{S}(\beta), \tag{31.10}$$

where

$$\hat{H}'(\beta) = \hat{\zeta}_0^{-1}(\beta)\,\hat{H}'\hat{\zeta}_0(\beta). \tag{31.11}$$

The analogy with the transformation to the normal interaction representation in quantum mechanics is particularly obvious here:

† It is only necessary to differentiate with respect to the parameter β occurring in the exponent. $\hat{H}_0$ and $\hat{H}'$ also depend on the actual inverse temperature of the system which we will denote by β_0 in the future.

$\hat{S}(\beta)$ is the analogue of the S-matrix, $\zeta_0(\beta)$ is the analogue of the transition operator $\exp(-i\hat{H}_0 t)$.

From the obvious boundary condition

$$\zeta(0) = \zeta_0(0) = 1$$

we have

$$\hat{S}(0) = 1. \tag{31.12}$$

We can thus say that $\hat{S}(\beta)$ represents the evolution of a system in terms of the temperature from the point $\beta = 0$, where interaction plays no part, up to the value β.†

Due to the formal identity (so far as the replacement $it \to \beta$ holds) of eqn. (31.10) and the equation for the normal S-matrix, we can immediately represent $\hat{S}(\beta)$ as a sum of T-products

$$\hat{S}(\beta_0) = T \exp\left\{-\int_0^{\beta_0} d\beta\, \hat{H}'(\beta)\right\} \tag{31.13}$$

or

$$\hat{S}(\beta_0) = \sum_{n=1}^{\infty} \frac{(-1)^n}{n!} \int_0^{\beta_0} d\beta_1 \ldots d\beta_n T[\hat{H}'(\beta_1) \ldots \hat{H}'(\beta_n)].$$

The time sequence must here be understood as an ordering in terms of the parameter β: all the factors under the sign of the T-product must follow the order in which β increases, from right to left.

If we know $\hat{S}(\beta_0)$, we can obtain complete thermodynamical information for a given system. Denoting by $Z_0 = \operatorname{Tr} \zeta_0$ the partition function for a system in the Hartree–Fock approximation, we can write

$$\frac{Z}{Z_0} = \frac{\operatorname{Tr} \zeta}{\operatorname{Tr} \zeta_0} \equiv \langle \hat{S}(\beta_0) \rangle_0. \tag{31.14}$$

The symbol $\langle \ldots \rangle_0$ here denotes an average over the statistical operator in the Hartree–Fock approximation.

† For an infinitely high temperature, all many-body systems (considered non-relativistically) transform to a perfect gas. This corresponds to the interaction being switched off for $t = \pm\infty$ in quantum mechanics, and is less formal than the adiabatic hypothesis.

Thus, to find the partition function and, consequently, the remaining thermodynamical quantities, we must know the average value of the S-matrix $\hat{S}(\beta_0)$. Equation (31.13) will serve as the initial basis for our further work.

31.3. We must now eliminate "virtual" operators, and reduce the T-products to N-products.

We will use the explicit form for the field operators in the "temperature-dependent" interaction representation

$$\left.\begin{aligned} \hat{\psi}(q,\beta) &= \zeta_0^{-1}\hat{\psi}(q)\,\zeta_0 = \sum_\nu [\hat{a}_\nu(1-\sqrt{n_\nu}) + \hat{b}_\nu^+\sqrt{n_\nu}] \\ &\quad \times \exp[-\beta(\varepsilon_\nu - \mu)], \\ \hat{\psi}^+(q,\beta) &= \sum_\nu [\hat{a}_\nu^+(1-\sqrt{n_\nu}) + \hat{b}_\nu\sqrt{n_\nu}]\exp[\beta(\varepsilon_\nu-\mu)]. \end{aligned}\right\} \tag{31.15}$$

We are using here the commutation rules

$$[\hat{H}_0, \hat{a}_\nu]_- = -\varepsilon_\nu\hat{a}_\nu; \quad [\hat{H}_0, \hat{a}_\nu^+]_- = \varepsilon_\nu\hat{a}_\nu^+,$$

etc. The operators $\hat{\psi}^+(q,\beta)$ and $\hat{\psi}(q,\beta)$ are not the Hermitian conjugates of each other. This fact, which is a consequence of the non-unitary character of the operator ζ_0, is, however, unimportant.

We can also introduce the concept of a normal product, defined as above, for the operators $\hat{\psi}(q,\beta)$ and $\hat{\psi}^+(q,\beta)$. The perturbation Hamiltonian $\hat{H}'(\beta)$, which enters into the relations of the preceding section, can then be written as

$$\hat{H}'(\beta) = \tfrac{1}{2}\int dq\,dq'\,N[\hat{\psi}^+(q,\beta)\,\hat{\psi}^+(q',\beta)\,\hat{V}\hat{\psi}(q',\beta)\,\hat{\psi}(q,\beta)]. \tag{31.16}$$

And here the parameter β is different from the actual temperature of the system appearing in n_ν.

It is important to find the average value of the normal product. This quantity goes to zero in quantum mechanics, when averaged over the ground state of the system, which makes the calculations far simpler. The present case is a good deal more complicated, since the averaging covers all states, including states where the number of particles and holes differs from zero. Nevertheless, we can put

$$\langle N[\hat{\psi}(\mathrm{I})\,\hat{\psi}^+(\mathrm{II})\ldots]\rangle_0 = 0, \tag{31.17}$$

where the roman numerals indicate a combination of the q-coordinates and the reciprocal temperature β.

We will consider first the N-product of a pair of operators. Since the states on the right and left are the same when the averaging takes place over each of the configurations Ψ_{0n}, it is sufficient if we consider only the expression $\langle N[\hat{\psi}^+(\mathrm{I})\,\hat{\psi}(\mathrm{II})]\rangle_0$; for the same reason, there are essentially only two terms

$$\langle[\hat{\psi}^+(\mathrm{I})_{(+)}\,\hat{\psi}(\mathrm{II})_{(-)} - \hat{\psi}(\mathrm{II})_{(+)}\,\hat{\psi}^+(\mathrm{I})_{(-)}]\rangle_0.$$

Substituting in the expansion (31.15), we find

$$\langle N[\hat{\psi}^+(\mathrm{I})\,\hat{\psi}(\mathrm{II})]\rangle_0 = \sum_\nu \chi_\nu^*(q_1)\,\chi_\nu(q_2)$$
$$\times \exp\left[(\beta_1 - \beta_2)(\varepsilon_\nu - \mu)\right]\{\langle\hat{a}_\nu^+\hat{a}_\nu\rangle_0 (1 - \sqrt{n_\nu})^2 - \langle\hat{b}_\nu^+\hat{b}_\nu\rangle_0\, n_\nu\}. \tag{31.18}$$

The quantity n_ν represents the mean value of the operator $\hat{A}_\nu^+\hat{A}^\nu$ (see § 3) or, after a transition to the hole description,

$$n_\nu = \langle\hat{A}_\nu^+\hat{A}_\nu\rangle = (1 - \sqrt{n_\nu})^2\,\langle\hat{a}_\nu^+\hat{a}_\nu\rangle_0 + n_\nu\langle\hat{b}_\nu\hat{b}_\nu^+\rangle_0.$$

Hence, with $\langle\hat{b}\hat{b}^+\rangle_0 = 1 - \langle\hat{b}^+\hat{b}\rangle_0$, we have

$$(1 - \sqrt{n_\nu})^2\,\langle\hat{a}_\nu^+\hat{a}_\nu\rangle_0 = n_\nu\langle\hat{b}_\nu^+\hat{b}_\nu\rangle_0,$$

which shows expression (31.18) to be equal to zero. From this, we can write

$$\langle N(\hat{A}_\nu^+\hat{A}_\nu)\rangle_0 = 0, \tag{31.17'}$$

i.e. the theorem of (31.17) is also true for individual terms corresponding to a given state ν.

We will expand each operator in the general expression

$$\langle N(\hat{\psi}^+(\mathrm{I})\,\hat{\psi}(\mathrm{II})\ldots)\rangle_0$$

in terms of the states ν. We obtain a series of terms; our future discussion will refer to any one of them. If one of the terms has at least a pair of identical operators, $\hat{A}_\nu\hat{A}_\nu$ or $\hat{A}_\nu^+\hat{A}_\nu^+$, then it goes to zero identically from the commutation properties of these operators. The case where the operators appear as pairs referring to different states is particularly important

$$\langle N(\hat{A}_\nu^+\hat{A}_\nu\hat{A}_{\nu'}^+\hat{A}_{\nu'}\ldots)\rangle_0.$$

But the statistical averaging in the Hartree–Fock approximation is performed independently over the different states. Since $\nu \neq \nu'$, we can place the operators of the different pairs arbitrarily relative to each other. Therefore, this last expression reduces to the product $\langle N(\hat{A}_\nu^+ \hat{A}_\nu)\rangle_0 \langle N(\hat{A}_{\nu'}^+ \hat{A}_{\nu'})\rangle_0 \ldots$, i.e. according to (31.17′), we obtain a zero result. We may thus assume the theorem to be proved.

31.4. We must next introduce the concept of a temperature contraction of the field operators in the interaction representation. We will define it in the usual way (T is the symbol of ordering in terms of β)

$$\overbrace{\hat{\psi}(\mathrm{I})\, \hat{\psi}^+}(\mathrm{II}) = T(\hat{\psi}\hat{\psi}^+) - N(\hat{\psi}\hat{\psi}^+) = \langle T(\hat{\psi}\hat{\psi}^+)\rangle_0. \tag{31.19}$$

We have here used the theorem proved above. The basic properties of the contraction, which were noted in § 10, remain true here. A straightforward calculation gives

$$\overbrace{\hat{\psi}(\mathrm{I})\, \hat{\psi}^+}(\mathrm{II}) = \sum_\nu \chi_\nu^*(q_2)\, \chi_\nu(q_1)$$
$$\times \exp\left[-(\varepsilon_\nu - \mu)(\beta_1 - \beta_2)\right] \begin{cases} 1 - n_\nu & \beta_1 > \beta_2 \\ -n_\nu & \beta_1 > \beta_2 \end{cases} \tag{31.20}$$

This expression can be obtained from (10.5) by the substitution

$$it_{1,2} \to \beta_{1,2};\ \varepsilon_\nu \to \varepsilon_\nu - \mu.\dagger$$

Wick's rules, which reflect the general algebraic properties of operators are also true in quantum statistics. We can thus exclude virtual operators, and so reduce the T-products in the $\hat{S}(\beta)$ expansion to normal products.

We can use all of the diagram technique in this temperature-dependent theory; this is an important advantage of the field theoretical approach over the old thermodynamical perturbation theory. The Feynman rules in the coordinate representation are changed slightly as follows:

† This would reduce simply to making the change $it_{1,2} \to \beta_{1,2}$ if we defined the operators in the interaction representation using the operator $\hat{U} = \exp[i(\hat{H}_0 - \mu\hat{N})\,t]$ as is often done in statistics. The operator $\hat{U} = \exp[i(\hat{H} - \mu\hat{N})\,t]$ would similarly serve for the transition to the Heisenberg representation. We note that it is usually the energy difference that is found in physical entities, so the actual difference is unimportant.

(a) the factor $(-i)^n$ is replaced by $(-1)^n$;
(b) the time integration from $-\infty$ to ∞ is replaced by an integration over β from 0 to β_0;
(c) the usual contraction of the operators is replaced by the temperature contraction.

The Feynman rules in the momentum representation (which are a good deal more convenient to apply to spatially uniform systems) will be considered in the next section.

We now return to the problem of calculating the partition function and consider the expression

$$\frac{Z}{Z_0} = \langle \hat{S}(\beta_0) \rangle_0 .$$

If we use the theorem that the average value of any N-product is equal to zero (apart, of course, from $N(1) = 1$), we see that only vacuum fluctuation diagrams give any contribution out of all the perturbation theory diagrams in Z. These diagrams can be written in exponential form (see § 14)

$$\frac{Z}{Z_0} = \exp L,$$

where L is the sum of all the connected vacuum-fluctuation diagrams. We can thus arrive at a simple expression for the thermodynamic potential Ω:

$$\Omega = \Omega_0 - \frac{1}{\beta} L, \tag{31.21}$$

where Ω_0 is the thermodynamic potential in the Hartree–Fock approximation.

31.5. The basic difference between the diagram technique in the momentum representation for $T \neq 0$, and the technique considered in § 13, is that the limits of the integration over β are finite. Otherwise, the corresponding Feynman rules hold as before.

We will consider the temperature-dependent free-particle Green function

$$G_0(\mathrm{I}, \mathrm{II}) = -i\overbrace{\hat{\psi}(\mathrm{I})\, \hat{\psi}^+(\mathrm{II})}. \tag{31.22}$$

This depends only on the difference $\beta = \beta_1 - \beta_2$, and is defined as a function of this difference in the interval from $-\beta_0$ to β_0.

It is expedient to continue G_0 in terms of β periodically along the entire β-axis. We can then expand $G_0(q_1, q_2, \beta)$ in a Fourier series in β

$$G_0(q_1, q_2, \beta) = \frac{1}{\beta_0} \sum_n G_0(q_1, q_2, \varepsilon_n) \exp(-i\varepsilon_n \beta), \qquad (31.23)$$

where

$$G_0(q_1, q_2, \varepsilon_n) = \frac{1}{2} \int_{-\beta_0}^{\beta_0} d\beta \, G_0(q_1, q_2, \beta) \exp(i\varepsilon_n \beta). \qquad (31.24)$$

To find the permissible values of ε_n, we can use the equality

$$G_0(q_1, q_2, \beta) = -G_0(q_1, q_2, \beta + \beta_0). \qquad (31.25)$$

The substitution of the expression (31.23) in this equality gives

$$\sum_n G_0(q_1, q_2, \varepsilon_n) [\exp(-i\varepsilon_n \beta_0) + 1)] \exp(-i\varepsilon_n \beta) = 0,$$

whence $\exp(-i\varepsilon_n \beta_0) = -1$, and

$$\varepsilon_n = \frac{(2n+1)\pi}{\beta_0}. \qquad (31.26)$$

This condition is in fact also true for the exact Green function.

We can use the relationships in (31.15) to prove (31.25). When $\beta_1 > \beta_2$, we can write (the denominator $\mathrm{Tr}\hat{\zeta}_0$ is omitted)

$$G_0(q_1, q_2, \beta) = \mathrm{Tr}\,[\hat{\zeta}_0(\beta_0)\hat{\zeta}_0^{-1}(\beta_1)\hat{\psi}(q_1)\hat{\zeta}_0(\beta_1)\hat{\zeta}_0^{-1}(\beta_2)\hat{\psi}^+(q_2)\hat{\zeta}_0(\beta_2)]$$

or, if we allow for the possibility of permutating the operators in the trance, and also the obvious relationships

$$\hat{\zeta}_0(\beta_1)\hat{\zeta}_0(\beta_2) = \hat{\zeta}_0(\beta_1 + \beta_2); \quad \hat{\zeta}_0(\beta)\hat{\zeta}_0(-\beta) = 1,$$

$$G_0(q_1, q_2, \beta) = \mathrm{Tr}\,[\hat{\zeta}_0(\beta_0)\hat{\zeta}_0^{-1}(\beta)\hat{\psi}(q_1)\hat{\zeta}_0(\beta)\hat{\psi}^+(q_2)].$$

Here $\beta = \beta_1 - \beta_2$. Transposing the last two factors to the left, we find

$$G_0(q_1, q_2, \beta) = \mathrm{Tr}\,[\hat{\zeta}_0(\beta_0)\hat{\zeta}_0^{-1}(-\beta + \beta_0)\hat{\psi}^+(q_2)\hat{\zeta}_0(-\beta + \beta_0)\hat{\psi}(q_1)].$$

When $\beta_1 < \beta_2$,

$$G_0(q_1, q_2, \beta) = -\mathrm{Tr}\,[\zeta_0(\beta_0)\,\zeta_0^{-1}(-\beta)\,\hat{\psi}^+(q_2)\,\zeta_0(-\beta)\,\hat{\psi}(q_1)].$$

Whence, for $\beta < 0$,

$$G_0(q_1, q_2, \beta) = -G_0(q_1, q_2, \beta + \beta_0).$$

A cyclic extension of this shows that it is true for any β.

If we go over to the momentum representation in the expression for the Green function, we have

$$G_0(q_1, q_2, \varepsilon_n) = \sum_\nu \frac{\chi_\nu^*(q_2)\,\chi_\nu(q_1)}{i\varepsilon_n + \mu - \varepsilon_\nu}. \tag{31.27}$$

All integrals over β which enter into the expressions for the matrix elements can be extended from $-\beta_0$ to β_0 by introducing the factor $\frac{1}{2}$. It appears, moreover, that the δ-function from the energy summation at each vertex of the diagram can be replaced by the Kronecker symbol $\delta_{0,\Sigma\varepsilon_n}$. This differs from zero only when the frequency sum, corresponding to the lines converging to this node, is zero [13].

We can say, in conclusion, that the expression for the matrix element $\hat{S}(\beta)$ is obtained from the analogous expression for $T = 0$ by replacing all the frequencies ε by $i\varepsilon_n$ [$\varepsilon_n = (2n + 1)\,\pi/\beta_0$], and by going from an integration over the frequencies to a summation over n

$$\int_{-\infty}^{\infty} \frac{d\omega}{2\pi} \to -\frac{i}{\beta_0}\sum_n .$$

The rules for computing the corresponding sums have been given elsewhere [26]. These rules depend basically on the fact that the summation over the frequencies can be replaced by the calculation of some contour integral into which we can introduce the function $\tan(\omega\beta_0/2)$ which has a pole at the necessary points. In this case the desired result is obtained as a sum of the residues of the expressions under the integral sign.

32. The Method of Green Functions in Quantum Statistics

32.1. The thermodynamical perturbation theory is not enough to solve all the types of problem that arise in quantum statistics.

On the one hand, the theory developed above is only suitable for providing information about purely thermodynamical characteristics of the system—the thermodynamical potentials, their derivatives, etc. These integral characteristics, although important, by no means exhaust all the information about the system that one would like to have. We can think, first of all, of local characteristics (the distribution of the average values of the dynamical variables, the correlation functions, the transport coefficients, etc.), and, secondly, of the characteristics of the excitation spectrum for the system.

On the other hand, even if we restrict ourselves to the purely thermodynamical characteristics, it may be insufficient to consider only some of the first perturbation-theory diagrams. We can carry out an analysis of the relative roles of perturbation-theory diagrams in quantum statistics, similar to that in § 16 for the zero-temperature case. If the condition

$$\frac{\beta\varrho^{2/3}}{M} \gtrsim 1$$

is satisfied, where ϱ is the average density of the substance, then the previous analysis still holds. For very high temperatures,

$$\frac{\beta\varrho^{2/3}}{M} \ll 1$$

and we should make the replacement $p_0 \to (M/\beta)^{1/2}$ in the interaction parameter. We note, by way of explanation, that the kinetic energy of a particle is determined by the larger of the quantities p_0^2/M and $1/\beta$.

For a non-zero temperature, one can have a situation that requires the effective summation of some infinite sequence of perturbation-theory diagrams. To solve the subsequent problems, we must introduce temperature-dependent and time- and temperature-dependent Green functions.

32.2. The temperature-dependent Green function can be introduced by generalizing the concept of a temperature-dependent contraction of operators (more accurately, of the free-particle temperature-dependent Green function) including the correlation interaction between particles

$$G(\mathrm{I}, \mathrm{II}) = -i \operatorname{Tr} \frac{\{\zeta T[\hat{\psi}_{\mathrm{H}}(\mathrm{I})\, \hat{\psi}_{\mathrm{H}}^{+}(\mathrm{II})]\}}{\operatorname{Tr}\zeta}. \tag{32.1}$$

We have introduced here the temperature-dependent Heisenberg operators

$$\hat{\psi}_{\mathrm{H}}(\mathrm{I}) = \exp[\beta_1(\hat{H} - \mu\hat{N})]\, \hat{\psi}(q_1) \exp[-\beta_1(\hat{H} - \mu\hat{N})].$$

The way in which they change as β varies is not determined by $\hat{H}_0$, as it was for the operator $\hat{\psi}(\mathrm{I})$, but by the total Hamiltonian $\hat{H}$.†

Using the definition (31.9), a straightforward calculation gives

$$G(\mathrm{I}, \mathrm{II}) = -i \frac{\langle T[\hat{\psi}(\mathrm{I})\, \hat{\psi}^{+}(\mathrm{II})\, \hat{S}(\beta_0)]\rangle_0}{\langle \hat{S}(\beta_0)\rangle_0}. \tag{32.2}$$

Here $\hat{S}(\beta_0)$ is the complete S-matrix, taken over the interval from 0 to β_0 — the actual reciprocal temperature of the system. The two-particle Green function can be introduced in a similar way

$$G(\mathrm{I}, \mathrm{II}, \mathrm{III}, \mathrm{IV}) = -\frac{\langle T[\hat{\psi}(\mathrm{I})\, \hat{\psi}(\mathrm{II})\, \hat{\psi}^{+}(\mathrm{III})\, \hat{\psi}^{+}(\mathrm{IV})\, \hat{S}]\rangle_0}{\langle \hat{S}\rangle_0}. \tag{32.3}$$

If we know the Green function, then we can obtain information on the probability distributions, the correlation properties, etc., in the usual way. In particular the single-particle density matrix for a system has the form

$$R(q_1, q_2) = -i \lim_{\beta_2 - \beta_1 \to +0} G(\mathrm{I}, \mathrm{II}). \tag{32.4}$$

We can obtain information about the excitation spectrum of a system in a somewhat more complicated way. For this, we need to have available the normal time-dependent Green function (see § 29). It turns out that we can construct the time-dependent Green function directly from the value of the temperature-dependent Green function. This can be done by continuing $G(\mathrm{I}, \mathrm{II})$ analytically into the complex plane in terms of the function $\beta_1 - \beta_2$.

† We should remark on the similarity between these latter relationships, and the general ones given in § 2. The additional factors $\exp(\pm \mu\beta\hat{N})$ here lead to a further factor $\exp[-\mu(\beta_1 - \beta_2)]$ appearing (see § 32.3).

We will compare the expressions for the temperature-dependent and the time-dependent Green functions

$$G_\beta(q_1, \beta_1, q_2, \beta_2) = -i\frac{\left\{\begin{matrix}\mathrm{Tr}\,\{\xi T\,[\exp(\hat{H}-\mu\hat{N})\beta_1]\,\hat{\psi}(q_1)\exp[-(\hat{H}-\mu\hat{N}) \\ \times(\beta_1-\beta_2)]\,\hat{\psi}^+(q_2)\exp[-(\hat{H}-\mu\hat{N})\beta_2]\}\end{matrix}\right\}}{\mathrm{Tr}\xi} \tag{32.5}$$

$$G_t(q_1, t_1, q_2, t_2) = -i\frac{\left\{\begin{matrix}\mathrm{Tr}\,\{\xi T\,[\exp(i\hat{H}t_1)\,\hat{\psi}(q_1)\exp[-i\hat{H}(t_1-t_2)] \\ \times\hat{\psi}^+(q_2)\exp(-i\hat{H}t_2)]\}\end{matrix}\right\}}{\mathrm{Tr}\xi}, \tag{32.6}$$

where we have introduced, for convenience, the suffixes β and t. We will transform the expression for G_t further, using the relationships

$$\left.\begin{aligned}\exp(-i\mu\hat{N}t)\,\hat{\psi}(q)\exp(i\mu\hat{N}t) &= \hat{\psi}(q)\exp(i\mu t),\\ \exp(-i\mu\hat{N}t)\,\hat{\psi}^+(q)\exp(i\mu\hat{N}t) &= \hat{\psi}^+(q)\exp(-i\mu t),\end{aligned}\right\} \tag{32.7}$$

which can be proved from the commutation rules (3.15) (see, also, Appendix B)

$$\exp(-i\mu\hat{N}t)\,\hat{\psi}(q)\exp(i\mu\hat{N}t) = \sum_{n=0}^{\infty}\frac{(i\mu t)^n}{n!}\,\hat{\psi}(q).$$

Then

$$G_t(q_1, t_1, q_2, t_2) = -i\frac{\left\{\begin{matrix}\mathrm{Tr}\,(\xi T\,\{\exp[i(\hat{H}-\mu\hat{N})t_1]\,\hat{\psi}(q_1)\exp[-i(\hat{H}-\mu\hat{N})(t_1-t_2)] \\ \times\,\hat{\psi}^+(q_2)\exp[-i(\hat{H}-\mu\hat{N})t_2]\}\end{matrix}\right\}}{\mathrm{Tr}\xi}.$$

It follows from these relationships that the analytical continuation of the function $G_\beta(\mathrm{I}, \mathrm{II})$ to the imaginary axis $\beta_1 \to it_1$, $\beta_2 \to it_2$, gives the function $G_t(1, 2)\exp[i\mu(t_1 - t_2)]$ with $t_1 > t_2$ for $\beta_1 > \beta_2$, and the function $G_t(1, 2)\exp[i\mu(t_1 - t_2)]$ with $t_1 < t_2$ for $\beta_1 < \beta_2$. The time-dependent Green function can thus be found from the temperature-dependent Green function in the coordinate representation.

We will now consider the corresponding transformation for the Fourier components of both the foregoing functions†

$$G_\beta(q_1, q_2, \varepsilon_n) = \frac{1}{2}\int_{-\beta_0}^{\beta_0} d(\beta_1 - \beta_2)\, G_\beta(\mathrm{I}, \mathrm{II}) \exp[i\varepsilon_n(\beta_1 - \beta_2)],$$

$$\{G_t(1, 2) \exp[i\mu(t_1 - t_2)]\}_\varepsilon = G_t(q_1, q_2, \varepsilon + \mu),$$

where $G_t(q_1, q_2, \varepsilon)$ is the Fourier transform of the Green function $G_t(1, 2)$. These relationships are not immediately suitable for an analytical continuation. It is convenient for this purpose to use the spectral representation.

32.3. We will now obtain the spectral function for the temperature-dependent Green function. Introducing a complete set of intermediate states into (32.1) we find

$$G_\beta(\mathrm{I}, \mathrm{II}) = -i \sum_{m,n} \exp[\beta_0(\Omega + \mu N_n - E_n)$$
$$+ (E_n - E_m + \mu)[(\beta_1 - \beta_2)] \langle n|\hat{\psi}(q_1)|m\rangle \langle m|\hat{\psi}^+(q_2)|n\rangle$$
$$\times \begin{cases} 1 & \beta_1 > \beta_2 \\ -\exp[\beta_0(E_n - E_m + \mu)] & \beta_1 < \beta_2 \end{cases}$$

Going over to an integration over $E = E_m - E_n$, and introducing $\beta = \beta_1 - \beta_2$, we can write

$$G_\beta(q_1, q_2, \beta) = \int_{-\infty}^{\infty} dE\, A(q_1, q_2, E)$$
$$\times \begin{cases} \exp[-\beta(E - \mu)] & \beta_0 > \beta > 0 \\ -\exp[-(\beta + \beta_0)(E - \mu)] & 0 > \beta > -\beta_0 \end{cases}$$

Replacing β by it ($t = t_1 - t_2$), we obtain an expression for the time-dependent Green function

$$G_t(q_1, q_2, t) \exp(i\mu t) = \int_{-\infty}^{\infty} dE\, A(q_1, q_2, E)$$
$$\times \begin{cases} \exp[-it(E - \mu)] & t > 0 \\ -\exp[-(it + \beta_0)(E - \mu)] & t < 0 \end{cases}$$

† Here, as before, $\varepsilon_n = (2n + 1)\pi/\beta$.

The spectral densities in these relations for G_β and G_t are equal. We now go over to the Fourier transforms of these functions. A straightforward calculation gives

$$G_\beta(q_1, q_2, \varepsilon_n) = \int_{-\infty}^{\infty} dE \frac{A(q_1, q_2, E)\{\exp[-\beta_0(E-\mu)]+1\}}{i(E-\mu)+\varepsilon_n},$$

$$G_t(q_1, q_2, \varepsilon) = \int_{-\infty}^{\infty} dE\, A(q_1, q_2, E)$$

$$\times\left\{\frac{1}{\varepsilon - E + i\delta} + \frac{\exp[-\beta_0(E-\mu)]}{\varepsilon - E - i\delta}\right\}.$$

It is evident from a comparison of these relationships that the singularities in the analytical continuation of the function $G_t(q_1, q_2, \varepsilon)$, which gives the characteristics of the quasi-particles, are determined by the singularities in the analytical continuation of the temperature-dependent function $G_\beta(q_1, q_2, \varepsilon_n)$ in terms of the variable ε_n.

We can, in fact, write the symbolic equality

$$G_t(q_1, q_2, \varepsilon) = -iG_\beta(q_1, q_2, \varepsilon_n)|_{\varepsilon_n \to i(\mu-\varepsilon)}, \tag{32.8}$$

which signifies that, if we wish to obtain the Fourier transform of the time-dependent Green function, it is sufficient to continue the Fourier transform of the temperature-dependent Green function to the point $\varepsilon_n = i(\mu - \varepsilon)$. The possibility of this continuation, and its uniqueness, is guaranteed by the presence of an infinite sequence of points which determine $G_\beta(\varepsilon_n)$, which has a point of condensation at infinity. The poles of the analytical continuation of the functions $G(\varepsilon_n)$ also provide the poles of the function $G(\varepsilon)$. Thus the temperature-dependent Green function contains all the essential information on the equilibrium state of a many-body system.

32.4. We now turn to the representation of the thermodynamic potential Ω using the temperature-dependent Green functions. A representation of this sort permits an effective summation of the perturbation-theory diagrams.

Our subsequent discussion coincides in its basic features with the calculations of §§ 15 and 19. We will replace the interaction

Hamiltonian $\hat{H}'$ by $\lambda\hat{H}'$ and differentiate

$$\Omega = \Omega_0 - \frac{1}{\beta_0}\ln\langle\hat{S}\rangle_0$$

with respect to λ, remembering that Ω_0 and the brackets of the operator $\hat{S}$ do not depend on λ. This gives

$$\frac{\partial\Omega}{\partial\lambda} = -\frac{\langle\partial\hat{S}/\partial\lambda\rangle_0}{\beta_0\langle\hat{S}\rangle_0}. \tag{32.9}$$

Moreover,

$$\frac{\partial\hat{S}}{\partial\lambda} = \frac{\partial}{\partial\lambda}T\left\{\exp\left[-\lambda\int_0^{\beta_0} d\beta\hat{H}'(\beta)\right]\right\} = -\int_0^{\beta_0} d\beta\, T\left[\hat{H}'(\beta)\,\hat{S}(\beta_0)\right].$$

It is evident that a supplementary time-ordering under the T-product sign does not alter the result. Thus with $\Omega|_{\lambda=0} = \Omega_0$, we can write

$$\Omega = \Omega_0 + \frac{1}{\beta_0}\int_0^1 d\lambda\int_0^{\beta_0} d\beta\,\frac{\langle T[\hat{H}'(\beta)\,\hat{S}(\beta_0)]\rangle_0}{\langle\hat{S}(\beta_0)\rangle_0}. \tag{32.10}$$

Denoting the expression under the integral sign by Q, and applying the method that led to eqn. (32.2), we obtain

$$Q(\beta) = \frac{\mathrm{Tr}\hat{\zeta}\hat{H}'_{\mathrm{H}}(\beta)}{\mathrm{Tr}\hat{\zeta}}.$$

Further discussion follows that in § 15.3. The sole difference is that the operator $i\partial/\partial t$ is replaced by $-\partial/\partial\beta - \mu$ as can be seen from a comparison of the equation of motion for the operator $\hat{\psi}_H(\mathrm{I})$

$$\frac{\partial}{\partial\beta_1}\hat{\psi}_H(\mathrm{I}) = [\hat{H} - \mu\hat{N}, \hat{\psi}_H(\mathrm{I})]_-$$

or $(\beta_1 = \beta_2)$

$$-\left(\frac{\partial}{\partial\beta_1} + \hat{T} + \mu\right)\hat{\psi}_{\mathrm{H}}(\mathrm{I}) = \lambda\int dq_2\hat{V}(q_1, q_2)\,\overset{+}{\hat{\psi}}_{\mathrm{H}}(\mathrm{II})\,\hat{\psi}_{\mathrm{H}}(\mathrm{II})\,\hat{\psi}_{\mathrm{H}}(\mathrm{I})$$

with eqn. (3.28). The quantity Q can therefore be expressed in terms of the single-particle Green function only

$$Q(\beta) = -\frac{i}{2\lambda}\int dq_1 \times$$

$$\underset{\mathrm{II}\to\mathrm{I}}{\mathrm{Lim}}\left[\frac{\partial}{\partial\beta_1} + \mu - \hat{T} - (1+\lambda)\,\hat{W}\right]\left[G(\mathrm{I},\mathrm{II}) - G_0(\mathrm{I},\mathrm{II})\right].$$

As in quantum mechanics, we can further expand $\hat{S}(\beta)$ in terms of normal products (§ 14), substitute this expansion in eqn. (32.2), and introduce the temperature-dependent mass operator $M(\mathrm{I},\mathrm{II})$, which is described by diagrams of the same type as the analogous quantity with $T = 0$ (§ 19). We arrive, as a result, at the temperature dependent analogue of Dyson's equation which can be written in differential form as

$$-\left(\frac{\partial}{\partial\beta} + \mu + \hat{T} + \hat{W}\right)G(\mathrm{I},\mathrm{II}) = \int d\mathrm{III}\, M(\mathrm{I},\mathrm{III})\, G(\mathrm{III},\mathrm{II}), \quad (32.11)$$

where we have put $\int d\mathrm{I} \equiv \int dq_1 \int_0^{\beta_0} d\beta_1$. Using this equation, we can write finally [132, 135]

$$\Omega = \Omega_0 + \frac{1}{\beta_0}\int_0^1 \frac{d\lambda}{\lambda}\left\{\int d\mathrm{I}\, d\mathrm{II}\, M(\mathrm{I},\mathrm{II})\, G(\mathrm{II},\mathrm{I}) - \int d\mathrm{I} \underset{\substack{\mathrm{II}\to\mathrm{I}\\ \beta_2-\beta_1\to+0}}{\mathrm{Lim}} \lambda\hat{W}[G(\mathrm{I},\mathrm{II}) - G_0(\mathrm{I},\mathrm{II})]\right\}. \quad (32.12)$$

33. Applications to Plasma Theory

33.1. We will now consider certain questions in the dynamical theory of plasmas (a completely ionized gas composed of electrons and, for definiteness, protons). Hydrogen can be converted to this state by various means—either by subjecting it to intense heating, or to high pressures (§ 6). We will consider the case of a hot plasma. The temperature of the plasma must be of at least the same order of magnitude as the ionisation potential of hydrogen

$$kT \sim e^2/a_0. \quad (33.1)$$

We will restrict ourselves to weak interactions, putting (§ 1)

$$\alpha \sim \frac{e^2}{dkT} \ll 1. \quad (33.2)$$

We will consequently be considering a dense system with Coulomb interactions. We have already emphasized in § 1 that such a system is dilute in the normal sense of the word.

It follows from (33.1) and (33.2) that the parameter $\varrho^{2/3}/MkT$ which represents the ratio of the Fermi energy of the electrons to their thermal energy, is small compared with unity, i.e.

$$\frac{\varrho^{2/3}}{MkT} \sim \left(\frac{a_0 kT}{e^2}\right)\left(\frac{e^2}{dkT}\right)^2 \ll 1.$$

This inequality holds even more strongly for the proton gas. We are therefore dealing with a system which obeys Boltzmann statistics, i.e. the occupation number of the levels is small compared with unity

$$n_\nu = \exp\left[\beta(\mu - \varepsilon_\nu)\right] \ll 1 \qquad (33.3)$$

and the chemical potential μ is large in absolute value and negative.

As the plasma is macroscopically uniform and electrically neutral, the direct self-consistent interaction term in the Hartree–Fock approximation disappears completely. So far as the small self-consistent exchange term is concerned, we may obtain it without difficulty from (30.10), if we use (33.3) to write

$$\hat{\varrho} = \exp\left[\beta(\mu - \hat{p}^2/2M)\right].$$

Whence,

$$C = -\frac{1}{2}\operatorname{Tr}\hat{W}\hat{\varrho}$$

$$= V\frac{e^2}{8\pi^2\beta^2}\left[M_{(e)}^2 \exp\left(2\beta\mu_{(e)}\right) + M_{(p)}^2 \exp\left(2\beta\mu_{(p)}\right)\right]$$

and

$$\Omega_0 = -\frac{2\varrho}{\beta}\left[1 - \frac{\pi}{8}\beta^2\varrho e^2\left(\frac{1}{M_{(p)}} + \frac{1}{M_{(e)}}\right)\right]V, \qquad (33.4)$$

where $\mu_{(p,e)}$ are the chemical potentials, $M_{(p,e)}$ are the masses, $\varrho_{(p)} = \varrho_{(e)}$ are the respective densities of the proton and the electron gases. We have taken into account here the fact that the exchange effects for the electron and the proton distributions must be considered separately.

The dynamical theory of plasmas with weak Coulomb interactions taking into account the correlations between the particles has previously been set up using the Debye–Hückel method [4]. This includes a self-consistent procedure which has, however, an uncertain region of applicability. A field theory of plasmas has also been developed [136].

The physical quantities—in particular, the thermodynamic potential—appear in this theory as expansions whose convergence improves as the parameter α gets smaller. The main term of the expansion coincides with the expression obtained by the Debye–Hückel method. Hence the field theory of plasmas generalises the foregoing results, increases their accuracy, and indicates the limits within which they are applicable.

33.2. We will now calculate the main term of the correlation part of the thermodynamic potential. We use as our starting point the fact that a plasma with weak interactions between the particles is a dense system. Hence, an analysis of the temperature-dependent diagrams of thermodynamical perturbation theory is basically similar to the analysis of § 16, and leads to the same results. As for any other dense system, the most important diagrams for a plasma are those which contain a maximum number of closed loops, and which have a single small momentum transfer. The requisite closed vacuum diagrams, which correspond to the thermodynamic potential, are shown in Fig. 67. These diagrams must be considered both for electrons and protons.

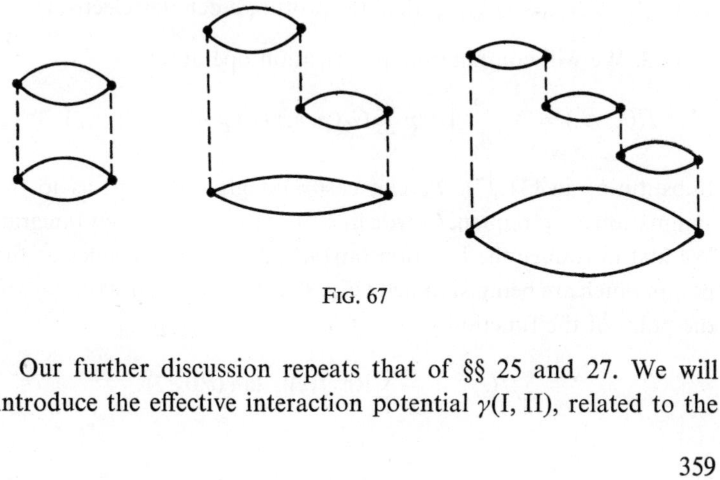

FIG. 67

Our further discussion repeats that of §§ 25 and 27. We will introduce the effective interaction potential $\gamma(\mathrm{I}, \mathrm{II})$, related to the true Coulomb potential by the equation

$$\gamma(\text{I, II}) = \frac{e^2}{r_{12}}\delta(\beta_1 - \beta_2) + \int d\text{III}\, d\text{IV} \frac{e^2}{r_{13}}\delta(\beta_1 - \beta_3) \times \Pi(\text{III, IV})\, \gamma(\text{IV, II}), \tag{33.5}$$

where the polarization operator Π is defined by the relation

$$\Pi(\text{I, II}) = iG_0(\text{I, II})\, G_0(\text{II, I}). \tag{33.6}$$

The mass operator of a dense system is related to the effective potential, γ, by the simple equation (see § 25).

$$M(\text{I, II}) = i\left[\gamma(\text{I, II}) - \frac{e^2}{r_{12}}\delta(\beta_1 - \beta_2)\right] G_0(\text{I, II}). \tag{33.7}$$

If this is subtituted in (32.12), we can use (33.5) to obtain the following expression for the thermodynamic potential:

$$\Omega - \Omega_0 = \frac{1}{\beta_0}\int_0^1 \frac{d\lambda}{\lambda}\int d\text{I}\, d\text{II}\left[\gamma(\text{I, II}) - \frac{e^2}{r_{12}}\delta(\beta_1 - \beta_2)\right]\Pi(\text{I, II}). \tag{33.8}$$

If we now transfer to the momentum representation, using the rules given in §§ 31 and 32, we obtain

$$\Omega - \Omega_0 = \frac{g^2 V}{\beta_0}\int_0^1 \frac{d\lambda}{\lambda}\int d^3\boldsymbol{k} \sum_n \left[\gamma(\boldsymbol{k}, \omega_n) - \frac{4\pi e^2}{k^2}\right]\Pi(\boldsymbol{k}, \omega_n). \tag{33.9}$$

Here, $\Pi = \Pi_{(p)} + \Pi_{(e)}$ (p denotes proton, e denotes electron).

33.3. We will compute the polarization operator

$$\Pi(\boldsymbol{k}, \omega_n) = -\frac{g}{\beta_0}\int d^3\boldsymbol{p} \sum_n G_0(\boldsymbol{p}, \varepsilon_n)\, G_0(\boldsymbol{p} + \boldsymbol{k}, \varepsilon_n + \omega_n).$$

Substituting in (31.27), we can easily integrate over $\boldsymbol{p}$; as to the summation over frequencies we can reduce this to a contour integral. We first introduce the function $\tan(\omega\beta_0/2)$ which has poles at the points which are being summed (Fig. 68). We can then write (ε_i are the poles of the function f)

$$\frac{1}{\beta_0}\sum_n f(\varepsilon_n) = \frac{1}{2}\sum_i \text{Res}\, f(\varepsilon)|_{\varepsilon_i} \tan(\varepsilon_i\beta_0/2).$$

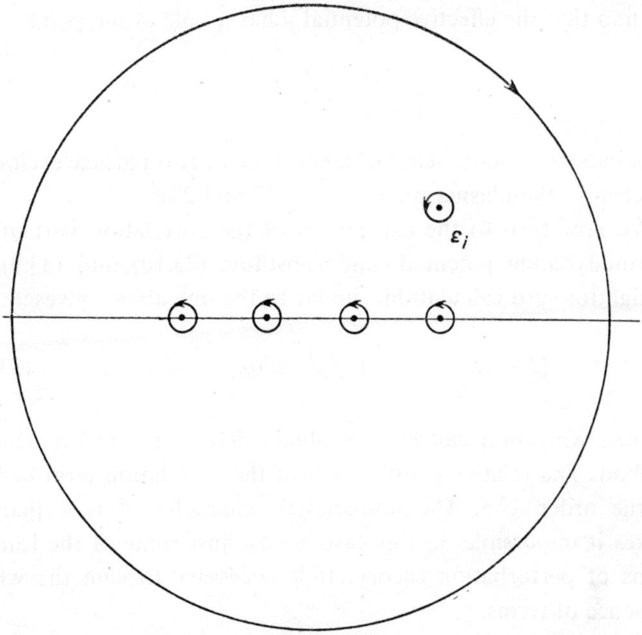

FIG. 68

We then obtain directly the following expression for the polarization operator:

$$\Pi(\boldsymbol{k}, \omega_n) = -\frac{9}{M}\int d^3\boldsymbol{p}\, n_p \frac{p^2 - (\boldsymbol{p}+\boldsymbol{k})^2}{\{[p^2 - (\boldsymbol{p}+\boldsymbol{k})^2]/2M\}^2 + \omega_n^2}. \quad (33.10)$$

For small values of $\boldsymbol{k}$, and $\omega/k \to 0$, we have

$$\Pi(0) = \Pi_{(e)}(0) + \Pi_{(p)}(0) = -\beta[\varrho_{(e)} + \varrho_{(p)}]. \quad (33.11)$$

From the results of § 27, this leads to Debye screening of the Coulomb potential. The corresponding radius R_0 is simply related to the polarization operator

$$\varkappa^2 \equiv \frac{1}{R_0^2} = -4\pi e^2 \Pi(0). \quad (33.12)$$

Putting (33.5) in the form

$$\gamma(\boldsymbol{k}, \omega_n) = \frac{4\pi e^2}{k^2}\left[1 - \frac{4\pi e^2}{k^2}\Pi(\boldsymbol{k}, \omega_n)\right]^{-1},$$

we find that the effective potential γ has a pole at the point

$$\omega^2 = 4\pi\left[\frac{\varrho_{(p)}}{M_{(p)}} + \frac{\varrho_{(e)}}{M_{(e)}}\right]e^2 \quad (k \to 0). \tag{33.13}$$

This indicates the presence of a new type of two-particle excitation spectrum—the plasma waves (see §§ 27 and 28).

We now turn to the calculation of the correlation part of the thermodynamic potential, and substitute (33.10) into (33.9). A straightforward calculation, similar to the one above, gives

$$\Omega - \Omega_0 = -\frac{2}{3}(\pi\beta)^{1/2} e^3(\varrho_{(p)} + \varrho_{(e)})^{3/2}. \tag{33.14}$$

This expression can also be obtained from the Debye–Hückel method. The relative contribution of the correlation term to Ω is of the order $\alpha^{3/2}$. The non-analytic character of this quantity makes it impossible, in this case, to use just some of the leading terms of perturbation theory. It is necessary to sum the whole sequence of terms.

The subsequent terms in the expansion of Ω, which are of order $\alpha^2 \ln \alpha$, have been calculated elsewhere [136]. We will simply give the corresponding expressions, when $kT < e^2/a_0$

$$\Omega - \Omega_0 = -\frac{2}{3}(\pi\beta)^{1/2} e^3(Z^2\varrho_{(i)} + \varrho_{(e)})^{3/2} + \frac{\pi}{3}\beta^2 e^6(Z^2\varrho_{(i)} - \varrho_{(e)})^2 \ln\left(\frac{1}{\beta e^2 R_0}\right) + O(\alpha^2), \tag{33.15}$$

and when $kT \gg Z^2e^2/a_0$

$$F - F_0 = -\frac{2}{3}(\pi\beta)^{1/2} e^3[Z(Z+1)]^{3/2} \varrho^{3/2} + \frac{\pi}{3} Z^2(Z^2 - 1)\beta^2 e^6 \varrho^2 \ln(m^{1/2}/\beta^{1/2}\varkappa) + O(\alpha^2). \tag{33.16}$$

Here Ω and F are the thermodynamic potential and the free energy respectively, $\varrho_{(i)}$ is the ion density and $\varrho_{(e)}$ the electron density.

Appendices

Appendix A. The Calculation of Mean Values of Operators

A.1. The problem of calculating the mean values of operators over the state Ψ_0, corresponding to the single-particle approximation, is much easier in the occupation-number representation. Unlike the configuration representation, where a solution may require tedious calculations using Slater–Fock determinants, in the occupation-number representation, the problem reduces to the application of simple permutation rules to the creation and annihilation operators.

We will begin by considering the n-particle density matrix, which is defined by the relationship

$$R(q_1 \cdots q_n, q_1' \cdots q_n') = \langle \Psi_0 | \hat{\psi}^+(q_n') \cdots \hat{\psi}^+(q_1')\, \hat{\psi}(q_1) \cdots \hat{\psi}(q_n) | \Psi_0 \rangle. \tag{A.1}$$

Substituting the expansion (3.9), we obtain

$$R = \sum_{\mu, \nu} \chi^*_{\mu_1}(q_1') \cdots \chi^*_{\mu_n}(q_n')\, \chi_{\nu_1}(q_1) \cdots \chi_{\nu_n}(q_n)\, A_{\mu_1 \ldots \mu_n, \nu_1 \ldots \nu_n}$$

where

$$A_{\mu_1 \ldots \mu_n, \nu_1 \ldots \nu_n} = \langle \Psi_0 | \hat{A}^+_{\mu_n} \cdots \hat{A}^+_{\mu_1} \hat{A}_{\nu_1} \cdots \hat{A}_{\nu_n} | \Psi_0 \rangle. \tag{A.2}$$

When calculating $A_{\mu\nu}$ we must remember that, since the right-hand and left-hand brackets of this expression coincide, the states of the annihilated particles $(\nu_1 \cdots \nu_n)$ must coincide in aggregate with the states of the created particles $(\mu_1 \cdots \mu_n)$. Since the state Ψ_0 corresponds to the single-particle approximation, and satisfies the condition (4.6), the occupation-number operators n_ν, which appear under the average sign, can be replaced by their eigenvalues n_ν.

We will compute some of the simplest functions $A_{\mu\nu}$. When $n = 1$, we have

$$A_{\mu\nu} = \langle \Psi_0 | \hat{n}_\nu | \Psi_0 \rangle\, \delta_{\mu\nu} = n_\mu \delta_{\mu\nu}. \tag{A.3}$$

When $n = 2$, we must allow for two possibilities ($\mu_1 = \nu_1, \mu_2 = \nu_2$ and $\mu_1 = \nu_2$, $\mu_2 = \nu_1$). Arranging the operators $\hat{A}$ and $\hat{A}^+$ in each case, so that the corresponding occupation-number operators appear, we find that

$$A_{\mu_1\mu_2, \nu_1\nu_2} = n_{\nu_1} n_{\nu_2} (\delta_{\mu_1\nu_1}\delta_{\mu_2\nu_2} - \delta_{\mu_1\nu_2}\delta_{\mu_2\nu_1}).$$

In the general case, as can be shown by induction, we have the following relationship:

$$A_{\mu_1 \ldots \mu_n, \nu_1 \ldots \nu_n} = n_{\nu_1} \cdots n_{\nu_n} \det |\delta_{\mu_i \nu_k}|, \tag{A.4}$$

where det represents the determinant constructed from δ-symbols ($i, k = 1, 2, \cdots, n$).

Returning to the expression for the density matrix, we can use (A.2) to write for $n = 1$,

$$R(q_1, q_1') = \sum_{\nu} n_\nu \chi_\nu^*(q_1') \chi_\nu(q_1), \tag{A.5}$$

for $n = 2$,

$$R(q_1, q_2, q_1', q_2') = R(q_1, q_1') R(q_2, q_2') - R(q_1, q_2') R(q_2, q_1'). \tag{A.6}$$

In the general case an n-particle density matrix can be represented as a determinant made up of single-particle matrices

$$R(q_1 \cdots q_n, q_1' \cdots q_n') = \det |R(q_i, q_k')|. \tag{A.7}$$

The possibility of reducing a higher-order density matrix to single-particle matrices is due entirely to our choice of the single-particle approximation.

The expression for a higher-order density matrix can be turned into a simple product of single-particle density matrices by introducing coordinate permutation operators $\hat{\mathscr{P}}$ (see § 1). For a two-particle density matrix, we obtain the expression

$$R(q_1, q_2, q_1', q_2') = (1 - \hat{\mathscr{P}}_{1,2}) R(q_1, q_1') R(q_2, q_2'), \tag{A.8}$$

where the suffixes on $\hat{\mathscr{P}}$ indicate the coordinates which are being transposed ($\hat{\mathscr{P}}$ acts only on the unprimed coordinates). For $n = 3$, we find

$$R(q_1, q_2, q_3, q_1', q_2', q_3')$$
$$= (1 - \hat{\mathscr{P}}_{1,2})(1 - \hat{\mathscr{P}}_{1,3} - \hat{\mathscr{P}}_{2,3}) R(q_1, q_1') R(q_2, q_2') R(q_3, q_3'). \tag{A.9}$$

We can represent higher-order density matrices in a similar way.

A.2. We now turn to the problem of calculating the average values of operators in the state Ψ_0. The average value of an n-particle operator $\hat{\alpha}_n$ in the occupation-number representation has the form

$$\langle \Psi_0 | \hat{\alpha}_n | \Psi_0 \rangle = \frac{1}{n!} \int dq_1 \cdots dq_n \langle \Psi_0 | \hat{\psi}^+(q_n) \cdots \hat{\psi}^+(q_1) \hat{a}_{q_1 \ldots q_n} \hat{\psi}(q_1) \cdots \hat{\psi}(q_n) | \Psi_0 \rangle,$$

where the suffixes on $\hat{a}$ indicate the variables on which it acts.

The operator can be removed from under the average sign, and the expression reduced to the density matrix. We will use here the identity

$$\hat{b}(q_1 \cdots q_n) \hat{a}_{q_1 \ldots q_n} \equiv \int dq_1' \cdots dq_n' \, \delta(q_1 - q_1') \cdots \delta(q_n - q_n') \hat{a}_{q_1 \ldots q_n} \hat{b}(q_1' \cdots q_n'),$$

which allows us to change the places of the field operator and $\hat{a}$. Hence

$$\langle \Psi_0 | \hat{\alpha}_n | \Psi_0 \rangle = \frac{1}{n!} \int dq_1 \cdots dq_n \, dq_1' \cdots dq_n'$$
$$\times \delta(q_1 - q_1') \cdots \delta(q_n - q_n') \hat{a}_{q_1 \ldots q_n} R(q_1 \cdots q_n, q_1' \cdots q_n'). \tag{A.10}$$

In particular we can write

$$\langle \Psi_0 | \hat{\alpha}_1 | \Psi_0 \rangle = \int dq_1 \, dq_1' \, \delta(q_1 - q_1') \hat{a}_{q_1} R(q_1, q_1') = \int dq_1 \lim_{q_1' \to q_1} \hat{a}_{q_1} R(q_1, q_1'). \tag{A.11}$$

In other words we must first act on the density matrix with the operator, and then put $q_1' = q_1$. We have, for a two-particle operator,

$$\langle \Psi_0 | \hat{\alpha}_2 | \Psi_0 \rangle = \tfrac{1}{2} \int dq_1\, dq_2 \lim_{q'_{1,2} \to q_{1,2}} \hat{a}_{q_1 q_2} (1 - \hat{\mathscr{P}}_{12})\, R(q_1, q_1')\, R(q_2, q_2'). \tag{A.11'}$$

We introduce the occupation operator $\hat{\varrho}$, defined by

$$R(q, q') = \hat{\varrho}_q \delta(q - q').$$

We can now rewrite (A.11) as

$$\langle \Psi_0 | \hat{\alpha}_1 | \Psi_0 \rangle = \int dq_1\, dq_1'\, \delta(q_1 - q_1')\, \hat{a}_{q_1}\, \hat{\varrho}_{q_1}\, \delta(q_1 - q_1').$$

But $\delta(q_1 - q_1')$ is the wave function in the coordinate representation of a state with a coordinate q_1'. The expression $\int dq_1\, \delta(q_1 - q_1')\, \hat{a}\hat{\varrho}\delta(q_1 - q_1')$ is therefore the diagonal matrix element of the operator $\hat{a}\hat{\varrho}$, and the subsequent integration over q_1' gives the sum of these matrix elements, i.e. the trace† of the operator $\hat{a}\hat{\varrho}$

$$\langle \Psi_0 | \hat{\alpha}_1 | \Psi_0 \rangle = \mathrm{Tr}\, \hat{a}\hat{\varrho}. \tag{A.12}$$

Similarly,

$$\langle \Psi_0 | \hat{\alpha}_2 | \Psi_0 \rangle = \tfrac{1}{2}\, \mathrm{Tr}\, [\hat{a}_{q_1 q_2} (1 - \hat{\mathscr{P}}_{1,2})\, \hat{\varrho}_{q_1} \hat{\varrho}_{q_2}], \tag{A.13}$$

etc.

A.3. Using the fact that the trace is independent of the set of functions selected, we can reduce (A.12) and (A.13) to a form suitable for application.

We will choose our set of functions to be the individual wave functions of the particles. From the relationship (4.9), and the definition of the matrix elements, we can write

$$\langle \Psi_0 | \hat{\alpha}_1 | \Psi_0 \rangle = \sum_\nu n_\nu \langle \nu | \hat{a}_1 | \nu \rangle, \tag{A.14}$$

$$\langle \Psi_0 | \hat{\alpha}_2 | \Psi_0 \rangle = \tfrac{1}{2} \sum_{\mu\nu} n_\mu n_\nu \langle \mu\nu | \hat{a}_2 (1 - \hat{\mathscr{P}}_{1,2}) | \mu\nu \rangle, \tag{A.15}$$

etc.; if we remember that a rearrangement of the arguments in the brackets of the latter matrix element is equivalent to a rearrangement of the state indices, we have

$$\langle \mu\nu | \hat{a}_2 (1 - \hat{\mathscr{P}}) | \mu\nu \rangle = \langle \mu\nu | \hat{a}_2 | \mu\nu \rangle - \langle \mu\nu | \hat{a}_2 | \nu\mu \rangle.$$

The other possible way of writing average values is obtained by using a system of plane waves (see (4.20)). Here

$$\langle \Psi_0 | \hat{\alpha}_1 | \Psi_0 \rangle = \mathrm{Tr}_{\sigma\tau} \int d^3\boldsymbol{x} \int d^3\boldsymbol{p} \langle \hat{a}_1 \hat{\varrho} \rangle_p \tag{A.16}$$

where the trace is taken in terms of the discrete indices, and the symbol $\langle \cdots \rangle_p$ signifies

$$\langle \ldots \rangle_p = \exp[-i(\boldsymbol{p} \cdot \boldsymbol{x})]\, (\ldots)\, \exp i(\boldsymbol{p} \cdot \boldsymbol{x}).$$

† We should remind ourselves of the basic properties of the trace of an operator. In the first place, its magnitude does not depend on the choice of the set functions used in its calculation. In the second place, a cyclic permutation of the operator factors is permissible in the trace.

If $\hat{a}_1$ is independent of the momentum operator, we can use (4.18) to write

$$\langle \Psi_0|\hat{\alpha}_1|\Psi_0\rangle = \mathrm{Tr}_{\sigma\tau} \int d^3x \int d^3p\, \hat{a}_1 f(x, p).$$

In the general case, putting $\hat{a}_1 = \hat{a}_1(x, \nabla_x)$, we have

$$\langle \hat{a}_1 \hat{\varrho}\rangle_p = \exp -i(p \cdot x)\, \hat{a}_1(x, \nabla_x) \exp i(p \cdot x) f(x, p),$$

whence†

$$\langle \Psi_0|\hat{\alpha}_1|\Psi_0\rangle = \mathrm{Tr}_{\sigma\tau} \int d^3x \int d^3p\, \hat{a}_1(x, \nabla_x + ip) f(x, p), \tag{A.16'}$$

where ∇_x acts on the distribution function. If, in particular, $\hat{a}_1$ is independent of x, we can replace $\nabla_x + ip$ by ip in this relationship (this may easily be confirmed by an integration by parts).

The corresponding expression for a two-particle average, when the operator $\hat{a}_2$ depends, so far as the momentum is concerned, only on the combination $(\nabla_{x_1} - \nabla_{x_2})/2$, is

$$\langle \Psi_0|\hat{\alpha}_2|\Psi_0\rangle = \tfrac{1}{2}\, \mathrm{Tr}_{\sigma_1\tau_1, \sigma_2\tau_2} \int d^3x_1\, d^3x_2 \int d^3p_1\, d^3p_2 \exp[-i(p_1 \cdot x_1) + i(p_2 \cdot x_2)]$$
$$\times\, \hat{a}_2 \hat{\varrho}_{q_1} \hat{\varrho}_{q_2} \{\exp[i(p_1 \cdot x_1) + i(p_2 \cdot x_2)] - \hat{\mathscr{P}}^{(\sigma)} \hat{\mathscr{P}}^{(\tau)} \exp[i(p_1 \cdot x_2) + i(p_2 \cdot x_1)]\}$$

or

$$\langle \Psi_0|\hat{\alpha}_2|\Psi_0\rangle = \frac{1}{2}\, \mathrm{Tr}_{\sigma_1\tau_1, \sigma_2\tau_2} \int d^3x_1\, d^3x_2 \int d^3p_1\, d^3p_2$$
$$\times \{1 - \hat{\mathscr{P}}^{(\sigma)} \hat{\mathscr{P}}^{(\tau)} \exp[i(p_1 - p_2 \cdot x_1 - x_2)]\}$$
$$\times\, \hat{a}_2\left(x_1, x_2\, \frac{i(p_1 - p_2) + \nabla_{x_1} - \nabla_{x_2}}{2}\right) f(x_1, p_1) f(x_2, p_2). \tag{A.17}$$

If $\hat{a}_2$ is simply a function of x_1 and x_2, then

$$\langle \Psi_0|\hat{\alpha}_2|\Psi_0\rangle = \tfrac{1}{2} \int d^3x_1\, d^3x_2 \int d^3p_1\, d^3p_2 \{g^2 - g \exp[i(p_1 - p_2 \cdot x_1 - x_2)]\}$$
$$\times\, \hat{a}_2(x_1, x_2) f(x_1, p_1) f(x_2, p_2), \tag{A.17'}$$

where $g = \mathrm{Tr}_{\sigma\tau} 1$ is the degeneracy factor.

Finally, if we take the set of functions to be the eigenfunction of the coordinates $\delta(q - q')$, then we return to the relationships (A.11) and (A.11').

We also need the trace of a two-particle operator which is taken with respect to only one of the sets of variables

$$\hat{S} = \mathrm{Tr}_{q_2} [\hat{a}_2(1 - \hat{\mathscr{P}}_{1,2})\, \hat{\varrho}_{q_2}] = \int dq_2\, dq_2'\, \delta(q_2 - q_2')\, \hat{a}_2(1 - \hat{\mathscr{P}}_{1,2})\, R(q_2, q_2'). \tag{A.18}$$

Using (A.5), we can write

$$\hat{S} = \sum_\nu n_\nu \int dq_2\, \chi_\nu^*(q_2)\, \hat{a}_2(1 - \hat{\mathscr{P}}_{1,2})\, \chi(q_2).$$

† We have here used Leibnitz' rule for differentiating a product

$$f(\nabla)\, a(x)\, b(x) = f(\nabla_a + \nabla_b)\, a(x)\, b(x),$$

where $\nabla_{a,b}$ acts, respectively, on $a(x)$ and $b(x)$.

If this operator acts on the function χ_μ, the result is

$$\hat{S}\chi_\mu(q_1) = \sum_\nu n_\nu \int dq_2\, \chi_\nu^*(q_2)\, \hat{a}_2\, \{\chi_\nu(q_2)\,\chi_\mu(q_1) - \chi_\nu(q_1)\,\chi_\mu(q_2)\}. \quad (A.19)$$

We will write the operator $\hat{S}$ in terms of the distribution function, retaining our previous assumptions concerning the operator $\hat{a}_2$. Since $\hat{S}$ can depend on the momentum operator, we will consider the result when it acts on $\exp i(\mathbf{p}_1 \mathbf{x}_1)$:

$$\hat{S}\exp(i\mathbf{p}_1\mathbf{x}_1) = \mathrm{Tr}_{\sigma_2\tau_2} \int d^3x_2 \int d^3p_2 \exp[\mathrm{t}] - i(\mathbf{p}_2 \cdot \mathbf{x}_2)$$
$$\times\, \hat{a}_2(1 - \mathscr{P}_{1,2})\,\hat{\varrho}_{q_2} \exp[i(\mathbf{p}_1 \cdot \mathbf{x}_1) + i(\mathbf{p}_2 \cdot \mathbf{x}_2)].$$

Carrying out the same calculations as for the average, we find

$$\hat{S}\exp i(\mathbf{p}_1 \cdot \mathbf{x}_1) = \mathrm{Tr}_{\sigma_2\tau_2} \int d^3x_2 \int d^3p_2\, \{\hat{a}_2[\mathbf{x}_1, \mathbf{x}_2, -\tfrac{1}{2}(\nabla_{x_2} + i\mathbf{p}_2)] f(\mathbf{x}_2, \mathbf{p}_1 + \mathbf{p}_2)$$
$$- \hat{\mathscr{P}}^{(\sigma)}\hat{\mathscr{P}}^{(\tau)} \exp i[(\mathbf{p}_2 \cdot \mathbf{x}_1 - \mathbf{x}_2)]\, \hat{a}_2[\mathbf{x}_1, \mathbf{x}_2, \tfrac{1}{2}(\nabla_{x_1} + i\mathbf{p}_2)] f(\mathbf{x}_1, \mathbf{p}_1 + \mathbf{p}_2)\} \exp i(\mathbf{p}_1 \cdot \mathbf{x}_1)$$

Finally, replacing $\mathbf{p}_1$ by the momentum operator $\hat{p}$, we can write

$$\hat{S} = \mathrm{Tr}_{\sigma_2\tau_2} \int d^3x_2 \int d^3p_2\, \{\hat{a}_2[\mathbf{x}_1, \mathbf{x}_2, -\tfrac{1}{2}(\nabla_{x_2} + i\mathbf{p}_2)] f(\mathbf{x}_2, \mathbf{p}_2 + \hat{\mathbf{p}})$$
$$- \mathscr{P}^{(\sigma)}\mathscr{P}^{(\tau)} \exp i[(\mathbf{p}_2 \cdot \mathbf{x}_1 - \mathbf{x}_2)]\, \hat{a}_2[\mathbf{x}_1, \mathbf{x}_2, \tfrac{1}{2}(\nabla_x + i\mathbf{p}_2)] f(\mathbf{x}_1, \mathbf{p}_2 + \hat{\mathbf{p}})\}.$$

A.4. It is a somewhat more complicated matter to calculate the average value of an operator product. We will consider the simplest examples of this kind.

The operator product $\hat{\alpha}\hat{\beta}$ can always be represented as a half-sum of the commutator $[\hat{\alpha}, \hat{\beta}]_-$ and anti-commutator $[\hat{\alpha}, \hat{\beta}]_+$. We will first calculate the average value of the commutator. The commutator of two single-particle operators $\hat{\alpha}_1$ and $\hat{\beta}_1$, can also be represented as a single-particle operator; this may be easily proved from the commutation relations (3.10):

$$[\hat{\alpha}_1, \hat{\beta}_1]_- = \int dq\, \hat{\psi}^+(q)\, [\hat{a}, \hat{b}]_-\, \hat{\psi}(q).$$

Whence we find, using (A.12), that

$$\langle \Psi_0 | [\hat{\alpha}_1, \hat{\beta}_1]_- | \Psi_0 \rangle = \mathrm{Tr}\,([\hat{a}, \hat{b}]_-\, \hat{\varrho}). \quad (A.21)$$

Since we can make a cyclic rearrangement of the factors in the spur, we can rewrite them as $\hat{a}\hat{b}\hat{\varrho} - \hat{a}\hat{\varrho}\hat{b} = \hat{a}[\hat{\varrho}, \hat{b}]_-$. If one of the operators, therefore, commutes with $\hat{\varrho}$, i.e. if the quantity corresponding to it is an integral of motion, then the corresponding average value goes to zero.

In a similar way, the commutator of one- and two-particle operators is a two-particle operator of the following type:

$$[\hat{\alpha}_1; \hat{\beta}_2]_- = \int dq_1\, dq_2\, \hat{\psi}^+(q_1)\, \hat{\psi}^+(q_2)\, [\hat{a}_{q_1}, \hat{b}_{q_1 q_2}]_-\, \hat{\psi}(q_2)\, \hat{\psi}(q_1),$$

whence

$$\langle \Psi_0 | [\hat{\alpha}_1, \hat{\beta}_2] | \Psi_0 \rangle = \mathrm{Tr}\,\{[\hat{a}_{q_1}, \hat{b}_{q_1 q_2}]_-\, (1 - \hat{\mathscr{P}}_{1,2})\, \hat{\varrho}_{q_1} \hat{\varrho}_{q_2}\}. \quad (A.22)$$

A.5. We will now find the average value of the anti-commutator of two operators. We will first consider the anti-commutator of two single-particle operators

$$\langle \Psi_0 | [\hat{\alpha}_1, \hat{\beta}_1]_+ | \Psi_0 \rangle = \int dq_1\, dq_2 \langle \Psi_0 | [\hat{\psi}^+(q_1)\, \hat{a}\hat{\psi}(q_1), \hat{\psi}^+(q_2)\, \hat{b}\hat{\psi}(q_2)]_+ | \Psi_0 \rangle.$$

Introducing the δ-function, we can reduce this expression to the form

$$\int dq_1\, dq_2\, dq_1'\, dq_2' \delta(q_1 - q_1')\, \delta(q_2 - q_2')\, (\hat{a}_{q_1}\hat{b}_{q_2} + \hat{a}_{q_2}\hat{b}_{q_1}) \times \langle \Psi_0 | \hat{\psi}^+(q_1')\, \hat{\psi}(q_1)\, \hat{\psi}^+(q_2')\, \hat{\psi}(q_2) | \Psi_0 \rangle .$$

The average value which figures here can be reduced to the density matrix. To do this, we will change the places of the second and third operators, using the permutation rules of (3.10). As a result, we obtain the average value in the form

$$R(q_1, q_2, q_1', q_2') + \delta(q_1 - q_2')\, R(q_1, q_2')\, R(q_2, q_1') = [(1 - \hat{\mathscr{P}}_{1,2})\, \hat{\varrho}_{q_1}\hat{\varrho}_{q_2} + \hat{\mathscr{P}}_{1,2}\hat{\varrho}^2_{q_1}]\, \delta(q_1 - q_1')\, \delta(q_1 - q_2').$$

We have introduced a transposition operator in the second term in the square brackets, so that the δ-function has the same form as in the previous expression. We have also used the relationship $\hat{\varrho}^2 = \hat{\varrho}$.

If we continue our discussion as in § A.2, we find, after the appropriate symmetrization in q_1 and q_2,

$$\langle \Psi_0 | [\hat{\alpha}_1, \hat{\beta}_1]_+ | \Psi_0 \rangle = 2\,\mathrm{Tr}\, \hat{a}\hat{\varrho}\ \mathrm{Tr}\, \hat{b}\hat{\varrho} + \mathrm{Tr}\, [\hat{a}_{q_1}\hat{b}_{q_2}(\hat{\varrho}_{q_1} - \hat{\varrho}_{q_2})^2\, \hat{\mathscr{P}}_{1,2}]. \tag{A.23}$$

We will finally calculate the average of the anti-commutator of one- and two-particle operators. We have

$$\langle \Psi_0 | [\hat{\alpha}_2, \hat{\beta}_1]_+ | \Psi_0 \rangle = \tfrac{1}{2} \int dq_1\, dq_2\, dq_3\, dq_1'\, dq_2'\, dq_3'\, \delta(q_1 - q_1')\, \delta(q_2 - q_2')\, \delta(q_3 - q_3')\, \hat{a}_{q_1 q_2}\hat{b}_{q_3}(S_1 + S_2).$$

The expression

$$S_1 = \langle \Psi_0 | \hat{\psi}^+(q_2')\, \hat{\psi}^+(q_1')\, \hat{\psi}(q_1)\, \hat{\psi}(q_2)\, \hat{\psi}^+(q_3)\, \hat{\psi}(q_3) | \Psi_0 \rangle$$

found here can be reduced to the density matrix by transposing the operator $\hat{\psi}^+(q_3')$ to the left:

$$S_1 = [\hat{\mathscr{P}}_{2,3}\delta(q_3 - q_3') + \hat{\mathscr{P}}_{1,3}\delta(q_3 - q_3')]\, R(q_1, q_2, q_1', q_2') + R(q_1, q_2, q_3, q_1', q_2', q_3').$$

Similarly,

$$S_2 = \langle \Psi_0 | \hat{\psi}^+(q_3')\, \hat{\psi}(q_3)\, \hat{\psi}^+(q_2')\, \hat{\psi}^+(q_1')\, \hat{\psi}(q_1)\, \hat{\psi}(q_2) | \Psi_0 \rangle$$
$$= [\hat{\mathscr{P}}_{2,3}\delta(q_2 - q_2')\, R(q_1, q_3, q_1', q_3') + \hat{\mathscr{P}}_{3,1}\delta(q_1 - q_1')\, R(q_2, q_3, q_2', q_3')] + R(q_1, q_2, q_3, q_1', q_2', q_3').$$

If we now go over to the trace, we obtain

$$\langle \Psi_0 | [\hat{\alpha}_2, \hat{\beta}_1]_+ | \Psi_0 \rangle = \tfrac{1}{2}\,\mathrm{Tr}\, \{\hat{a}_{q_1 q_2}\hat{b}_{q_3}[2((1 - \hat{\mathscr{P}}_{1,2})\, (1 - \hat{\mathscr{P}}_{2,3}) - \hat{\mathscr{P}}_{1,3})\, \hat{\varrho}_{q_1}\hat{\varrho}_{q_2}\hat{\varrho}_{q_3}(\hat{\mathscr{P}}_{2,3} + \hat{\mathscr{P}}_{1,3}) \times (1 - \hat{\mathscr{P}}_{1,2})\hat{\varrho}^2_{q_1}\hat{\varrho}_{q_2} + \hat{\mathscr{P}}_{2,3}(1 - \hat{\mathscr{P}}_{1,3})\hat{\varrho}^2_{q_1}\hat{\varrho}_{q_3} + \hat{\mathscr{P}}_{1,3}(1 - \hat{\mathscr{P}}_{2,3})\hat{\varrho}^2_{q_3}\hat{\varrho}_{q_2}]\}.$$

If, in the last two terms, we make the replacement

$$\hat{\mathscr{P}}_{2,3}(1 - \hat{\mathscr{P}}_{1,3}) \rightarrow (1 - \hat{\mathscr{P}}_{1,2})\, \hat{\mathscr{P}}_{2,3}, \quad \hat{\mathscr{P}}_{1,3}(1 - \hat{\mathscr{P}}_{2,3}) \rightarrow (1 - \hat{\mathscr{P}}_{1,2})\, \hat{\mathscr{P}}_{1,3},$$

we finally obtain

$$\langle\Psi_0|[\hat{\alpha}_2, \hat{\beta}_1]_+|\Psi_0\rangle = \mathrm{Tr}\,[\hat{a}_2(1-\hat{\mathscr{P}}_{1,2})\,\hat{\varrho}_{q_1}\hat{\varrho}_{q_2}]\,\mathrm{Tr}\,(\hat{b}\hat{\varrho})$$
$$+\,\mathrm{Tr}\,[\hat{a}_{q_1q_2}\hat{b}_{q_3}(1-\hat{\mathscr{P}}_{1,2})\hat{\varrho}_{q_2}(\hat{\varrho}_{q_1}-\hat{\varrho}_{q_3})^2\,\hat{\mathscr{P}}_{1,3}]. \tag{A.24}$$

As is evident from eqns. (A.21) to (A.24), the average of an operator product is by no means equal to the corresponding product of the averages. The supplementary terms correspond to the presence of an exchange (or statistical) correlation between particles, due to the anti-symmetry of the wave functions, i.e. to Pauli's principle.

Appendix B. The Basic Equations of Operator Calculus

B.1. Throughout the development of quantum theory, there has been a tendency for theoretical entities to be represented in an explicit operator form. We might, give as examples, the partition function in quantum statistics [137], the Lippman–Schwinger scattering theory [138], *S*-matrix theory [139] and statistical many-body theory.

The reason for changing to an operator formulation is, on the one hand, to obtain greater simplicity and compactness. On the other hand, these expressions describe the corresponding physical entity in a form which is independent of the type of representation. Finally, the operator formulation explicitly reflects the fact that the basic difference between quantum and classical methods of description lies in the operator character of the quantum approach. The operator formulation clearly reflects the implications of the Heisenberg relations. Also the problem of going over to the classical (or quasi-classical) case is much simplified, and its solution merely amounts to ignoring the commutators of the corresponding operators.

The simplicity of the operator formulation is to some extent compensated by the fact that operator expressions contain much more mathematically complicated quantities. As a rule, operator expressions are functions of several arguments which do not commute with one another. The rules for calculating these functions are very complicated: we will consider them in the next section. We will follow a simpler method, suggested by Bloch, which has an obvious physical interpretation [140]. Short descriptions and a bibliography of other methods of operator calculus are given elsewhere [51].

B.2. From now on, we will consider functions of sums of operators.

Suppose we have some function $f(\hat{a}+\hat{b})$ which is the sum of two non-commuting arguments $\hat{a}$ and $\hat{b}$. According to the definition, two operators $\hat{A}$ and $\hat{B}$ are considered to be functionally related $[\hat{A}=f(\hat{B})]$ if their eigenfunctions coincide, and their eigenvalues have the same functional relationship as the operators

$$\hat{B}\psi_n = B_n\psi_n;\quad \hat{A}\psi_n = f(B_n)\,\psi_n.$$

A function of non-commuting arguments is mathematically much more complicated than an ordinary function. Thus computations with certain functions do not obey the elementary mathematical rules. For example, $(\hat{a}+\hat{b})^2$ and $\exp(\hat{a}+\hat{b})$ are not equal to $\hat{a}^2+2\hat{a}\hat{b}+\hat{b}^2$ and $\exp\hat{a}\exp\hat{b}$, respectively; they depend in a very involved manner on the commutators of the operators $\hat{a}$ and $\hat{b}$.

The aim of operator calculus is to reformulate the rules so that an appropriate dependence on the commutators appears.

Appendix B

For a wide range of assumptions about the form of the function f, it is possible to expand it in a Fourier (or Laplace) integral

$$f(\hat{a}+\hat{b}) = \int d\tau f_\tau \exp[\tau(\hat{a}+\hat{b})], \tag{B.1}$$

where τ is a real or imaginary parameter. We may then restrict our consideration to the exponential function only

$$\hat{E}(\tau) = \exp[\tau(\hat{a}+\hat{b})].$$

If the operators $\hat{a}$ and $\hat{b}$ commuted, we would have simply $\hat{E}(\tau) = \exp(\tau\hat{b})\exp(\tau\hat{a})$. The additional dependence on the commutators can be included by introducing a supplementary factor $\hat{K}$

$$\hat{E}(\tau) = \exp(\tau\hat{b})\,\hat{K}(\tau)\exp(\tau\hat{a}). \tag{B.2}$$

Each of the factors in an operator product acts on all quantities to the right of it.

The order of the factors in (B.2) may vary (if another order is chosen, different expressions for $\hat{K}$ will, of course, be obtained). The one taken is suitable for the case where $f(\hat{a}+\hat{b})$, acts on ψ_a—the eigenfunction $\hat{a}$ with eigenvalue a. Here,

$$\hat{E}(\tau)\,\psi_a = \exp(\tau\hat{b})\,\hat{K}(\tau)\exp(\tau a)\,\psi_a = \exp[\tau(a+\hat{b})]\,\hat{K}(\tau)\,\psi_a.$$

The quantities in the exponential index now commute with one another, and the difference from ordinary exponents appears only in the factor $\hat{K}$. The function ψ_a, on which the given operator acts, will be omitted from our future discussion for the sake of brevity.

We will now find an equation for $\hat{K}$. Differentiating (B.2) with respect to τ, we obtain

$$\frac{\partial\hat{K}}{\partial\tau} = \exp(-\tau\hat{b})\,\hat{a}\exp(\tau\hat{b})\,\hat{K} - \hat{K}\hat{a}. \tag{B.3}$$

We note that the combination

$$\exp(-\tau\hat{b})\,\hat{a}\exp(\tau\hat{b}) = \sum_{n=0}^{\infty}\frac{(-\tau)^n}{n!}\overbrace{[\hat{b}[\hat{b}\ldots[\hat{b},\hat{a}]_-\ldots]_-]_-}^{n}$$

is expressed only in terms of the commutators of $\hat{a}$ and $\hat{b}$, as can be easily verified by expanding in a series in τ.

Putting $\tau = 0$ in (B.2), we obtain the boundary condition for $\hat{K}(\tau)$:

$$\hat{K}(0) = 1$$

B.3. If the operators $\hat{a}$ and $\hat{b}$ are such that the majority of their commutators are equal to zero, then eqn. (B.3) can have simple solutions.

1. Suppose, for example, that the commutator $[\hat{a},\hat{b}]_-$ is a c-number; this will be the only non-zero commutator. Then

$$\exp(-\tau\hat{b})\,\hat{a}\exp(\tau\hat{b}) = \hat{a} - \tau[\hat{b},\hat{a}]_-$$

and we obtain Glauber's equation [141]

$$\hat{E}(\tau) = \exp[\tau(a+\hat{b})]\exp\left\{-\frac{\tau^2}{2}[\hat{b},\hat{a}]_-\right\}. \tag{B.4}$$

For the more complicated case, where the commutators $[\hat{b}[\hat{b},\hat{a}]_-]_-$ and $[[\hat{b},\hat{a}]_-\hat{a}]_-$ are c-numbers, we have the equation

$$\hat{E}(\tau) = \exp[\tau(a+\hat{b})]\exp\left\{-\frac{\tau^2}{2}[\hat{b},\hat{a}]_-\right\}\exp\left(\frac{\tau^3}{6}\{[\hat{b}[\hat{b},\hat{a}]_-]_- + [[\hat{b},\hat{a}]_-\hat{a}]_-\}\right).$$

Similarly, we can separate out factors, corresponding to higher-order commutators, from $\hat{E}(\tau)$.

2. Suppose only commutators of the type $[\hat{b}[\hat{b} \ldots [\hat{b}, \hat{a}]_- \ldots]_-]_-$ are non-zero, i.e. they all commute with one another and with $\hat{a}$, then $\hat{K}$ also commutes with $\hat{a}$, and we can solve the equation for $\hat{K}$ easily

$$\hat{E}(\tau) = \exp(\tau\hat{b}) \exp\left\{\int_0^\tau dt \exp(-t\hat{b})\, \hat{a} \exp(t\hat{b})\right\}. \tag{B.5}$$

For example, if $\hat{a} = F(x)$ and $\hat{b} = \partial/\partial x$, then, from the properties of the displacement operator,† we obtain

$$\exp\left\{\tau\left[\frac{\partial}{\partial x} + F(x)\right]\right\} = \exp\left(\tau\frac{\partial}{\partial x}\right) \exp\left[\int_{x-\tau}^{x} F(\xi)\, d\xi\right]$$

$$= \exp\left[\int_x^{x+\tau} F(\xi)\, d\xi\right] \exp\left(\tau\frac{\partial}{\partial x}\right). \tag{B.6}$$

Another particular case corresponds to the commutator relations

$$[\hat{b}, \hat{a}]_- = \lambda\hat{a},$$

where λ is a c-number.

Here,

$$\hat{E}(\tau) = \exp(\tau\hat{b}) \exp\left[\frac{1 - \exp(-\tau\lambda)}{\lambda} a\right]. \tag{B.7}$$

3. A more complicated case occurs when commutators of the type $[\hat{a}\,[\hat{a} \ldots [\hat{a}, \hat{b}]_- \ldots]_-]_-$ are non-zero (commute with one another).

Then,

$$\frac{\partial \hat{K}}{\partial \tau} = [\hat{a}, \hat{K}]_- + \tau[\hat{a}, \hat{b}]_- \hat{K}.$$

Putting $\hat{K} = \exp(\hat{a}\tau)\, \hat{L} \exp(-\hat{a}\tau)$, we find

$$\hat{L} = \exp\left\{\int_0^\tau dt\, t \exp(-\hat{a}t)\, [\hat{a}, \hat{b}]_- \exp(\hat{a}t)\right\}.$$

In particular, when $[\hat{a}, \hat{b}]_- = \lambda\hat{b}$, we obtain

$$\hat{L} = \exp\left\{\frac{1 - \exp(-\lambda\tau)\,(1 + \lambda\tau)}{\lambda}\hat{b}\right\}.$$

Allowing for the relationship $\exp(\hat{a}\tau) f(\hat{b}) \exp(-\hat{a}\tau) = f[\exp(\lambda\tau)\, \hat{b}]$, which may easily be proved by expanding as a series in b, we finally find

$$\hat{E}(\tau) = \exp\left[\frac{\exp(\lambda\tau) - 1}{\lambda}\hat{b}\right] \exp(\tau a). \tag{B.8}$$

† The displacement operator is the quantity $\exp(a\nabla)$

$$\exp(a\nabla)\, \varphi(x) = \varphi(x + a) \exp(a\nabla).$$

B.4. Even if all the commutators are non-zero there is a case which gives a simple solution.

Commutation can be thought of as an operation acting on a set which includes $\hat{a}$ and $\hat{b}$ and all their commutators. If this set is actually finite, then further commutation will take us back to elements we already have. The operation will have a cyclic character.

We may naturally expand the required function in terms of the elements of the set, and the problem reduces to finding the finite number of the appropriate coefficients of the expansion [51, 142].

Suppose, for example, that the following relationships hold†

$$[\hat{a}, \hat{b}]_- = \hat{c}; \quad [\hat{a}, \hat{c}]_- = -\lambda\hat{a}; \quad [\hat{b}, \hat{c}]_- = \lambda\hat{b}.$$

The set consists of three elements: $\hat{a}$, $\hat{b}$ and $\hat{c}$. Our solution, correspondingly, has the form

$$\hat{E}(\tau) = \exp[\alpha(\tau)\,\hat{b}]\exp[\beta(\tau)\,\hat{c}]\exp[\gamma(\tau)\,\hat{a}], \tag{B.9}$$

where α, β and γ are unknown functions of τ. Differentiating this expression with respect to τ, we obtain

$$\hat{a} + \hat{b} = \alpha'(\tau)\,\hat{b} + \beta'(\tau)\exp(\alpha\hat{b})\,\hat{c}\exp(-\alpha\hat{b})$$
$$+ \gamma'(\tau)\exp(\alpha\hat{b})\exp(\beta\hat{c})\,\hat{a}\exp(-\beta\hat{c})\exp(-\alpha\hat{b}).$$

If we now use the commutation relations, and compare the coefficients of identical operators, we find

$$\gamma'(\tau) = \exp(-\lambda\beta); \quad \beta'(\tau) = \alpha; \quad \alpha'(\tau) = 1 - \lambda\alpha^2/2.$$

The solutions of these equations, which go to zero for $\tau = 0$, have the form

$$\alpha = \gamma = \left(\frac{2}{\lambda}\right)^{\frac{1}{2}} \tanh\left[\left(\frac{\lambda}{2}\right)^{\frac{1}{2}}\tau\right]; \quad \beta = \frac{2}{\lambda}\ln\cosh\left[\left(\frac{\lambda}{2}\right)^{\frac{1}{2}}\tau\right]. \tag{B.10}$$

The relationships (B.9) and (B.10) solve our problem.

This example corresponds to the quantum-mechanical oscillator problem. We have, in fact, in this case

$$\hat{a} = \hat{p}^2/2M; \quad \hat{b} = \frac{M\omega^2x^2}{2}; \quad \hat{c} = -\frac{\hbar^2\omega^2}{2}\left(1 + 2x\frac{\partial}{\partial x}\right); \quad \lambda = 2\hbar^2\omega^2.$$

From the equations we have obtained, we will calculate the following quantity

$$\exp\left[\tau\left((\hat{p}^2/2M) + \frac{M\omega^2x^2}{2}\right)\right]\exp(ipx)$$
$$= [\cosh(\hbar\omega\tau)]^{-1/2}\exp\left[\frac{\tanh(\hbar\omega\tau)}{\hbar\omega}\left(\frac{p^2}{2M} + \frac{M\omega^2x^2}{2}\right)\right]\exp\left[\frac{ipx}{\cosh(\hbar\omega\tau)}\right]. \tag{B.11}$$

This can be used to solve numerous problems concerning the oscillator.

† The commutators of three operators cannot be combined independently in pairs, since the Jacobi identity must be satisfied:

$$[[\hat{a}, \hat{b}]_-, \hat{c}]_- + [[\hat{b}, \hat{c}]_-, \hat{a}]_- + [[\hat{c}, \hat{a}]_-, \hat{b}]_- = 0.$$

When computing (B.11), we had to deal with functions of the operator product $\exp(kx\,\partial/\partial x)$. In principle we can apply the theory developed to this case, too. It is simpler, however, to use the following approach. Introducing a new variable $\ln x$, we reduce the operator to the normal displacement operator

$$\exp\left(kx\frac{\partial}{\partial x}\right)f(x) = \exp\left[k\frac{\partial}{\partial(\ln x)}\right]f[\exp(\ln x)] = f[\exp(k)\,x]. \tag{B.12}$$

Thus the operator $\exp(kx\,\partial/\partial x)$ acts as a scale-change operator.

B.5. In the general case all the commutators of $\hat{a}$ and $\hat{b}$ are non-zero, and form an infinite set. The solution of the problem now becomes extremely complicated. The situation simplifies if there are small parameters present.

Suppose, for example, that the role of the higher-order commutators decreases as their complexity increases. Then we can look for an expansion of the function concerned as a series in commutators of growing complexity. Each term of the series is, in the general case, a product of separate commutators. We will call the total number of commutations in a term the order of that term in the series. So, $[\hat{b}, [\hat{b}, \hat{a}]_-]_-$ and $[\hat{b}, \hat{a}]_-^2$ are of order two.

If we denote by $\hat{Q}_n^m$ the commutator of the $(n+m)^{\text{th}}$ order,

$$\hat{Q}_n^m = \underbrace{[\hat{a}[\hat{a} \dots [\hat{a}}_{m}\,\underbrace{[\hat{b} \dots [\hat{b}}_{n}, \hat{a}]_- \dots]_-]_- \dots]_-]_-$$

and by $\hat{K}_n$ the aggregate of all the nth order terms in $\hat{K}$, we can obtain the following recurrent relation from (B.3):

$$\frac{\partial \hat{K}_n}{\partial \tau} = [\hat{a}, \hat{K}_{n-1}]_- + \sum_{k=1}^{n} \frac{(-\tau)^k}{k!}\hat{Q}_k^0\hat{K}_{n-k}.$$

A straightforward calculation with $\hat{K}_0 = 1$ gives [54]

$$\hat{K}_1 = -\frac{\tau^2}{2}\hat{Q}_1^0, \tag{B.13}$$

$$\hat{K}_2 = \frac{\tau^3}{6}(\hat{Q}_2^0 - \hat{Q}_1^1) + \frac{\tau^4}{8}(\hat{Q}_1^0)^2, \tag{B.14}$$

$$\hat{K}_3 = -\frac{\tau^4}{24}(\hat{Q}_3^0 - \hat{Q}_2^1 + \hat{Q}_1^2) + \frac{\tau^5}{120}(3\hat{Q}_1^1\hat{Q}_1^0 + 7\hat{Q}_1^0\hat{Q}_1^1 - 4\hat{Q}_1^0\hat{Q}_2^0 - 6\hat{Q}_2^0\hat{Q}_1^0)$$
$$- \frac{\tau^6}{48}(\hat{Q}_1^0)^3, \tag{B.15}$$

$$\hat{K}_4 = \frac{\tau^5}{120}(\hat{Q}_2^2 - \hat{Q}_1^3 - \hat{Q}_3^1 + \hat{Q}_4^0) + \frac{\tau^6}{720}[3\hat{Q}_1^2\hat{Q}_1^0 + 10(\hat{Q}_1^1)^2 + 12\hat{Q}_1^0\hat{Q}_1^2$$
$$- 4\hat{Q}_1^1\hat{Q}_2^0 - 9\hat{Q}_1^0\hat{Q}_2^1 - 6\hat{Q}_2^1\hat{Q}_1^0 - 16\hat{Q}_2^0\hat{Q}_1^1 + 5\hat{Q}_1^0\hat{Q}_3^0 + 10(\hat{Q}_2^0)^2 + 10\hat{Q}_3^0\hat{Q}_1^0]$$
$$+ \frac{\tau^7}{1680}[15\hat{Q}_2^0(\hat{Q}_1^0)^2 - 11\hat{Q}_1^0\hat{Q}_1^1\hat{Q}_1^0 - 19(\hat{Q}_1^0)^2\hat{Q}_1^1 + 8(\hat{Q}_1^0)^2\hat{Q}_2^0 + 12\hat{Q}_1^0\hat{Q}_2^0\hat{Q}_1^0$$
$$- 5\hat{Q}_1^1(\hat{Q}_1^0)^2] + \frac{\tau^8}{384}(\hat{Q}_1^0)^4, \quad \text{etc.} \tag{B.16}$$

If we wish to change to an expansion in terms of the commutators of the function $f(\hat{a}+\hat{b})$ it is sufficient to note that, according to (B.1), the parameter τ acts as an operator for differentiating the function f in terms of its argument. Therefore, if we restrict ourselves to second-order terms for simplicity, we have

$$f(\hat{a}+\hat{b}) = f(a+\hat{b}) - \tfrac{1}{2}f''(a+\hat{b})\,[\hat{b},\hat{a}]_- + \tfrac{1}{6}f'''(a+\hat{b})\,\{[\hat{b}[\hat{b},\hat{a}]_-]_- - [\hat{a}[\hat{b},\hat{a}]_-]_-\} + \tfrac{1}{8}f^{\mathrm{IV}}(a+\hat{b})\,[\hat{b},\hat{a}]_-^2 + \dots \tag{B.17}$$

Here, as before, the number a in the argument of the function f and its derivatives is the eigenvalue of the operator $\hat{a}$, corresponding to the function on which $f(\hat{a}+\hat{b})$ acts.

B.6. If one of the arguments of our functions is small compared with the other, then the problem reduces to the operator analogue of perturbation theory, and consists of the expansion of $f(\hat{a}+\hat{b})$ in terms of the small argument. The equation thus obtained is distinguished from the Taylor series by terms which depend on the commutators.

We will first consider the expansion in terms of $\hat{b}$. Putting

$$\hat{E}(\tau) = \hat{L}(\tau)\exp(\tau\hat{a}),$$

differentiating this relationship with respect to τ, and denoting the nth order term in $\hat{b}$ by $\hat{L}_n$, we obtain

$$\frac{\partial \hat{L}_n}{\partial\tau} = \hat{L}_{n-1}\exp(\tau\hat{a})\,\hat{b}\exp(-\tau\hat{a});\quad \hat{L}_n = \int_0^\tau dt\hat{L}_{n-1}\exp(t\hat{a})\,\hat{b}\exp(-t\hat{a}).$$

Remembering that $\hat{L}_0 = 1$, we find [139]

$$\hat{E}(\tau) = \exp(\tau\hat{a}) + \int_0^\tau dt\exp(t\hat{a})\,\hat{b}\exp[(\tau-t)\,\hat{a}] + \dots \tag{B.18}$$

This expression can be put in another form

$$\hat{E}(\tau) = \left\{1 + \tau\hat{b} + \sum_{n=1}^{\infty}\frac{\tau^{n+1}}{(n+1)!}\underbrace{[\hat{a}[\hat{a}\dots[\hat{a},\hat{b}]_-\dots]_-]_-}_{n}\right\}\exp(\tau\hat{a}).$$

From these relationships, we can also find higher-order terms. Similarly, we can obtain an expansion in terms of $\hat{a}$

$$\hat{E}(\tau) = \exp(\tau\hat{b}) + \int_0^\tau dt\exp[(\tau-t)\,\hat{b}]\,\hat{a}\exp(t\hat{b}) + \dots \tag{B.18'}$$

The inverse function is given by the simple equation

$$(\hat{a}+\hat{b})^{-1} = (1+\hat{a}^{-1}\hat{b})^{-1}\,\hat{a}^{-1} = \sum_{n=0}^{\infty}(-1)^n(\hat{a}^{-1}\hat{b})^n\,\hat{a}^{-1}. \tag{B.19}$$

We will now consider the differentiation of the function $\hat{E}_\lambda(\tau) = \exp[\tau(\hat{a}+\lambda\hat{b})]$ in terms of the parameter λ. We start with the definition

$$\partial\hat{E}_\lambda(\tau)/\partial\lambda = \frac{1}{\delta\lambda}\,[\hat{E}_{\lambda+\delta\lambda}(\tau) - \hat{E}_\lambda(\tau)]$$

and expand the exponent as a series in $\delta\lambda$. We obtain as a result

$$\partial \hat{E}_\lambda(\tau)/\partial\lambda = \int_0^\tau dt \hat{E}_\lambda(t)\, \hat{b} \hat{E}_\lambda(\tau - t). \tag{B.20}$$

Since, moreover, $\hat{E}_\lambda(\tau - t) = \hat{E}_\lambda(-t)\, \hat{E}_\lambda(\tau)$, and

$$\hat{E}_\lambda(t)\, \hat{b} \hat{E}_\lambda(-t) = \sum_{n=0}^{\infty} \frac{t^n}{n!} \left[\overbrace{[\hat{a} + \lambda\hat{b}[\hat{a} + \lambda\hat{b} \ldots [\hat{a} + \lambda\hat{b}, \hat{b}]_- \ldots]_-}^{n} \right]_-,$$

we find that

$$\frac{\partial \hat{E}_\lambda(\tau)}{\partial\lambda} = \left\{ \tau\hat{b} + \sum_{n=1}^{\infty} \frac{\tau^{n+1}}{(n+1)!} \left[\overbrace{[\hat{a} + \lambda\hat{b}[\hat{a} + \lambda\hat{b} \ldots [\hat{a} + \lambda\hat{b}, \hat{b}]_- \ldots]_-]_-}^{n} \right]_- \right\} \hat{E}_\lambda(\tau). \tag{B.21}$$

This is called the left-hand derivative. If we make the replacement in (B.20) $t \to t + \tau$, we also determine the right-hand derivative

$$\frac{\partial \hat{E}_\lambda(\tau)}{\partial\lambda} = \hat{E}_\lambda(\tau) \left\{ \tau\hat{b} - \sum_{n=1}^{\infty} \frac{(-\tau)^{n+1}}{(n+1)!} \left[\overbrace{[\hat{a} + \lambda\hat{b}[\hat{a} + \lambda\hat{b} \ldots [\hat{a} + \lambda\hat{b}, \hat{b}]_- \ldots]_-}^{n} \right]_- \right\}. \tag{B.22}$$

More complicated rules, referring to the relativistic case, can be found elsewhere [51].

Appendix C. Singular Integrals

C.1. In many-body theory we frequently encounter integrations of singular functions, which go to infinity in the region of integration. It is convenient to represent such integrals in the form

$$I = \int_{-\infty}^{\infty} dx \frac{F(x)}{\varphi(x)}, \tag{C.1}$$

where $F(x)$ is a function which is regular over the region of integration, and $\varphi(x_i) = 0$, $(i = 1, 2, \cdots)$. The x_i are the singular points of the given function.

This type of integral is not determinate; we need special, supplementary conditions to tell us how we should define the integration around the pole (that is to say, how we should enclose it).

The integrals of singular functions are indeterminate because the singular function $1/\varphi(x)$ is itself indeterminate at the points x_i, where there is no unique value of the function. We can put

$$\frac{1}{\varphi(x)} = \begin{cases} \dfrac{1}{\varphi(x)} & x \neq x_i \\ A_i & x = x_i \end{cases}$$

where the A_i are certain completely arbitrary quantities. This expression can be written as a sum of the principal value of the singular function

$$\mathrm{P}\frac{1}{\varphi(x)} = \begin{cases} \dfrac{1}{\varphi(x)} & x \neq x_i \\ 0 & x = x_i \end{cases} \tag{C.2}$$

and of the function

$$\mathrm{D}\frac{1}{\varphi(x)} = \begin{cases} 0 & x \neq x_i \\ A_i & x = x_i . \end{cases} \tag{C.3}$$

We can consider the singular function as a distribution, i.e. we are not interested in the value of this function, but in the result of integrating it with arbitrary weight functions $F(x)$. Hence if the A_i are finite, the contribution of the function $\mathrm{D}\,(1/\varphi)$ to the integral is zero, and the singular function does not, in fact, differ from its principal value. The case of real interest is when the A_i are infinite: the corresponding integral over a small region round a point x_i now gives a finite contribution.

If we introduce the δ-function $\delta(x-x_i)$, which is equal to zero for $x \neq x_i$, and which has an integral which is equal to one for any small region round the point x_i, we can write the following for (C.3):

$$\mathrm{D}\frac{1}{\varphi(x)} = \sum_i a_i \delta(x-x_i), \tag{C.4}$$

where the a_i are arbitrary quantities, as before.†

The most frequent case is when the singular function has a first-order pole, i.e. $\varphi'(x_i) \neq 0$. The last equation can then be rewritten as

$$\mathrm{D}\frac{1}{\varphi(x)} = A(x)\,\delta[\varphi(x)],$$

where A is an arbitrary function which does not have zeroes or poles at the points x_i.

In fact,

$$A(x)\,\delta[\varphi(x)] = \sum_i \frac{A(x_i)}{|\varphi'(x_i)|}\delta(x-x_i);$$

identifying $A/|\varphi'|$ with a_i, we get back to (C.4).

Thus we can obtain the final most general form for the singular function $1/\varphi(x)$ by putting

$$\frac{1}{\varphi(x)} = \mathrm{P}\frac{1}{\varphi(x)} + A(x)\,\delta[\varphi(x)]. \tag{C.5}$$

If the function under consideration has no poles, we obtain the identity $1/\varphi = 1/\varphi$.

† Although it might appear at first glance that we could also introduce derivatives of the δ-function, the differentiation can, in fact, be transferred to the function F.

C.2. The functions $P(1/\varphi)$ and $\delta(\varphi)$ can be written explicitly using some limiting process

$$P\frac{1}{\varphi} = \lim_{\delta \to 0} \frac{\varphi}{\varphi^2 + \delta^2}, \tag{C.6}$$

$$\delta(\varphi) = \lim_{\delta \to 0} \frac{1}{\pi} \frac{\delta}{\varphi^2 + \delta^2}. \tag{C.7}$$

To prove this, we will consider the values of both sides of these relationships, first of all for the points $x \neq x_i$. Where $\varphi \neq 0$, (C.6) gives $1/\varphi$ in the limit and (C.7) zero, in agreement with the definitions of these functions. So far as the points x_i are concerned, where $\varphi = 0$, then, from (C.2), $P\,1/\varphi(x_i) = 0$, and $\delta[\varphi(x_i)] = 1/\pi\delta \to \infty$.

The integral†

$$\int_{x_i-\alpha}^{x_i+\beta} dx\, \delta[\varphi(x)] = \frac{1}{\pi} \int_{x_i-\alpha}^{x_i+\beta} dx\, \frac{\delta}{\varphi^2(x) + \delta^2}$$

is now equal to $1/|\varphi'(x_i)|$ (as it ought to be), as can be easily confirmed by putting $\varphi(x_i) = \varphi'(x_i)(x - x_i)$, and making the replacement $x = x_i + \delta t/|\varphi'(x_i)|$.

We see from (C.6) and (C.7) that $P(1/\varphi)$ is an odd function of φ, while $\delta(\varphi)$ is an even function. These properties could, in principle, break down near the points x_i. In this case the results of computing the corresponding integrals might be different. It is therefore usual to require that $P(1/\varphi)$ is the limit of the odd function, and $\delta(\varphi)$ that of the even function, i.e. the expressions conserve these properties as we go to the limit.

Expanding (C.6) and (C.7) as Fourier integrals in φ, we can write down an integral representation of the functions

$$P\frac{1}{\varphi} = \frac{1}{2i} \lim_{\delta \to 0} \int_{-\infty}^{\infty} dt\, \mathrm{sign}\,(t) \exp\,(it\varphi - \delta|t|), \tag{C.8}$$

$$\delta(\varphi) = \frac{1}{2\pi} \lim_{\delta \to 0} \int_{-\infty}^{\infty} dt\, \exp\,(it\varphi - \delta|t|). \tag{C.9}$$

We can show that an integral of the type (C.1) taken near one of the points x_i, is completely determinate in magnitude when substituted for $1/\varphi$ in (C.5). We can, in fact, make a replacement similar to the one above, near x_i. If we use (C.6) and (C.7), this gives

$$\int_{x_i-\alpha}^{x_i+\beta} dx\, \frac{F(x)}{\varphi(x)} = \frac{F(x_i)}{\varphi'(x_i)} \{\ln\,(\beta/\alpha) + A(x_i)\}.$$

It is characteristic for a logarithm to appear in the integral of a principal value of a function.

Hence an integral containing the principal value of a singular function,‡ or a δ-function, has a completely determinate value.

† The region of integration is chosen in such a way that only one pole x_i occurs within it.

‡ It is called a principal value integral, and is denoted by

$$\int dx\, P\frac{1}{\varphi} F = P \int dx\, \frac{F}{\varphi}.$$

C.3. The following linear combination of the functions P $1/\varphi$ and $\delta(\varphi)$ plays a special part in many-body theory:

$$\frac{1}{\varphi_{(\pm)}} = \mathrm{P}\frac{1}{\varphi} \pm i\pi\delta(\varphi). \tag{C.10}$$

The general expression (C.5) can be expressed in terms of these by

$$\frac{1}{\varphi} = \frac{1}{2}\left(1 - \frac{iA}{\pi}\right)\frac{1}{\varphi_{(+)}} + \frac{1}{2}\left(1 + \frac{iA}{\pi}\right)\frac{1}{\varphi_{(-)}}. \tag{C.11}$$

We can use the limit process to express the functions $1/\varphi_{(\pm)}$

$$\frac{1}{\varphi_{(\pm)}} = \lim_{\delta \to 0} \frac{1}{\varphi \mp i\delta}. \tag{C.12}$$

This can be proved by transferring the imaginary term to the numerator, and using (C.6) and (C.7).

From the integral representations (C.8) and (C.9), we can express the functions $1/\varphi_{(\pm)}$ in integral form

$$\frac{1}{\varphi_{(\pm)}} = \pm i \int_{-\infty}^{\infty} dt \, \frac{1 \mp \operatorname{sign}(t)}{2} \exp(it\varphi - \delta|t|) \tag{C.13}$$

or

$$\frac{1}{\varphi + i\delta} = -i \int_{0}^{\infty} dt \exp(it\varphi - \delta|t|), \tag{C.14}$$

$$\frac{1}{\varphi - i\delta} = i \int_{-\infty}^{0} dt \exp(it\varphi - \delta|t|). \tag{C.15}$$

C.4. It is very convenient in calculating integrals of the singular-function type (C.1), to use the methods of complex-variable theory, i.e. analytically continue F/φ into the complex plane of x.

We usually have to deal in practice with the case where the analytical continuation of the function F/φ into the upper, or lower, half-planes gives a function with the following two properties. Firstly, the function decreases sufficiently rapidly on the arc of a large circle† and, secondly, the poles are the only singularities in the half-plane concerned. Denoting by the suffixes $\pm$ the integral for which these properties are satisfied in the upper and lower half-plane, we can complete the integration contour round a large arc

$$I_{\pm} = \int_{C\pm} dx \, \frac{F(x)}{\varphi(x)}$$

(the contours $C_{\pm}$ are shown in Fig. 69). Using the residue theorem, we can write

$$I_{\pm} = \pm 2\pi i \sum \operatorname{Res}\left(\frac{F}{\varphi}\right), \tag{C.16}$$

† We will suppose that the integral of F/φ round this arc gives a vanishingly small contribution.

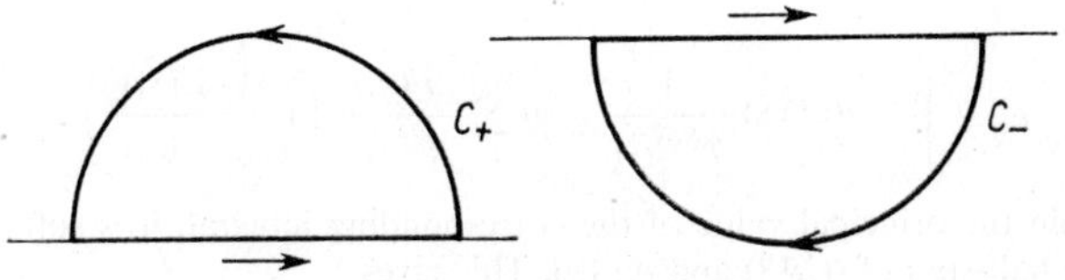

FIG. 69

where the summation is taken over all the poles in the expression under the integral sign, and within the contour.

Some of these poles are those at the points x_i. It is evident from (C.11) that it is sufficient to investigate only the poles of the function $1/\varphi_{(\pm)}$. These poles are near the points x_i, but are situated in the complex plane. Supposing that the pole $\bar{x}_i$, for which $\varphi(\bar{x}_i) \pm i\delta = 0$, can be put in the form $\bar{x}_i = x_i + \delta x_i (\delta x_i \ll x_i)$, we have $\varphi'(x_i)\,\delta x_i \pm i\delta = 0$, whence

$$\bar{x}_i = x_i \mp i\frac{\delta}{\varphi'(x_i)}.$$

Thus the pole of the function $1/\varphi_{(+)}$ lies in the lower (for $\varphi' < 0$), or upper (for $\varphi' > 0$), half-plane; the converse is true for $1/\varphi_{(-)}$ (Fig. 70). The residue has the form

$$\frac{F(x_i)}{\varphi'(x_i)}\frac{1}{2}\left(1 \mp \frac{iA(x_i)}{\pi}\right)$$

for the first and second term in (C.11) respectively.

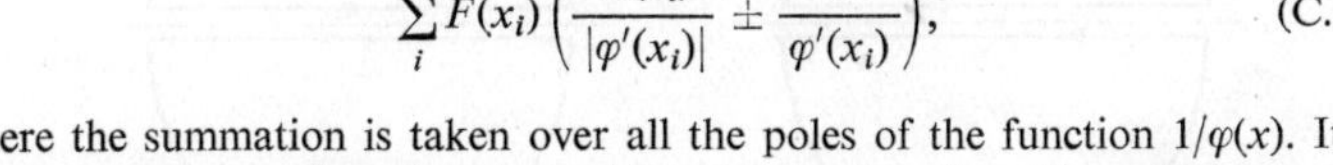

FIG. 70

We can therefore write the contribution of the poles x_i to $I_\pm$ in the form

$$\sum_i F(x_i)\left(\frac{A(x_i)}{|\varphi'(x_i)|} \pm \frac{\pi i}{\varphi'(x_i)}\right), \tag{C.17}$$

where the summation is taken over all the poles of the function $1/\varphi(x)$. It is also necessary to allow for the residues of the function F/φ, which are not connected with the poles of $1/\varphi$.

We will consider some particular cases of importance. Suppose the expression under the integral sign contains only poles at the point x_i. Then $I_\pm$ coincides with the expression (C.17). Putting $A(x) = -i\pi$, we can write

$$I_\pm = \int_{-\infty}^{\infty} dx F(x)\frac{1}{\varphi(x)_{(-)}} = -\pi i \sum_i \frac{F(x_i)}{|\varphi'(x_i)|}\left(1 \mp \frac{\varphi'(x_i)}{|\varphi'(x_i)|}\right). \tag{C.18}$$

In other words, only those poles of $1/\varphi(x)$ for which the derivative $\varphi'(x_i)$ is negative (or positive) contribute to this integral. Similarly, putting $A(x) = i\pi$, we find

$$I_{\pm} = \int_{-\infty}^{\infty} dx F(x) \frac{1}{\varphi(x)_{(+)}} = \pi i \sum_i \frac{F(x_i)}{|\varphi'(x_i)|}\left(1 \pm \frac{\varphi'(x_i)}{|\varphi'(x_i)|}\right). \qquad \text{(C.19)}$$

To obtain the principal value of the corresponding integral, it is sufficient to take the half-sum of (C.18) and (C.19). This gives

$$I_{\pm} = \mathrm{P}\int_{-\infty}^{\infty} dx \frac{F(x)}{\varphi(x)} = \pm\pi i \sum_i \frac{F(x_i)}{\varphi'(x_i)}. \qquad \text{(C.20)}$$

If one pole has a positive derivative, then

$$\left.\begin{aligned}
&\int_{-\infty}^{\infty} dx F(x) \frac{1}{\varphi(x)_{(-)}} = \begin{cases} 0 \\ -2\pi i \dfrac{F(x_i)}{\varphi'(x_i)}, \end{cases} \\
&\int_{-\infty}^{\infty} dx F(x) \frac{1}{\varphi(x)_{(+)}} = \begin{cases} 2\pi i \dfrac{F(x_i)}{\varphi'(x_i)}, \\ 0 \end{cases} \\
&\mathrm{P}\int_{-\infty}^{\infty} dx F(x) \frac{1}{\varphi(x)} = \begin{cases} \pi i \dfrac{F(x_i)}{\varphi'(x_i)} \\ -\pi i\, F(x_i)/\varphi'(x_i), \end{cases}
\end{aligned}\right\} \qquad \text{(C.21)}$$

where the upper value corresponds to closure of the integral in the upper half-plane, and the lower value correspondingly in the lower half-plane. The relationships of (C.21) are illustrated in Figs. 71a and 71b, where the poles

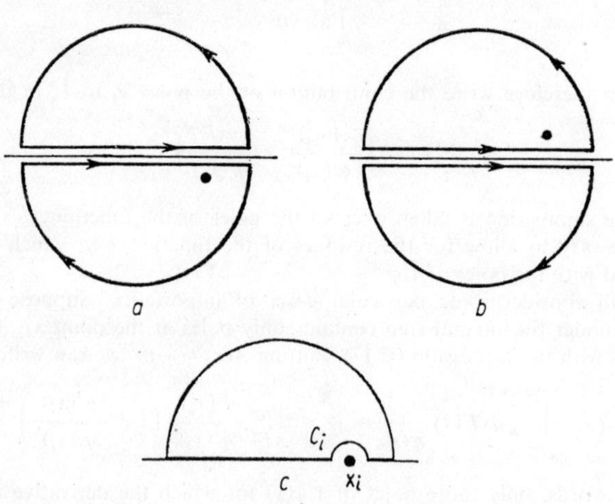

FIG. 71

of the expression in the first two integrals are indicated. A zero value for these integrals corresponds to Cauchy's theorem, since under these circumstances there are no poles inside the integration contour.

The principal-value integral is taken over the region indicated in Fig. 71c. If the contour is closed by a large arc, this integral is equal to the integral round the small contour C_i, i.e. the half-residue of the expression under the integral sign at the point x_i.

C.5. The fact that certain of the integrals in (C.21) vanish is related to the question of causality. Let us consider the integral over ε which enters into the propagation function (see § 10):

$$I = \int_{-\infty}^{\infty} \frac{d\varepsilon}{2\pi} \frac{\exp(-i\varepsilon\tau)}{\varepsilon - \varepsilon_\nu + i\delta \operatorname{sign}(\varepsilon_\nu - \varepsilon_F)}, \tag{C.22}$$

where $\tau = t_1 - t_2$. When $\varepsilon_\nu < \varepsilon_F$, the function is $1/(\varepsilon - \varepsilon_\nu)_{(+)}$, when $\varepsilon_\nu > \varepsilon_F$, the function is $1/(\varepsilon - \varepsilon_\nu)_{(-)}$. Introducing the occupation number $n_\nu = \theta(\varepsilon_F - \varepsilon_\nu)$, we can write

$$I = \int_{-\infty}^{\infty} \frac{d\varepsilon}{2\pi} \exp(-i\varepsilon\tau) \left\{ \frac{n_\nu}{(\varepsilon - \varepsilon_\nu)_{(+)}} + \frac{1 - n_\nu}{(\varepsilon - \varepsilon_\nu)_{(-)}} \right\}.$$

The possibility of closing the integral in the upper, or lower, half-plane is closely connected with the sign of the time difference τ. Writing the complex ε as $\alpha + i\beta$, we have

$$\exp(-i\varepsilon\tau) = \exp(-i\alpha\tau) \exp(\beta\tau).$$

As we go away from the real axis, i.e. as $|\beta| \to \infty$, $\exp(\beta\tau)$ grows without limit for $\beta\tau > 0$, and decreases for $\beta\tau < 0$. Closure is only possible in the latter case. When $\tau > 0$, the closure is in the lower half-plane ($\beta < 0$), when $\tau < 0$, it is in the upper half-plane ($\beta > 0$).

Using (C.21), we have

$$I = \begin{cases} -i(1 - n_\nu) \exp(-i\varepsilon_\nu\tau) & \tau > 0 \\ i n_\nu \exp(-i\varepsilon_\nu\tau) & \tau < 0 \end{cases} \tag{C.22'}$$

Thus particles are propagated forward in time ($\tau < 0$), holes are propagated backward ($\tau < 0$). This is a result of causality. The corresponding method of integrating around a pole (the function $i\delta \operatorname{sign}(\varepsilon_\nu - \varepsilon_F)$ in the denominator) is called the causal (or Feynman) method.

C.6. We will now integrate certain products of Green functions $G_\mu(\varepsilon)$ over the energy ε.

We will first consider the product of two Green functions.

$$I = \int_{-\infty}^{\infty} \frac{d\varepsilon}{2\pi} \{[\varepsilon - \varepsilon_{\mu_1} + i\delta \operatorname{sign}(\varepsilon_{\mu_1} - \varepsilon_F)] [\varepsilon - \varepsilon_{\mu_2} + i\delta \operatorname{sign}(\varepsilon_{\mu_2} - \varepsilon_F)]\}^{-1}. \tag{C.23}$$

In this case the contour can be closed both above and below. From the disposition of the poles in this expression

$$\varepsilon_{1,2} = \begin{cases} \varepsilon_{\mu_{1,2}} - i\delta & \varepsilon_{\mu_{1,2}} > \varepsilon_F \\ \varepsilon_{\mu_{1,2}} + i\delta & \varepsilon_{\mu_{1,2}} < \varepsilon_F \end{cases}$$

it is evident that the integral goes identically to zero if both energies $\varepsilon_{\mu_{1,2}}$ are to one side of the Fermi surface. Both the poles here are on one side of the real axis. When the contour in the other half-plane is closed, a zero result is obtained from Cauchy's theorem. We must therefore consider two cases:

$$\varepsilon_{\mu_1} > \varepsilon_F \quad \text{and} \quad \varepsilon_{\mu_2} < \varepsilon_F; \quad \text{or} \quad \varepsilon_{\mu_1} < \varepsilon_F \quad \text{and} \quad \varepsilon_{\mu_2} > \varepsilon_F.$$

Suppose we close the contour in the upper half-plane. In the first case the contour of integration encloses the pole $\varepsilon = \varepsilon_{\mu_2} + i\delta$. The residue is equal to $(\varepsilon_{\mu_2} - \varepsilon_{\mu_1} + i\delta)^{-1}$, that is†

$$I = \frac{i}{\varepsilon_{\mu_2} - \varepsilon_{\mu_1} + i\delta}.$$

In the second case we must exchange the indices $1 \rightleftarrows 2$. Summing the results obtained, and introducing occupation numbers, we can write

$$I = i\left[\frac{(1-n_{\mu_1})\,n_{\mu_2}}{\varepsilon_{\mu_2} - \varepsilon_{\mu_1} + i\delta} + \frac{(1-n_{\mu_2})\,n_{\mu_1}}{\varepsilon_{\mu_1} - \varepsilon_{\mu_2} + i\delta}\right]. \tag{C.24}$$

We next consider integrating the product of three Green functions

$$I = \int \frac{d\varepsilon_1\, d\varepsilon_2}{(2\pi)^2} \{[\varepsilon_1 - \varepsilon_{\mu_1} + i\delta \operatorname{sign}(\varepsilon_{\mu_1} - \varepsilon_F)]\,[\varepsilon_2 - \varepsilon_{\mu_2} + i\delta \operatorname{sign}(\varepsilon_{\mu_2} - \varepsilon_F)]$$

$$\times\, [\varepsilon_1 + \varepsilon_2 - \varepsilon_{\mu_3} + i\delta \operatorname{sign}(\varepsilon_{\mu_3} - \varepsilon_F)]\}^{-1}. \tag{C.25}$$

We will first integrate over ε_2. The corresponding integral reduces to (C.18), if we make the replacement $\varepsilon_{\mu_3} \to \varepsilon_{\mu_3} - \varepsilon_1$. We obtain as a result

$$i\left[\frac{(1-n_{\mu 3})\,n_{\mu 2}}{\varepsilon_{\mu 2} - \varepsilon_{\mu 3} + \varepsilon_1 + i\delta} + \frac{(1-n_{\mu 2})\,n_{\mu 3}}{\varepsilon_{\mu 3} - \varepsilon_{\mu 2} - \varepsilon_1 + i\delta}\right].$$

Hence the repeated integral can be separated into a linear combination of two integrals

$$I_\pm = \int \frac{d\varepsilon_1}{2\pi} \{[\varepsilon_1 - \varepsilon_{\mu 1} + i\delta \operatorname{sign}(\varepsilon_{\mu 1} - \varepsilon_F)]\,[\varepsilon_1 + \varepsilon_{\mu 2} - \varepsilon_{\mu 3} \pm i\delta]\}^{-1}.$$

It is convenient to make the closure in the upper half-plane for the integral I_+. The single pole that must be allowed for, when $\varepsilon_{\mu_1} < \varepsilon_F$, is at the point $\varepsilon_{\mu_1} + i\delta$. When $\varepsilon_{\mu_1} > \varepsilon_F$ all the poles are on one side of the real axis, and give a zero contribution. Thus

$$I_+ = \frac{i n_{\mu 1}}{\varepsilon_{\mu 1} + \varepsilon_{\mu 2} - \varepsilon_{\mu 3} + i\delta}.$$

† There is, of course, no difference between $2i\delta$ and $i\delta$ as $\delta \to 0$.

Similarly,

$$I_- = \frac{-i(1-n_{\mu_1})}{\varepsilon_{\mu_1}+\varepsilon_{\mu_2}-\varepsilon_{\mu_3}-i\delta}.$$

Finally,

$$I = -\frac{(1-n_{\mu_1})(1-n_{\mu_2})\,n_{\mu_3}}{\varepsilon_{\mu_1}+\varepsilon_{\mu_2}-\varepsilon_{\mu_3}-i\delta} - \frac{n_{\mu_1}n_{\mu_2}(1-n_{\mu_3})}{\varepsilon_{\mu_1}+\varepsilon_{\mu_2}-\varepsilon_{\mu_3}+i\delta}. \qquad \text{(C.26)}$$

In some cases we have to deal with multiple poles. Suppose we consider, for example, the integral

$$I = \int_{-\infty}^{\infty} \frac{d\varepsilon}{2\pi}\,\frac{F(\varepsilon)}{[\varepsilon-\varepsilon_\nu+i\delta\,\mathrm{sign}\,(\varepsilon_\nu-\varepsilon_F)]^2}, \qquad \text{(C.27)}$$

where the function $F(\varepsilon)$ does not have a singularity in the upper (or lower) half-plane, and is bounded on the arc of a large circle. Closing the contour in the corresponding half-plane, and working out the residue at the multiple pole, we have

$$I_- = i(1-n_\nu)\,F'(\varepsilon_\nu). \qquad \text{(C.28)}$$

Similarly,

$$I_+ = -in_\nu F'(\varepsilon_\nu). \qquad \text{(C.29)}$$

Here the $I_\pm$ are the integrals corresponding to closure in the upper (or lower) half-plane.

C.7. We obtain a characteristic example of a singular function when we consider the quantum-mechanical scattering problem. Schrödinger's equation in the centre-of-mass system, has the form

$$(\nabla^2+k^2)\,\psi(\boldsymbol{r}) = MV\psi(\boldsymbol{r}).$$

If we change to the momentum representation, and divide both sides of the equation by k^2-p^2, we obtain

$$\psi_p = M(V\psi)_p/(k^2-p^2).$$

This expression is indeterminate unless we take into account the boundary conditions which come from the physical picture of the scattering. The solution must contain, as one term, a plane incident wave,† $\psi_{0p} = \delta(p-k)$; the second term ψ_p must describe a diverging wave which has the asymptotic form $\exp(ikr)/r$ for $r\to\infty$ in the coordinate representation.

These two conditions completely determine the singular function $(k^2-p^2)^{-1}$:

$$\frac{1}{k^2-p^2} = C\delta(k^2-p^2) + \frac{1}{k^2-p^2+i\delta}. \qquad \text{(C.30)}$$

where $C(V\psi)_p = 1/4\pi Mk$. The expression (C.30) corresponds to the general relationship (C.5), with $A = C - i\pi$.

It is simple to confirm that the first term gives the required contribution, if we remember that

$$\delta(k^2-p^2) = \frac{\delta(k-p)}{2k},$$

† We consider, for simplicity, only s-scattering.

where $k > 0$ and $p > 0$. The contribution of the second term to ψ_p is given by $M(V\psi)_p/(k^2 - p^2 + i\delta)$. In the coordinate representation this is:

$$M \int \frac{dp^3 (V\psi)_p \exp i(\boldsymbol{p} \cdot \boldsymbol{r})}{k^2 - p^2 + i\delta}.$$

For large r, p must be small. Assuming short-range forces, finite in magnitude, we find

$$\frac{M(V\psi)_0}{r} \int_0^\infty \frac{\sin(pr)\, p\, dp}{k^2 - p^2 + i\delta} = -\frac{M(V\psi)_0}{r} \frac{\partial}{\partial r} \int_0^\infty \frac{\cos(pr)\, dp}{k^2 - p^2 + i\delta}.$$

The integral in this case can be written as

$$\frac{1}{2} \int_{-\infty}^{\infty} \frac{dp \exp(ipr)}{k^2 - p^2 + i\delta} \sim \exp(ikr).$$

We have here closed the contour in the upper half-plane ($r > 0$), and used the relationships of C.4. We thus obtain a diverging wave. If we changed the sign of $i\delta$, the wave would be converging.

Bibliography

1. TER HAAR, D., *Introduction to the Physics of Many-body Systems*, Interscience (1958).
2. BRUECKNER, K., and GAMMEL, J., *Phys. Rev.* **109**, 1023 (1958); *Rev. Mod. Phys.* **30**, 561 (1958).
3. BOGOLIUBOV, N. N., *Lectures on Quantum Statistics*, Kiev, Sovietskaia Shkola, 1949 (in Ukrainian). English translation to be published by Gordon & Breach, New York.
4. LANDAU, L. D., and LIFSHITZ, E. M., *Statistical Physics*, Pergamon Press 1958.
5. BONCH-BRUEVICH, V. L., and TYABLIKOV, S. V., *Green Function Methods in Statistical Mechanics*, North Holland Publishing Co. (1962).
6. ABRIKOSOV, A. A., GORKOV, L. P., and DZIALOSHINSKII, I. E., *Methods of Quantum Field Theory in Statistical Physics*, 2nd ed. Pergamon Press 1965.
7. MIGDAL, A. B., *Soviet Phys, JETP* **13**, 478 (1961); **16**, 1366 (1962). *Theory of Finite Fermi-Systems and the Properties of Atomic Nuclei*, Moscow (1965). English translation to be published by Interscience.
8. BOGOLIUBOV, N. N., TOLMACHEV, V. V., and SHIRKOV, D. V., *A New Method in Superconductivity Theory*, Consultants Bureau (1959).
9. THOULESS, D. J., *The Quantum Mechanics of Many-body Systems*, Academic Press, N. Y., London 1961.
10. FRADKIN, E. S., *Soviet Phys. JETP* **9**, 912 (1959); **11**, 114 (1960); *Nucl. Phys.* **12**, 465 (1959).
11. BONCH-BRUEVICH, V. L., and KOGAN, SH. M., *Ann. of Physics* **9**, 125 (1960).
12. ZUBAREV, D. N., *Soviet Phys. Uspekhi* **3**, 320 (1960).
13. ALEKSEEV, A. I., *Soviet Phys. Uspekhi* **4**, 23 (1961).
14. *The Many-body Problem* (Cours donnés à l'école d'été de physique théorique, Les Houches, 1958), Methuen, London 1959.
15. Proceedings of the International Congress on Many-particle Problems (Utrecht, 1960), Physica Supplement (1962).
16. MARTIN, P., and SCHWINGER, J., *Phys. Rev.* **115**, 1342 (1959); KADANOFF, L., and MARTIN, P., *Phys. Rev.* **124**, 670 (1960).
17. BOGOLIUBOV, N. N., *J. Phys. U.S.S.R.* **11**, 23 (1947).
18. BELIAEV, S. T., *Soviet Phys. JETP* **7**, 289 (1958).
19. BRUECKNER, K., and SAWADA, K., *Phys. Rev.* **106**, 1117, 1128 (1957).
20. KONSTANTINOV, O. V., and PEREL, V. I., *Soviet Phys. JETP* **12**, 142 (1960); DZIALOSHINSKII, I. E., *Soviet Phys. JETP* **15**, 778 (1962). L. V. KELDYSH, *Soviet Phys. JETP* **20**, 1018 (1964).
21. *The Application of Quantum Field Methods to Many-body Theory* (Ed. by ALEKSEEV, A. I.), Gosatomizdat, 1963 (in Russian).

22. Pines, D., *The Many-body Problem*, Benjamin (1961).
23. Tolmachev, V. V., *The Field Theoretical Form of Perturbation Theory for the Many-electron Problem in Atoms and Molecules*, Tartu University Press 1963 (in Russian).
24. Landau, L. D., *Soviet Phys. JETP* **5**, 101 (1957); **8**, 70 (1958); *Collected Papers*, Pergamon Press 1965, pp. 701, 752.
25. Abrikosov, A. A., and Khalatnikov, I. M., *Rep. Progr. Phys.* **22**, 329 (1959).
26. Fradkin, E. S., Dissertation Inst. Teor. i Eksperim. Fiz. (1960); *Nucl. Phys.* **12**, 465 (1959); *Soviet Phys. JETP* **11**, 114 (1960).
27. Bogoliubov, N. N., *J. Phys. U.S.S.R.* **11**, 23 (1947).
28. Beliaev, S. T., *Soviet Phys. JETP* **7**, 289, 299 (1958).
29. Abrikosov, A. A., and Khalatnikov, I. M., *Adv. Phys.* **8,** 45 (1958).
30. Dirac, P. A. M., *Principles of Quantum Mechanics*, 4th edn., Oxford, University Press 1958.
31. Landau, L. D., and Lifshitz, E. M., *Quantum Mechanics*, Pergamon Press 1958.
32. Pauli, W., *Handb. Phys.* **241**, 83 (1933).
33. Blokhintsev, D. I., *Foundations of Quantum Mechanics*, Reidel Dordrecht (1964).
34. Schweber, S., Bethe, H., and de Hofmann, F., *Mesons and Fields*, Vol. I.
35. Akhiezer, A. I., and Berestetskii, V. B., *Quantum Electrodynamics*, Interscience (1965).
36. Bogoliubov, N. N., and Shirkov, D. V., *Introduction to Quantum Field Theory*, Interscience (1959).
37. Davydov, A. S., *Theory of the Atomic Nucleus*, Prentice Hall.
38. Bethe, H., and Salpeter, E., *Quantum Mechanics of One- and Two-electron Atoms*, Academic Press 1957.
39. Gomes, L., Walecka, I., and Weisskopf, V., *Ann. of Phys.* **3**, 241 (1958).
40. Akhiezer, A. I., and Pomeranchuk, I. Ia., *Some Problems of Nuclear Theory*, Moscow, Gostekhizdat (1950).
41. Phillips, R., *Rep. Prog. Phys.* **22**, 562 (1959).
42. Gammel, J., and Thaller, R., *Prog. Elementary Particle and Cosmic Ray Physics* **5**, 99 (1960).
43. Jastrow, R., *Phys. Rev.* **81**, 636 (1951).
44. Huang, K., and Yang, C., *Phys. Rev.* **105**, 367 (1957); Martin, P., and de Dominicis, C., *Phys. Rev.* **105**, 1417 (1957).
45. Abrikosov, A. A., and Khalatnikov, I. M., *Soviet Phys. JETP* **6**, 883 (1957).
46. Vagradov, G. M., and Kirzhnits, D. A., *Soviet Phys. JETP* **11**, 1082 (1960).
47. Silin, V. P., and Rukhadze, A. A., *Electromagnetic Properties of Plasma and Plasma-like Media*, Gordon and Breach (1965).
48. Zeldovich, Ia. B., *Soviet Phys. JETP* **11**, 812 (1960).
49. Wigner, E., *Trans. Faraday Soc. JETP* **34**, 678 (1938).
50. Kirzhnits, D. A., *Soviet Phys. JETP* **11**, 365 (1960).
51. Kirzhnits, D. A., *Tr. Fiz. Inst. Ak. Nauk U.S.S.R.* **16**, 3 (1961).
52. Kompaneiets, A. S., *Zh. Eksperim. i Teor. Fiz.* **25**, 540 (1953); **26**, 153 (1954).

53. Hartree, D., *Calculation of Atomic Structures*, Wiley (1957); Fock, V. A., *Jubilee Symposium Ak. Nauk U.S.S.R.* **1**, 255 (1947).
54. Kirzhnits, D. A., *Soviet Phys. JETP* **5**, 64 (1957).
55. Golden, S., *Phys. Rev.* **105**, 604 (1957).
56. Dirac, P. A. M., *Proc. Camb. Phil. Soc.* **26**, 376 (1930).
57. Kirzhnits, D. A., *Soviet Phys. JETP* **7**, 1113 (1958).
58. Kompaneiets, A. S., and Pavlovskii, E. S., *Soviet Phys. JETP* **4**, 328 (1957).
59. Gombás, P., *Statistical Theory of the Atom and its Applications*, Springer, Berlin.
60. Zeldovich, Ia. B., and Rabinovich, E. M., *Soviet Phys. JETP* **10**, 924 (1959).
61. Alfred, L., *Phys. Rev.*, **121**, 1275 (1961); Baraff, G., and Borowitz, S., *Phys. Rev.* **121**, 1704 (1961); Baraff, G., *Phys. Rev.* **123**, 2087 (1961); Levine, P., and von Roos, O., *Phys. Rev.* **125**, 207 (1962).
62. Weizsäcker, C., *Z. Phys.* **96**, 431 (1935).
63. Abrikosov, A. A., *Soviet Phys. JETP* **12**, 1254 (1960); **14**, 408 (1961).
64. Born, M., and Huang, K., *Dynamical Theory of Crystal Lattices*, Oxford, University Press 1954.
65. Seitz, F., *Modern Theory of Solids*, McGraw-Hill (1940).
66. Abrikosov, A. A., *Questions of Cosmogony*, Vol. III, p. 11, Moscow (Ak. Nauk U.S.S.R.), 1954.
67. Gandelman, G. M., and Pavlovskii, E. S., *Soviet Phys. JETP*, **11**, 851 (1960); Gandelman, G. M., *Soviet Phys. JETP*, **16**, 94 (1962).
68. Latter, R., *Phys. Rev.* **99**, 1854 (1955).
69. Feynman, R., Metropolis, N., and Teller, E., *Phys. Rev.* **75**, 1561 (1949); Latter, R., *J. Chem. Phys.* **24**, 280 (1956); Cowan, K., and Ashkin, T., *Phys. Rev.* **105**, 144 (1957).
70. Kirzhnits, D. A., *Soviet Phys. JETP* **8**, 1081 (1958).
71. Kirzhnits, D. A., *Soviet Phys. JETP* **8**, 1081 (1958); **11**, 1106 (1960).
72. Altshuller, L. V., Krupnikov, K. K., Ledenev, B. N., Zhuchikhin, V. I., and Brazhnik, M. I., *Soviet Phys. JETP* **7**, 606 (1958).
73. *Handbuch der Physik* Vol. 39, Springer, Berlin (1957).
74. Nemirovskii, P. E., *Contemporary Models of the Atomic Nucleus*, Moscow, Gosatomizdat (1960).
75. *Rev. Mod. Phys.* **30**, 1 (1958).
76. Bell, J., and Squires, E., *Advances in Phys.* **10**, 39 (1961).
77. Brueckner, K., Gammel, J., and Kabis, J., *Phys. Rev.* **118**, 1095 (1960); Sood, P., and Moszkowski, S., *Nucl. Phys.* **21**, 582 (1960); Salpeter, E., *Ann. of Physics* **11**, 393 (1960); Cameron, E., *Astrophys. J.* **130**, 884 (1959); Levinger, L., and Simmons, L., *Phys. Rev.* **124**, 916 (1961).
78. Gombás, P., *Uspekhi Fiz. Nauk* **49**, 385 (1953); *Fortschr. Phys.* **5**, 159 (1957).
79. Hara, Y., *Prog. Theor. Phys.* **24**, 1179 (1960).
80. Vagradov, G. M., and Kirzhnits, D. A., *Soviet Phys. JETP* **16**, 923 (1962).
81. Kirzhnits, D. A., *Soviet Phys. JETP* **10**, 414 (1959).
82. Goldstone, J., *Proc. Roy. Soc.* **239**, 267 (1957).
83. Galitskii, V. M., *Soviet Phys. JETP* **7**, 104 (1958).
84. Hubbard, J., *Proc. Roy. Soc.* **240**, 539 (1957).
85. Klein, A., and Prange, R., *Phys. Rev.* **112**, 994, 1008 (1958).

86. GELL-MANN, M., and LOW, F., *Phys. Rev.* **84**, 350 (1951).
87. KIRZHNITS, D. A., *Soviet Phys. JETP* **8**, 835 (1958).
88. BETHE, H., *Phys. Rev.* **103**, 1353 (1956).
89. KIRZHNITS, D. A., *Optika i Spektroskopia* **5**, 485 (1958).
90. HARTREE, D., and HARTREE, W., *Proc. Roy. Soc.* A **150**, 9 (1935).
91. HYLLERAAS, E., *Z. Phys.* **65**, 209 (1930).
92. ZELDOVICH, IA. B., *Soviet Phys. JETP* **4**, 942 (1957).
93. GREEN, A., *Rev. Mod. Phys.* **30**, 569 (1958).
94. GALITSKII, V. M., and MIGDAL, A. B., *Soviet Phys. JETP* **7**, 96 (1957).
95. LANDAU, L. D., *Soviet Phys. JETP* **3**, 930 (1957); **5**, 101 (1957) **8**, 70 (1959); *Collected Papers*, Pergamon Press 1965, pp. 723, 731, 752.
96. PITAIEVSKII, L. P., *Soviet Phys. JETP* **10**, 1267 (1959).
97. AMUSIA, M. IA., *Soviet Phys. JETP* **16**, 667 (1963).
98. LOW, F., *Phys. Rev.* **97**, 1392 (1955).
99. IOFFE, B. L., In the symposium *Questions in the Theory of Strong and Weak Interactions of Elementary Particles*, Erevan, Izd. Ak. Nauk Arm. S.S.R. (1962).
100. ALIAMOVSKII, V. N., and KIRZHNITS, D. A., *Proceedings of the Second All-Union Conference on Quantum Chemistry. Lithuanian Phys. Trans.*, Vol. III, Nos. 1–2, Gos. Izd. Polit. i Nauchn. Lit. Lithuanian S.S.R. (1963).
101. SAWADA, K., *Phys. Rev.* **106**, 372 (1957); SAWADA, K., BRUECKNER, K., FUKUDA, N., and BROUT, R., *Phys. Rev.* **108**, 507 (1957).
102. BONCH-BRUEVICH, V. L., *Soviet Phys. JETP* **4**, 457 (1956).
103. KALLÉN, G., *Det. Kgl. Danske. Mat.-fys.* **27**, 12 (1953); LEHMANN, H., *Nuovo Cimento* **11**, 342 (1954).
104. LAVRENTEV, M. A., and SHABAT, B. V., *Methods in the Theory of Functions of a Complex Variable*, Moscow, Gostekhizdat (1951).
105. KIRZHNITS, D. A., FAINBERG, V. IA., and FRADKIN, E. S., *Soviet Phys. JETP* **11**, 174 (1960).
106. AGRANOVICH, V. M., and GINZBURG, V. L., *Soviet Phys. Uspekhi* **5**, 323, 675 (1962).
107. GINZBURG, V. L., RUKHADZE, A. A., and SILIN, V. P., *Soviet Phys.–Solid State*, **3**, 1337 (1961).
108. PEIERLS, R. E., *Quantum Theory of Solids*, Oxford, University Press 1955.
109. KOSIEVICH, A. M. and LIFSHITZ, I. M., *Soviet Phys. JETP*, **2**, 646 (1956).
110. PEKAR, S. I., *Investigations into the Electron Theory of Crystals*, Moscow, Gostekhizdat (1951).
111. MIGDAL, A. B., In the symposium *Questions in the Theory of Strong and Weak Interactions of Elementary Particles*, Erevan, Izd. Ak. Nauk Arm. S.S.R. (1962).
112. HUGENHOLTZ, N., and VAN HOVE, L., *Physica* **24**, 363 (1958).
113. KARPMAN, V. I., *Soviet Phys. JETP* **12**, 133 (1960).
114. GORKOV, L. P., *Soviet Phys. JETP* **7**, 505 (1958).
115. MIGDAL, A. B., *Soviet Phys. JETP* **5**, 333 (1957).
116. FRÖHLICH, H., *Proc. Roy. Soc.* A **215**, 291 (1952).
117. BONCH-BRUEVICH, V. L., *Soviet Phys. JETP* **3**, 278 (1956).
118. TOLMACHEV, V. V., *Proceedings of the Second All-Union Conference on Quantum Chemistry. Litovskii Fizicheskii Sbornik*, Vol. III, Nos. 1–2, Gos. Izd. Polit. i Nauchn. Lit. Lithuanian S.S.R. (1963).

119. Klimontovich, Iu. L., and Silin, V. P., *Zh. Eksperim. i Teor. Fiz.* **23**, 151 (1952); Silin, V. P., *Zh. Eksperim. i Teor. Fiz.* **23**, 641 (1952).
120. Ginzburg, V. L., *Soviet Phys. JETP* **7**, 1096 (1958).
121. Kulik, I. O., *Soviet Phys. JETP* **13**, 946 (1961).
122. Amusia, M. Ia., *Soviet Phys. JETP* **14**, 309 (1962); **16**, 205 (1963).
123. Gell-Mann, M., and Brueckner, K., *Phys. Rev.* **106**, 364 (1957).
124. Gombás, P., *Acta phys. Hungarica* **13**, 233 (1961).
125. Lewis, H., *Phys. Rev.* **111**, 1554 (1958).
126. Ambartsumian, V. A., and Saakian, G. S., *Soviet Astron.* **4**, 187 (1960).
127. Bloch, F., *Z. Phys.* **81**, 363 (1933); Feinberg, E. L., *Soviet Phys. JETP* **7**, 780 (1958). Sobelman, I. I., and Feinberg, E. L., *Soviet Phys. JETP* 7, 339 (1958).
128. Fox, R., *Advances in Mass Spectrometry*, Pergamon Press 1959; *J. Chem. Phys.* **33**, 200 (1960).
129. Aliamovskii, V. N., *Soviet Phys. JETP* **15**, 1067 (1962).
130. Landau, L. D., *Soviet Phys. JETP* **7**, 182 (1958).
131. Bogoliubov, N. N., and Tyablikov, S. V., *Soviet Phys. Doklady* **4**, 589 (1960).
132. Abrikosov, A. A., Gorkov, L. P., and Dzialoshinskii, I. E., *Soviet Phys. JETP* **9**, 636 (1959).
133. McDougall, D., and Stoner, E., *Phil. Trans.* **237**, 67 (1938).
134. Matsubara, T., *Prog. Theor. Phys.* **14**, 351 (1955).
135. Fradkin, E. S., *Soviet Phys. JETP* **9**, 672 (1959).
136. Vedenov, A. A., and Larkin, A. I., *Soviet Phys. JETP* **9**, 806 (1959).
137. Khalatnikov, I. M., *Dokl. Ak. Nauk U.S.S.R.* **87**, 539 (1952).
138. Lippman, B., and Schwinger, *J.*, *Phys. Rev.* **79**, 469 (1950).
139. Feynman, R., *Phys. Rev.* **84**, 395 (1951).
140. Bloch, F., *Z. Phys.* **74**, 295 (1932).
141. Glauber, R., *Phys. Rev.* **84**, 395 (1951).
142. Fujiwara, I., *Prog. Theor. Phys.* **7**, 433 (1952); Tani, S., *Prog. Theor. Phys.* **11**, 190 (1954).
143. Brink, D. M. *Nuclear Forces*, Pergamon Press (1965).
144. ter Haar, D., *Elements of Thermostatistics*, Holt, Rinehart & Winston, New York (1966).

Index

Index

Index

MADE IN GREAT BRITAIN

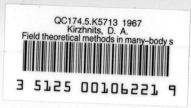